Fire Stream Management Handbook
Second Edition

FIRE STREAM MANAGEMENT HANDBOOK

SECOND EDITION

DAVID P. FORNELL

FIRE ENGINEERING • BOOKS •

Disclaimer

The recommendations, advice, descriptions, and methods in this book are presented solely for educational purposes. Photos are for instructional purposes only. Always wear the proper level of approved PPE when conducting training drills and operating at incidents. The author and publisher assume no liability whatsoever for any loss or damage that results from the use of any of the material in this book. Use of the material in this book is solely at the risk of the user.

Copyright © 2025 by
Fire Engineering Books
110 S. Hartford Ave., Suite 200
Tulsa, Oklahoma 74120 USA

800.752.9764
+1.918.831.9421
info@fireengineeringbooks.com
www.FireEngineeringBooks.com

Executive Vice President: Eric Schlett
Vice President, Group Publishing: Amanda Champion
Vice President of Content Operations: Starlet Franz
Sales and Customer Service Manager: Lane Nash
Managing Editors: Diane Rothschild and David Rhodes
Production Manager: Tony Quinn
Book Development Editor: Daniel Edward Petrino
Book Designer: Robert Kern, TIPS Publishing Services, Carrboro, NC
Cover Designer: Brandon Ash

ISBN: 9781593706098

Library of Congress Cataloging-in-Publication Data Available on Request

All rights reserved. No part of this book may be reproduced, stored in a retrieval system, or transcribed in any form or by any means, electronic or mechanical, including photocopying and recording, without the prior written permission of the publisher.

Printed in the United States of America

1 2 3 4 5   28 27 26 25

This book is dedicated to Becky, my lovely wife, who has had to endure my endless hours at the firehouse, long nights in front of the computer, and my traveling to endless interviews and demonstrations. Most appreciated is her uncanny ability to read hundreds of manuscript pages of unfamiliar technical data, then help me put them into proper order with just the right perspective to make the words readable. My heartfelt thanks for her love, understanding, and help.

Contents

Foreword by Andrew O'Donnell...xiii
Acknowledgments..xv
Introduction..xvii

1. **Fire Stream Basics**..1
    Basic Hydraulics 2
    Introduction To Water Movement 2
    Factors Affecting Friction Loss...3
    Water Movement Terms 4
    Fire Pump Operations..5
    Onboard Water Tank Operations...5
    Drafting 6
    Supply Line Hydraulics..8
    Forward Lay Supply Operations...8
    Determining Supply Line Hose Sizes..9
    Ways of Increasing Supply Line Flows.....................................10
    Laying Additional Lines 10
    Increasing Hoseline Diameter 12
    Increasing Pressure to Increase Flow 14
    Combining Procedures 16
    Managing Water Supplies 16
    Get Water Out of the Pump 17
    Factors Affecting Water Movement...19
    Simplifying Water Delivery Calculations..................................22
    A Brief Review...23
    Chapter 1 Review Questions...24

2. **Strategic Considerations**...25
    Setting Goals 26
    Legal Challenges 28
    Extinguishing More Fire with Fewer Personnel 29

Response District Evaluation ... 30
How Much Flow Do You Apply? ... 36
Personnel Requirements ... 36
Establishing Flow Rates ... 39
Fire Flow Formulas ... 40
Individual Line Flows ... 42
Unwritten Planning ... 44
Evaluating Rates of Flow and Areas of Application ... 45
Evaluation of Line Sizes ... 48
Reaching the Fire ... 53
High-Volume Handlines ... 54
High-Rise Buildings ... 54
Review ... 56
Chapter 2 Review Questions ... 57

3. Nozzle Theory ... 59

Basic Methods ... 59
Direct Attack Method ... 60
Indirect Attack Method ... 60
Indirect Attack Theory ... 61
Iowa Research ... 66
Evaluating Indirect Attack Tactics ... 70
Underwriters Laboratories Fire Safety Research Institute Research ... 74
Stream Selection for Penetration ... 76
Life Safety During Firefighting Operations ... 76
Other Testing ... 77
Indirect Approach Applications ... 79
Development Of Sensible Tactics ... 79
Preventing and Mitigating the Effects of Flashover ... 81
Chapter 3 Review Questions ... 82

4. Nozzle Basics ... 83

Basic Nozzle Design ... 83
Nozzle Components ... 84
Nozzle Types ... 85
Nozzle Basics ... 85
World War II ... 86
The Early Days of Fog for Structural Use ... 87
Stream Pattern Control ... 88
Controlling Gallonage ... 89
Shutoff Devices ... 92
Flush Features ... 93
Teeth ... 94
Protection Myth ... 95
Combination Nozzle Types ... 96
Using the Adjustable Gallonage Nozzle in Place of the Automatic Nozzle ... 100

Distributor and Piercing Nozzles 100
Opposed Discharge Nozzle 103
Bent Discharge Nozzle 105
High-Pressure Delivery 106
Conserving Water versus Extinguishing the Fire 107
Fog Pressure Apparatus in Chicago 108
Modern Experience with High-Pressure Delivery 110
U.S. Air Force High-Pressure Operation 111
Chapter 4 Review Questions 111

5. Automatic Nozzles 113
Automatic Nozzle Design 114
High-End Performance 116
The Automatic Nozzle's Ability to Increase Flow 118
Automatic Nozzle Operation at Lower Flows 118
Utilizing Automatics 120
Automatic Nozzle Master Stream Operation 120
Handline Operation Using Automatic Nozzles 122
High-Pressure/Maximum Volume Method of Operation 123
Fixed-Gallonage Method of Operation 125
Predetermined Pressure Method of Operation 125
Nozzle Evaluation 126
Handline Nozzle Reaction 126
Automatic Nozzle Flow Ranges 128
Pump Operation for Automatic Nozzles 129
Operational Hints 130
Attempting To Evaluate Amount of Flow 130
Chapter 5 Review Questions 132

6. Smoothbore Nozzles 133
Nozzle Comparison 136
Smoothbore Nozzle Tips 137
Effect of Tip Size Selection 140
Smoothbore Firefighting Tactics 142
The Smoothbore Nozzle's Operational Advantages 142
Penetration 144
Managing Unwanted Steam 145
Smoothbore Nozzle Use with Standpipes 148
Reducing Stress 149
Training 150
Combination Stream Attack 151
Twin Tips 151
Threaded Tip Styles 155
Variations on Common Smoothbore Nozzles 156
Conclusion 158
Chapter 6 Review Questions 159

7. Low-Pressure Nozzles 161
- High-Flow Handlines 164
- The History of Low-Pressure Combination Nozzles 166
- The Low-Pressure Age Begins 167
- Safety In Handling 169
- Low-Pressure Nozzle Evaluations 169
- Converting Present Nozzles 172
- Explanation of Nozzle Reaction Forces 173
- Review 175
- Chapter 7 Review Questions 175

8. Heavy Streams 177
- Heavy Stream Equipment 178
- Combination Aerial and Pumper Apparatus 179
- Heavy Stream Nozzles 179
- Using Heavy Streams Effectively 180
- Blitz Attack Tactics 182
- Mounting 183
- Large Fire Combat Tactics 190
- Staging Area 194
- Considering the Effectiveness of Smoothbore Streams 197
- Ground-Attack Monitors 201
- Fighting a Losing Battle 203
- Protecting Exposures 204
- Directing Heavy Streams 206
- Heavy Stream Safety 208
- Chapter 8 Review Questions 210

9. Fire Hose Basics 211
- Hose Construction 212
- Fire Hose Yarns 214
- Fire Hose Sizes 217
- Fire Hose Pressure Ratings 218
- Countering the Effects of Heat and Wear 219
- Rubber-Covered Hose 220
- Other Methods of Protecting Hose 223
- Forestry Hose 223
- A Discussion About Lightweight Lined Hose 224
- Hose for High-Rise Use 226
- Large-Diameter Hose 228
- Large-Diameter Flow Systems 230
- Inspection and Maintenance 232
- Chapter 9 Review Questions 234

Contents

10. Class A Foam 235
    History 237
    Development of a Suitable Agent 238
    What Are Class A Agents? 239
    Putting Class A Agents to Work 240
    Premix or Batch Mixing 241
    Discharge-Side Eductors and Proportioners 243
    Balanced- and Positive-Pressure Injection Metering Systems .. 243
    Compressed Air Injection 244
    Using Class A Extinguishing Agents for Structural Firefighting .. 248
    Increasing Firefighter Safety by Using Class A Agents 248
    Class A Agents on Class B Fires 249
    The Future of Class A Agents 250
    Chapter 10 Review Questions 251

11. Tactical Fire Attack 253
    Engine Company Size-Up 253
    The Art of Applying the Water 263
    Ventilation 271
    Overhaul 272
    Water Damage 274
    What Happens When Unwanted Steam Is Generated? 274
    Preventing Fire Spread by Stream Placement 278
    Ventilation in Conjunction with Engine Work 278
    Consider Water Supply 281
    Summary 282
    Chapter 11 Review Questions 284

12. Managing Hose 285
    Attack Line Hose Loads 285
    Preconnected Beds 286
    Bulk Hosebeds 293
    Reverse Lays 294
    Forward Lays 300
    Flying Lay 305
    Handling the Line 306
    Summary 312
    Chapter 12 Review Questions 313

13. Evaluation and Maintenance 315
    Nozzle Operation Review 316
    Performing Evaluations 316
    Equipment For Evaluation 322

Purchasing Equipment 325
Pricing Factors..326
Nozzle Maintenance 327
Maintenance Procedures.......................................328
Repair Parts..329
Chapter 13 Review Questions...................................330

Answers for Chapter Review Questions..............................331
Index..337
About the Author...355

Foreword

When asked to write this foreword for my good friend, Dave Fornell, I was honored. Dave and I have been good friends since the early 70s—when he started riding with us—while I was the lieutenant of the second busiest engine company in Chicago. We had fires. That was during the time that even the federal government wrote a report called "America Burning." Vandalism, civil unrest, and arson for profit were running rampant, especially in larger urban cities.

We went to fires. We talked about what we did and how it worked out and what we could have done better. Of course, engine operations were always at the top of our list and were discussed extensively.

As time went by, Dave and I, along with other enthusiastic firefighters, taught together for many years at several larger annual fire schools in Illinois, Indiana, and Wisconsin, promoting and improving flows and water hardware.

One thing we agreed on was the fact that fire suppression efforts consider engine company operations to be the most important part of extinguishing a fire. Sometimes things don't go exactly right, but without water, nothing goes right! Everybody else gladly supports the engine's efforts.

Ladder and squad companies are not to be disregarded in their efforts in subduing a fire. They greatly assist suppression efforts in performing other necessary tasks such as forced entry, ventilation, search and rescue, and overhaul operations. A group effort is what puts fires out. Those thoughts were best spoken by a Fire Department of New York truck captain, who often used the phrase, "The quicker you get water on the fire, a lot of your other problems start to go away." Good, trained people working together give you those needed results.

This book explains the history of different efforts in developing water flow hardware and tactics over the years, the improvements that were made, and what worked and didn't work well. The choices are astronomical, and this book brings all of them into their proper perspective

Dave's book should be on the shelves of every fire station and reviewed by the on-duty shift, chapter by chapter, as a daily training session. Fire duty is down. That is why daily training and educating ourselves is more important than ever. I've always been into training. When we are called upon to go to work, we should be more prepared than ever before because we are trained. We don't reach that high level of experience by watching television or playing games on our mobile phones.

—Andrew O'Donnell
Retired District Chief and Director of Training, Chicago (IL) Fire Department
Retired Chief Lemont, IL Fire Protection District

Acknowledgments

The first edition of this book, as with firefighting operations, was truly a team effort. No single person can possibly know the answers to all the questions about the science and art of firefighting.

I would like to thank the following people who contributed to the success of the first edition text:

Chief Andrew O'Donnell, Supervisor of Training, Chicago (IL) Fire Department
Chief Claude Creasey, Los Angeles (CA) Fire Department
Keith Royer, Iowa State University
Bob Barraclough, Span Instruments
Chief Clarence Dixon, Chicago (IL) Fire Department
Chief Martin Pierce, Boston (MA) Fire Department
Chief Timothy O'Neill, Delavan (WI) Fire Department
Chief Neil Svetanics, St. Louis (MO) Fire Department
Battalion Chief Jeff Coffman, Fairfax County (VA) Fire and Rescue
Dave Clark and Dan Smith, University of Illinois
Lee Prazer and his engineering department at Akron Brass Manufacturing
Captain Don Lansu, Elmhurst (IL) Fire Department
Chief Larry Davis, President of International Society of Fire Service Instructors, ISFSI
Bruce Guard, Gordon Harris, and the rest of the wonderful gang at Elkhart Brass
    Manufacturing
George Hughes, National Fire Hose
Chief Bill Treadway, Little Rock (AR)
Chief Ed Phipps and Captain Gil Moreno, San Francisco (CA) Fire Department
Lieutenant Ted Aff, Oakland (CA) Fire Department
Paul Blankenship, California Department of Forestry and Fire Protection
Jill Nelson, Hypro Manufacturing
Ron Rockna, Department of the Interior
Clarence Grady, Odin Corp.

Acknowledgments

Keith Olson, Ansul Chemical Co.
Mark Conran, Neidner Ltd.
Hurley Matthews, Humat Inc.
Ed Sutch, Snap Tite
Bill Darley, W.S. Darley & Co.
Gary Rawlings, Monsanto

Most of all, I would like to thank my former fellow officers and members of Beckerle & Company, Hose Company, Engine 9 of the Danbury (CT) Fire Department for the endless evaluations, photo sessions, hose packings, and repetitive training they endured on behalf of this book. It never would have happened or been more authoritative without their support.

For helping with the second edition, I would like to thank

Captain Dennis LeGear, Oakland (CA) Fire Department (RET) and Fire Service Guru Extraordinaire
Andy Plofkin, Chris Martin, and Jerry Herbst of Elkhart Brass
Richard Riley, Prince Georges (VA) County Fire Department
Marshall Brian Kazmierzak, Benton Harbor (MI) Fire Department
Chief Gordy Nord, and Assistant Chief Ken Wojtecki, Lyons (IL) Fire Department
James Regan, Fire Protection Engineer, (RET). Special thanks for his marathon and helpful review of the galley proofs.
Steve Redeck, Chicago (IL) Fire Department (RET)
Staff Chief W. Parker Browne (RET), Local and National Fire Service Educator, Instructor and Fire Service Consultant, Foremost Water Supply Specialist and life-long Friend.

—David P. Fornell

Introduction

A disturbing trend has been surfacing for years in fire departments across the country: the lessening of emphasis on basic firefighting operations and tactical skills. Budget cutting, coupled with a demand for departments to deliver high-quality medical care and hazardous material incident response (for example, high angle and trench rescues), has tended to push basic firefighting more into the background of the emergency service picture. Certainly, fire protection is still the primary duty of fire departments, and fires continue to be fought, but more and more precious training and evaluation time is being spent studying and planning operations rather than firefighting. These attempts to broaden fire department responsibilities are beginning to divert the energy and time formerly devoted to research, evaluation, and training in firefighting operations.

Recent national fire trade publications include articles about management, coping with stress, hazardous material response, the crisis in trauma care delivery, and using computers more effectively. In two publications, columnists write monthly about truck and rescue work, when to open the roof, how the outside vent person is the most important individual on the fireground, hydraulic rescue tool operation, and planning for cave-in rescues. Since most firefighters in this country rarely encounter such operations, this information can be extremely helpful and certainly its presentation should continue to be closely studied.

What concerns me is the growing perception that engine company work is somehow no longer as important as other fireground operations. It lacks the glamour of rescue or ladder company duty. The searches, forcible entries, rescues, and roof work somehow seem more exciting than standing over a hydrant waiting for the order to charge the supply line. The lack of information presented in trade publications is one symptom. Another is the widespread attitude that we have progressed as far as possible in the science of applying water.

While researching material for this book, both the first and second editions, I reviewed a number of training videos. In most, the recommended method of attack was to set an automatic nozzle on a wide spray pattern, then to duck-walk into a burning building with no concern given to flow rate, water distribution, and the effect of the water pattern on the safety of the firefighters. The saddest part was that many of the videos did not even describe the water application tactics being used; they simply implied that the tactics were correct.

There has been much time and effort wisely expended in the development of practical, protective turnout gear. These improved protective clothing and breathing apparatus provide a high level of protection for the firefighter engaged in interior firefighting operations. But firefighters are still burned despite being wrapped in the best protective envelope money can buy. Many injuries occur during supervised training sessions where proper firefighting tactics are being taught. There is a message here: it's time we take a close look at the damage our water application techniques are causing firefighters and victims alike.

This book started out as a short and simple guide to nozzles. As research began, I noticed that not only was a nozzle guide needed by the fire service, but also some type of handbook on the application of water in structural firefighting. As more and more people were interviewed, it was found that most firefighters had little knowledge of how engine company operations affect total fireground operation. Sure, they knew how to put the pump in gear and how to hold a 1¾" line, but few had knowledge of how water application rates can be calculated before the fire, how water supply operations affect attack effectiveness, or how much more efficient a large-diameter hose is over multiple smaller lines. Most importantly, I realized how rarely most firefighters actually look a structure fire in the face from the business end of a hoseline.

If we consider the fact that every fire service organization in this country has some type of pumping apparatus, we can begin to see the scope of the problem. Comparatively few have a ladder or rescue apparatus. In effect, most firefighters across the nation are directly involved with the supply and application of water. Relatively few exclusively perform search and ventilation duties.

Fire stream management can be defined as the systematic application of water streams to a fire for the purpose of extinguishment. Managing fire streams includes the following points:

- Flow rate
- Reach and penetration of stream
- Form of water discharge
- Effective tactics
- Water supply
- Safety
- Integration with other functions such as ventilation and search and rescue

This book has a rather narrow focus-water movement and application for structural firefighting. It addresses and explains that basic aspect of firefighting that most of us thought we had mastered early in our careers. Reason tells us something must be amiss since we continue to burn firefighters and buildings despite aggressive interior attacks. Something is not working properly in many of the tactics we use for fire extinguishment.

What started out to be a simple explanation of nozzle functions and operations guide eventually grew into a comprehensive handbook describing the basic understanding of water movement, scientific principles of fire extinguishment, and the practical means of applying the "wet stuff to the red stuff." When the means of extinguishment fundamentals presented within these pages are understood and practiced, improvements in those means with techniques—such as ventilation and Class A agents—can become more effective.

Introduction

It must be kept in mind that no one part of the fire stream management system, such as a single nozzle style, can be taken by itself as a complete answer to fire extinguishment. This book presents many ideas and concepts which, to function effectively, must be integrated into a total extinguishment plan.

No one person could ever know all there is to know about putting out fires. One of the unique aspects of the fire service is the universal willingness to share knowledge, to pass it on. Though most of the contents of this book result from personal experience, there are many others who filled in the gaps and smoothed over the rough spots with their desire to make the fire service stronger, safer, and more efficient.

This book is a result of years of study, discussion, and many, many long, and sometimes frightening, hours of slinking down smoke-filled and heat-saturated hallways, calling for water, and attacking fire. It is also a result of critical evaluations after firefighting operations and untold hours of testing, training, and planning.

The techniques and procedures described in this book have been effectively used in actual fire combat. These techniques and procedures should be carefully evaluated by each individual organization to make sure they will integrate properly and safely with that organization's particular operation. What works well in a five-story frame apartment building in Boston may not be as effective or safe when used on a hay-filled barn in Bullet Hole, WY.

It is also assumed that any department wishing to utilize the information presented in this book mandates the use of full protective clothing and breathing apparatus in accordance with National Fire Protection Agency (NFPA) standards, including NFPA 1500. For these techniques to be safely implemented on the fireground, an incident command system must be in operation, which includes the supervision by a safety officer.

Anyone in the business of putting out fires needs this book. If you have ever dragged a line down a living-hell hallway only to be pushed back by fire lapping over your stream, you need to know why it happened and how to prevent it in the future.

A good hydraulics book will tell you how to make fire streams.

This book will tell you how to use them.

A Note on the Photographs

There are many photographs used in this book to illustrate important points and to clarify certain fireground operations. Some of the photographs show departments using improper or unsafe tactics or becoming involved in questionable operations. It is not the purpose of this text to criticize these departments or to ridicule their operations. The photos are presented in the interest of improving everyone's tactical firefighting operations and making readers more aware of firefighter safety. To these departments, the author apologizes for any embarrassment suffered and offers his thanks for their understanding that the fire service will be strengthened as a result.

Yes, many photographs show scenes from many years ago, but they are included because they illustrate a point. Rather than complaining that a firefighter is shown wearing hip boots or not wearing a self-contained breathing apparatus (SCBA), take a look to consider the point the photograph is trying to make.

Unless otherwise indicated, all photos are by the author.

Fire Stream Basics

To effectively generate and apply fire streams, all firefighters involved in producing them must have a basic understanding of how and why water performs as it does when it's moved through and out a hoseline. Hydraulics textbooks and classroom instruction are excellent ways of learning how to move water and create fire streams efficiently and safely. However, many firefighters have difficulty understanding and computing the algebraic formulas necessary to solve hydraulics problems. While many municipal departments require pump operators and officers to be proficient in computing friction loss, engine pressures, and other hydraulic problems, the majority of fire service personnel have a very limited working knowledge of what factors affect water movement on the fire scene.

In practical terms, it certainly is difficult for many of us to perform the needed calculations on the fireground, especially when fire stream overspray keeps shorting out your electronic hand calculator. A full knowledge of hydraulics and its mathematics is desirable for all pump operators. A thorough understanding of some basic principles concerning hydraulics and water application can lay the foundation for more efficient supply and pumping operations and be augmented by further study. All personnel involved with water supply and application should be capable of determining when the nozzle pressure is too low or too high to provide an effective fire stream, as well as what steps are needed to correct the situation. All firefighters should also be able to recognize the limitations of single or small-diameter supply lines and should be able to quickly determine solutions to problems of insufficient water supply.

While it seems deceptively simple and basic, the information provided in this chapter must be completely understood if any of the information presented in the subsequent chapters is to make any sense. For firefighters with little experience moving water, these paragraphs will serve as an introduction to engine company operations. For those with experience, it will be a good review.

Basic Hydraulics

Water is moved through a hose from one point to another by pressure forcing the liquid through the line. The science of moving water by pressure is called hydraulics. The subject of hydraulics and its accompanying mathematical computations can become rather complicated. Its use on the fireground is certainly more involved than simply twisting the throttle until the hoselines feel hard. In addition to studying hydraulics in formal fire school classes, one of the best ways to achieve an understanding of its practices is to read one of the many books devoted exclusively to hydraulic science and computations. Anyone involved with water movement and application should seriously consider the detailed study of hydraulics to become more familiar with its mathematics.

Many find hydraulics and its many numbers intimidating, and some even attempt to ignore its existence in actual use because it is thought that the theories are too complicated for practical application. This practice is neither wise nor safe. To ensure safe and efficient operation, firefighters, pump operators, and fire officers must know and understand the basic principles that govern how the water gets from the water source to the nozzle and what happens after it passes through the device on its way to the fire.

Introduction to Water Movement

Since every firefighter is familiar with nozzles, a description of how a simple, smoothbore nozzle operates is an excellent place to start explaining water movement. The nozzle's name can also serve as its description: a round piece of metal with a smooth hole bored through the center. This hole has sides that taper from the large opening on its inlet end to a smaller hole on the outlet end. Since the water flowing through the nozzle has to pass from a larger to a smaller opening, it builds up pressure. As with all liquids, water cannot be compressed; it has to go somewhere. If there is an opening available, it will attempt to force its way through. Remember that the nozzle, restricting the flow, is what generates working fire stream pressures; the pump only puts the water in motion.

This can be demonstrated by supplying a pumper with water from a hydrant or draft and discharging through a fully opened discharge valve with no hose attached. The discharge piping and valves will create some restriction in the flow, generating some pressure. However, it will be difficult for the pump to build up normal working pressures discharging through the large opening.

If a nozzle is attached to each discharge and the valves opened, the pump discharge pressure will show an increase over the open outlet pressure.

It should be remembered that the pump cannot generate working pressure by itself. It must have some type of restriction at the end of the line to harness the force of moving water, which in turn provides effective streams.

The effect of restricting the force to provide pressure can be demonstrated by placing a thumb over the open end of a garden hose. The thumb constricts the outlet flow which will build up nozzle pressure. This pressure causes the water to increase in speed, which in turn forces the water to pick up distance after it exits through the hole. This speed is referred to as *velocity*. The amount of constriction, or tip size, determines how much water will flow at a certain pressure and how much reach the water will obtain after the stream exits the tip.

If the spigot can be opened to supply more water to the hose, and if the thumb keeps the same size hole at the end of the hose, an increase in spigot pressure will cause the water to move faster through the hose, with more water being forced out the same size hole and more reach being provided.

In the aforementioned example, the thumb kept the opening at the end of the hose a consistent size. With fire nozzles, a smoothbore tip keeps the size constant as well. They are manufactured in many sizes, and the selection of the proper size tip for the job at hand depends on hose size, nozzle pressure available, the desired amount of water to be applied to the fire, and the staff available to handle the line. To obtain a certain flow at a certain pressure, there is only one size tip which will do the job.

For example, to flow 210 gallons per minute (GPM) at 50 pounds per square inch (PSI) nozzle pressure, the only size tip which can be used is 1". Other size tips can also flow 210 GPM, but the flow will be achieved at different nozzle pressures.

If the 1" tip is removed and a larger tip substituted—let's say one with an opening of 1¼"—at the same engine pressure, more water will flow because the hole is bigger. But because the constriction is now larger, less pressure is being generated, so the stream exiting the tip will not have as much reach as the stream from the 1" tip. This is because the water is moving slower, creating less velocity.

Keeping the engine pressure the same, if a smaller tip than 1" is used—say ¾"—the flow will be less because the hole is smaller, but the nozzle pressure will be higher because of the increase in constriction. The stream exiting the tip should have more reach because the water is moving faster.

The pressure available at the nozzle to create velocity and flow is dependent on two factors: how hard it is pushed at the starting end of the line and how much of this push force is used up by water traveling to the nozzle. As water flows through a hose, it rubs the interior hose walls. This rubbing action causes friction that takes energy away from the moving water. This loss of energy due to friction is cleverly termed *friction loss*. The amount of friction loss is the difference in pressure between the beginning of the line and the end of the line.

Factors Affecting Friction Loss

Water flow is described as *laminar* when all the water moving through a hoseline is traveling in a relatively straight, forward line. Because it is rubbing on the inside of the hose walls, the water at the edges will flow a little slower than the water near the center. If the hose is kinked or formed into a tight loop, if the interior hose surface is rough, or if the pressure pushing the water is excessive, the water inside the hose begins to swirl. This swirling effect is called *turbulence*. Turbulence increases friction loss.

If an attempt is made to move more water through a hoseline, more pressure is needed to push the water faster. The faster the water moves, the more turbulence and friction loss is generated.

Friction loss can be affected by the following:

- Hose diameter: The larger the diameter, the less the friction loss; the smaller the diameter the more the friction loss.

- Length of hoselines: The longer the hoselines, the more the friction loss.

- The amount of water flowing in a hoseline of a given size means that the more water flowing, the harder it rubs on the hose walls, creating turbulence, causing increased friction loss.

- Other factors that can also increase friction loss are kinks in the line, use of wyes or shutoffs which have smaller interior waterway diameters than the hose to which they are attached, or hose with a rough interior surface.

Water Movement Terms

All firefighters should be able to make sense of the basic factors affecting water movement. Some basic terms are described as follows:

Pump Pressure: The amount of force registered at the pump outlet that is available to move water through a hoseline. Measured in PSI.

Nozzle Pressure: The actual amount of force, measured at the end of the hose, that is available to make the nozzle work. Also measured in PSI.

Nozzle Operating Pressure: The optimum amount of force needed at the nozzle inlet to make the nozzle operate most efficiently, as designed by the manufacturer. As a general rule, smoothbore handline nozzles are designed to operate between 40 PSI and 60 PSI, combination automatic nozzles at 70 PSI to 100 PSI, and the newer, low-pressure combination nozzles at 50 PSI to 75 PSI. The nozzle pressure is what is actually available on the fireground to make the nozzle work. The nozzle operating pressure is what should be available to make the nozzle work according to the manufacturer.

Nozzle Reaction: The amount of pushing force generated at the nozzle in the opposite direction of stream travel. It is measured in pounds force and is caused by a combination of the volume of water flowing and the nozzle pressure at which the water is discharged.

Friction Loss: The amount of energy lost by the water traveling through the hose. The amount is measured by the PSI difference between pump pressure and nozzle pressure.

Flow: The amount of water moving through a hoseline or out a nozzle. Measured in GPM.

Reach: The distance water travels from the nozzle until it hits the ground. Effective reach is considered as the distance in which 85% of the actual water flow traveling from the nozzle hits the ground.

Attack Line: The hoseline stretched from the pumper to the fire. Attack lines usually have nozzles or portable master stream devices attached to the working end.

Supply Line: The hoseline stretched from a water source that supplies the apparatus with water.

Fire Pump Operations

Except in some special cases, all fire department streams are supplied from pumps mounted on fire apparatus or from fire department portable pumps. While probably millions of words have been written and spoken about running fire pumps, their basic operation can be described in three sentences:

- Put the pump in gear.
- Get water into the pump.
- Get water out of the pump.

These three operations must be performed in exact sequence in order for the pump to perform its intended job.

Water can be supplied to the pump in one of four ways:

- From the onboard water tank
- From the water tanks of other vehicles
- From a supply line laid from a hydrant or another pumper
- By drafting from a source of water or from a portable tank kept full by tanker apparatus

Onboard Water Tank Operations

Flowing water to the pump from the water tank is the simplest and quickest method of pump supply. It is frequently used along with one of the other methods to rapidly provide water for attack lines while a second water source is being secured. While most fires can be extinguished using tank water alone, another source of supply should be established when encountering any working fire in case the tank runs dry before the fire is knocked down. As soon as a second source begins to supply water, a wise pump operator will refill the pumper's own tank while supplying the attack lines from the second water source. If the second source fails due to a burst hose or mechanical breakdown, the pump operator can immediately switch back to the tank supply while steps are taken to warn the nozzle crew of the loss of the secondary water source.

Supplying water from other apparatus tanks is simply a matter of the second apparatus pumping into the first. While widely practiced in rural areas, I have also witnessed this operation in large cities where hydrants are frequently found frozen in the winter months.

Drafting

Drafting water from a source requires much practice if the department desires a consistently effective operation. Improper equipment is one of the common problems with the operation. Old-style, unwieldly, rubber-covered suction hose is being replaced with lightweight hose specifically designed for rural drafting operations. Likewise, the old brass barrel strainer, which, if not kept off the bottom of the water source, is likely to intake as much silt and mud as it is water, is being replaced with floating strainers that allow the unit to float on the water's surface, skimming debris-free water from near the top of the source (fig. 1–1).

When drafting, the pumper does not suck in water. The pump creates a low-pressure area in the pump casing and the water is pushed in by atmospheric pressure. We all know from our high school science classes that the weight of air pushes down on the earth's surface. How much it pushes depends on altitude. At sea level, the atmospheric pressure is 14.7 PSI. At altitudes above sea level, the weight is less; below sea level it is more. During priming, the pump creates an area of low pressure within itself and its supply suction hose. Atmospheric pressure pushing down on the water source forces water through the suction hose and into the pump. If there are leaks in the suction hose, the pump seals, or in any connection, air, rather than water, will attempt to fill the low-pressure area in the pump. The more air that seeps into the pump, the less chance water will have to fill the area. One of the most common causes for not being able to obtain a draft is that air is leaking into the system from somewhere, most likely through a loose coupling.

Figure 1–1. A supply pumper drafting from a dry hydrant to supply tankers during a water shuttle operation. Dry hydrants can be located in areas that provide quick and easy access for pumpers wanting to obtain a water supply from a pond or lake.

Stories have been told for years about pumpers connecting to a hydrant with a hard suction and then "drafting" from the water main system. Since atmospheric pressure must be present for a draft to be obtained, it can be reasonably said that it is impossible to draft from a sealed water source. What does happen when hard suction is connected to a hydrant is that kinks (common when soft suction lines are used and the pumper is not properly positioned) cannot be made in the hard suction line, creating a full-flow waterway from the hydrant to the pumper.

If a hydrant is flowing at its maximum capacity and the water mains, valves, or hydrants are in poor repair, the very low pressure and higher velocity flows created by the pump could damage mains, cause valves to leak or fail, or cause hydrants to become disconnected from the main and pushed out of the ground. These incidents are due to deteriorated fittings and mains failing because of lowered internal pressure, not because the pumper sucked them out of the ground.

In areas where drafting may be frequently encountered, dry hydrants should be installed. These devices contain a strainer, large-diameter plastic pipes and fittings, and standard fire department suction adapters.

When installing dry hydrants, the strainer is placed in a supported position about halfway between the bottom and the surface of the water source. A supply pipe is buried in the bank area and the pumper suction fittings positioned so suction hose can be easily connected (fig. 1–2).

When in use, the dry hydrant acts as rigid suction hose, making the drafting operation much quicker to set up while allowing the apparatus to remain on firm ground.

The secret to quick and efficient use of drafting sources is training. As with any operation, the more a department trains, the more second nature the operation becomes.

Figure 1–2. An example of an assembled dry hydrant. The perforated strainer has a hinged end cap that will open under pressure so the device can be backflushed for cleaning. The height of the riser above the lower elbow and amount of extension from the elbow into the water source will depend on each individual application.

Supply Line Hydraulics

On the fireground, discharging water from a pumper to the fire through attack hoselines is the easiest pumping operation to manage because the pump operator has complete control of output pressures and flows. Managing the water being supplied to a pumper can be more difficult because unless water is supplied by another pumper, the pump operator has to simply take what they can get and usually has little influence on the incoming flow or pressure.

There are two basic supply hose lays used when operating from a hydrant system or from a pumper at a drafting source. The most commonly used lay is the straight or forward lay, where a supply line is stretched from the hydrant or other water source to the fire. The other general type of hose lay is the reverse lay, where a line is laid from the fire to the water source. The reverse lay can be a combination of both supply and attack lines if the line supplies attack lines through wyes or other distribution devices. The reverse lay is useful if only a limited number of personnel are available or if maximum use of the water source is desired. Its pumping operation is similar to that of attack lines and is covered later in the chapter (fig. 1–3).

Forward Lay Supply Operations

Most departments utilize some form of the forward lay, also called a straight lay, for supply operations. Its deployment is usually quicker than other methods, and its use will place the pumper near the fire building where hoselines and needed tools are readily available. The amount of water the pumper will receive through the forward lay line will depend on the following factors:

- Pressure and volume available from the hydrant or supply pumper
- Size of hose laid
- Number of hoselines laid

The pressure available at the hydrant or supplied by the drafting pumper will determine how much energy is available to force the water through the supply hose to the engine. Supply hoselines are subject to the same friction loss rules as attack lines. Pressures available at the hydrant cannot be increased unless a pumper is connected to the water source and pumps water to the receiving engine at a pressure higher than that supplied by the hydrant alone. Although in some cities the water department can increase water main pressure on request, the only practical method of increasing flow without using another pumper is to increase the size or number (or both) of the supply lines.

By using larger diameter hose or laying multiple lines, a supply pumper can move more water over longer distances because friction loss is reduced.

Figure 1–3. A once common hosebed configuration throughout the country for many years were two 1½" preconnected hoselines outboard and 1,200' to 1,500' of 2½" supply hose. With this load, limited flow rates will be obtained especially if a good amount of the supply hose is laid as a single line. The simple replacement of the supply line with larger diameter hose, higher flow nozzles, and the addition of a 2½" preconnected hoseline will greatly increase this unit's firepower.

Determining Supply Line Hose Sizes

For many years, the largest hose considered for practical supply line use was 2½" hose. Bulky, woven, cotton double jacket hose construction, as well as the heavy weight of brass couplings (common in older days before the 1960s), limited the useful size of supply hose because of the difficulty in handling the weight and bulk of large-diameter hose sizes. Before the Second World War, a few big city fire departments carried 3" or 3½" supply hose on standard pumpers in addition to 2½" line. Many of these departments used the larger diameter hose to supply water from high pressure hydrant systems. While larger hose was available in the marketplace, it was not commonly found outside of large municipal departments. The 3" hose gained popularity after the Second World War and is frequently found in many departments, carried in a second bed alongside the standard 2½" line. The 3" hose is usually supplied with 2½" thread couplings to make deployment less complicated because adapters are not needed to make up connections with the pumpers' standard 2½" fittings.

The small restriction created by the 2½" coupling tends to be minimized by a small increase in water velocity as the water passes through the coupling, as well as the extremely short distance (about 1" to 2") the water has to travel. This makes the 2½" coupling's overall effect on friction loss very small, about 0.5 PSI per 100'.

While larger diameter 4", 4½", 5", and 6" double jacketed hoses have been available for many years, their use was generally limited to short lengths of soft suction line because of weight and expense. America's introduction to the large-diameter, 4" and 5" rubber-covered hose from Europe in 1968 finally provided the fire service with a practical method of moving great quantities of water long distances through what actually can be considered an above-ground water main.

Ways of Increasing Supply Line Flows

Let's assume a pumper, first arriving on the scene of a fire in a two-story frame dwelling, forward lays a 3" supply line from a hydrant 400' away from the structure. Assuming a hydrant pressure of 60 PSI is available, a flow of about 400 GPM can be expected at the pumper, which should be adequate to handle most fires (unless the building becomes fully involved, numerous exposures are threatened, or both). If the fire does increase in size, more water will be needed on the fireground. This can be accomplished in a number of ways. The most common method is to have another pumper lay additional lines to a different hydrant. With proper planning, however, the first pumper could have increased the available flow while laying the first supply line through the application of certain steps. These steps would be to

- Lay additional lines
- Increase the diameter of the supply line
- Provide hydrant relay valves so a second-due pumper can boost the pressure in the initial line laid from the hydrant
- Perform any combination of these steps

Laying Additional Lines

Many departments that still use 2½" or 3" diameter hose for supply lines have divided beds so two lines can be laid at the same time. While this action increases waterflow if dual lines are laid from the same source, it also doubles the amount of hose placed in the street and, in the case of long lays where the beds are coupled together and only a single line is laid, reduces the volume available (fig. 1–4).

For example, if a pumper carries 1,400' of 2½" line in two beds, it can lay only one 1,400' line or 700' of dual lines. This decrease in operating range may or may not be a factor depending on hydrant spacing and building location within the department's response area.

For example, if a pumper needs to be supplied with 750 GPM and the hydrant or drafting source is 300' away, the following pressure will be needed to move the flow to the pumper and maintain a minimum of 10 PSI at the pumper inlet:

Hose Size	Friction Loss Per 100'	Pump Pressure
2½"	110 PSI	340 PSI
3"	45 PSI	145 PSI
3½"	19 PSI	61 PSI
4"	11 PSI	43 PSI
5"	4.5 PSI	23.5 PSI
Two 2½"	28 PSI	94 PSI
Two 3"	11.5 PSI	44.5 PSI

Note. The 340-PSI shown in the chart is for illustration only

From this example, it can be seen that two 2½" lines are certainly more efficient than one 2½" or one 3" line. However, when utilizing a dual lay, the total friction loss can still be high, and 600' of hose will now be needed on the street rather than 300'.

Laying dual lines is the least expensive way of increasing supply line effectiveness without purchasing new hose or equipment. However, when determining personnel and apparatus needs, remember to consider the problems of decreased available hose lay distance and greater effort needed during hose pickup that the dual lay requires.

Figure 1–4. Long stretches of 2½" or 3" line become inefficient for moving large quantities of water because of excessive friction loss. Laying multiple lines of the same size hose helps reduce friction loss moving the same amount of water over that of a single line; however, multiple lines require more personnel for pickup.

Increasing Hoseline Diameter

The most common solution to the problem of moving increased amounts of water within the past 10 to 15 years has been the purchase and utilization of large-diameter supply hoses. Friction loss comparisons illustrate the large-diameter hose's effectiveness (fig. 1–5).

Let us assume that a hydrant has 65 PSI available when flowing water and the pumper will be operating 300' from the hydrant. The following is the maximum amount of water that each line can supply at that pressure and distance. Since 10 PSI is desired on the pumper inlet, 55 PSI is left to move the water.

Hose Size	GPM at 300'
2½"	300 GPM
3"	470 GPM
3½"	665 GPM
4"	470 GPM
5"	1,500 GPM

If the hydrant pressure remains the same but distance to the fire is doubled to 600', there will still be only 55 PSI to move the water twice the distance.

Figure 1–5. This pumper has two forward lay beds: a 3" main bed on the left and a 5" main bed on the right. The 5" line will supply as much water as three 3" lines at the same pressure.

Hose Size	GPM at 600'
2½"	210 GPM
3"	330 GPM
3½"	520 GPM
4"	670 GPM
5"	1,250 GPM

By laying a single line of 5" large-diameter hose, the amount of water capable of being delivered to the scene is increased almost six times over what is available from a single line of 2½" hose.

In many departments, large-diameter supply hose is rapidly replacing the smaller hoselines previously used for supply lines because the former can move great amounts of water at normal pressures, is easier to deploy and pick up than multiple lines of smaller hose, is relatively maintenance free, and costs less than multiple lines of smaller hose needed to move the same amount of water. Manufacturers report the most popular size for large-diameter hose is 5", followed by 4".

The 6" and larger size supply hose is rarely found in municipal fire departments, although it now is regularly used by the petrochemical industries for fire protection in refineries and storage areas. Sizes in these applications are available up to 12".

A 3½" supply hose, while once a popular large-diameter size in double jacket, in many departments, is usually replaced with 5" or 4" as it wears out.

The major reason for the popularity of 5" hose is that its large diameter can move a great amount of water while the hose's bulk and weight remain reasonably manageable, striking a successful balance between flow efficiency and handling ease. While 4" hose handles almost the same and takes up almost the same amount of space in a hosebed, it can only move approximately 66% of the water of the 5".

The 6" hose can move more water, but it takes up more storage space and is unwieldy to deploy and pick up on the fireground. The 5" size seems the most practical if maximum water movement combined with a reasonable managed amount of deployment effort is desirable.

Some departments are utilizing 4" split beds of 800'–1,000' which allow fire departments to essentially deploy dual 4" lays in the commercial zones of their jurisdictions, yielding a 6" hose hydraulic equivalent friction loss (as hydrants are connected to mains by 6" service, this makes complete sense operationally). This also allows up to 1,000' lays in residential areas where required suppression flows are lower. This dual 4" setup takes advantage of different requirements for fire flow based on zoning, water main designs, and hydrant-spacing-based water works standards in the same or less space than 5" hose.

To get an idea of the amount of water 5" hose can move, the following list compares the number of smaller lines needed to move the same amount of water as a single 5" line at equal pressures.

Hose Size	Number of Hoses Needed
2½"	6
3"	4
3½"	2
4"	1½

The advantages of laying a single 5" supply line in place of multiple lines of 3½" lines (or smaller) are that it provides more water on the fire scene more quickly, can move more water over longer distances, and is easier to place back in service, especially if only a limited number of personnel are available on scene. If 5" line is laid from a hydrant, it can probably supply most of the hydrant's capacity at some distance without requiring a pumper at the supply end. However, at larger water-use fires where multiple hydrants are used, zone residual pressure can drop to 20 PSI (or less), and only pumps on hydrants with adequate supply hose sizes between supply and attack pumpers yield the full water flow for the area being used. The distance depends on hydrant pressure. Remember if a 3½" or 4" supply hose is used, maximum waterflow capabilities are available from the hydrant only if the line is supplied by another pumper or if additional lines are laid.

The practical effects of switching to large-diameter hose were demonstrated when a fire company purchased a new pumper equipped with 5" hose. As a general rule of thumb, the company expected an average available flow of about 600 GPM when they were using two 3" lines for a standard hydrant-to-fire lay given the average pressures available throughout the wat er system and average lengths of the lays. Since the company placed 5" hose in service, the average available flow increased to 1,250 GPM.

Of course, these figures will vary depending on water main size, length of hose lay, and elevation between the water source and the fire. Actual experience, however, indicates that the deployment of 5" hose alone has easily doubled the company's fireground water-moving ability.

In actual practice, the company has found 5" hose can be laid directly from a hydrant and can supply just about all the hydrant is capable of flowing at lengths of up to 800'. The friction loss for 5" hose is so low that, up to a flow rate of 1,000 GPM, it actually does not have to be figured as a general rule, except at heavy stream fires where the water supply in the area around the fire reduces the rate of maximum fire flow to 20 PSI at the hydrant outlet.

Likewise, when relaying water from a drafting source, the pumper can supply more water over longer distances when large-diameter hose is used in place of 2½" or 3" lines. In areas where large-diameter hose is commonly used, it is not unusual to see two pumpers moving 1,000 GPM a mile or more.

There is a trend in the fire service toward widespread use of large-diameter hose—typically 4" and 5" hose—for supply and relay pumping operations in both city and rural areas. It has proven to be cost-effective and, as any department using the hose will testify, its low friction loss makes moving a high volume of water over long or short distances much more efficient. Its purchase is certainly a much better investment in firefighting power than gold leaf lettering or aluminum wheels—neither of which will extinguish much fire.

Increasing Pressure to Increase Flow

Practical experience is finding that 2½" to 4" hose obtains its maximum water-moving efficiency when supplied by a pumper on the supply end, rather than fed directly off the water system. Supply lines of these diameters can certainly be laid and initially supplied by hydrant pressure, and in most cases, the resulting water flow will prove sufficient for common one or two room dwelling fire.

However, if the fire increases in size, or if more water is needed to protect exposures, departments utilizing 2½" to 4" supply hose can consider the use of a special hydrant relay valve when making a hydrant-to-fire lay.

These hydrant relay valves, sometimes called four-way, Hydrassist, or Humat valves, offer an efficient way of quickly boosting pressure on the initially laid line without interrupting the flow. When using these valves, the first engine lays a supply line and receives water through the valve directly from the hydrant using hydrant pressure. If more water is needed, another engine, using the intake and discharge ports on the valve, can make the appropriate connections and pump into the initial line without having to disconnect any hose or fittings or shut down the first line (fig. 1–6).

While some departments utilize four-way valves in their supply operations, other large departments—such as Boston and Baltimore—have found their use particularly successful. If additional water is needed, the valve is connected to a supply pumper with short sections of 4" or 5" hose.

This operation allows first-due engine companies to be utilized to full capacity, reduces fireground hose clutter, and increases water availability much faster than if the extra-alarm engines were to lay their own lines and secure hydrants farther away from the fire scene.

Four-way valves can also have other uses. Some California counties have utilized four-way valves when fighting brush fires. When stretching the low flow and low-pressure lines common in combating running wildland blazes, these lines can be stretched directly from a hydrant using a four-way valve and are easily supplied only by the use of hydrant pressure.

Figure 1–6. The use of a hydrant relay valve allows this aerial to lay its own 3½" supply line and obtain immediate water in the initial stages of operation after arrival on the fireground. If more water is needed, another company simply makes its connections to the valve and increases waterflow by stepping up the pressure.

This action can free up pumper apparatus in case it is needed elsewhere. Additionally, if a fast-spreading fire calls for rapid evacuation, the pumper will not have to disconnect from a hydrant when making a hasty exit.

If the fire threatens a structure or if more hose is needed to reach a remote area, a pumper can then connect to the hydrant and begin to supply higher pressures to the lines that are already in place.

Combining Procedures

The amount of water moved through any size supply can be greatly increased if a pumper is placed at the supply source to boost the pressure. Paul Shapiro of the Las Vegas Fire Department conducted a number of tests in 1989 and 1990. He found that by combining large-diameter hose and a source pumper, flows of over 2,000 GPM were not uncommon using a single 4" or 5" line. He also found that supplying a 1,500 GPM pumper from two different hydrants using large-diameter hose increased its capacity to almost 3,000 GPM.

Managing Water Supplies

The cause of poor streams being delivered on the fire can be traced in almost every case to insufficient incoming water supply. The water supply may be available in sufficient volume but may not be received by a pumper at the fireground because of poor water supply tactics.

The photograph in figure 1–7 shows a pumper in a medium-size city that had laid 1,500' of 2½" supply line and was unable to deliver an effective stream on the fire. The attempted

Figure 1–7. At this fire, the second-alarm company at left laid 1,500' of 2½" line in a procedure normally used for short lays to dwelling fires. As can be seen by the kink in the line, the incoming pressure was not sufficient to supply even one large stream. The pumper on the right, connected to the hydrant with a large suction sleeve, has more than enough water to pump its full capacity. The pumper on the left is located on the wrong end of the hose lay.

solution at the time was to have the water department increase the pressure, but that measure did not solve the problem. Note the pumper on the right, connected to a hydrant with large-diameter suction hose. This unit was showing 20 PSI on the intake gauge while all discharge ports were in use. The engine on the left could have easily provided a useful stream with the equipment available if the pumper was placed on the supply end of the line and delivered the pressure needed to properly operate the nozzle. The tactical means of providing an effective stream was available, but it simply was not used.

It is common to hear arguments from fire officers against the use of large-diameter hose on poor water systems. "Our system is so old that it won't support the use of large-diameter hose," one chief said.

If the water is in the ground at almost any pressure or volume, doesn't it make sense to create the least amount of friction loss possible when delivering it to a fire? Maybe the reason the chief thought their water system was so bad was that they had never taken the time or effort to utilize it properly.

Sure, if a dead-end hydrant will only flow 300 GPM at 30 PSI, it may not be worth much at a large fire, but if it must be used, why not make it as effective as possible by using a hose size which has almost no friction loss at that flow, rather than a 2½" smaller line that will deliver practically no water 250' away because of friction loss? Effective management of water supply is simply the practice of getting the most water into the pump with the least resistance possible.

Get Water Out of the Pump

Getting sufficient water into the pump is more difficult than getting water out of the pump. Once attack lines are laid from the pumper to the fire, it is a relatively simple matter to slowly open the proper discharge gate and throttle up to the desired pressure. The only difficult part about getting water out of the pump is determining the desired output pressure.

One simple way to determine output or pump pressures, especially with preconnected hoselines, is to use the slide-rule-type calculator or the printed charts that come with certain styles of nozzles. Simply determine the type of nozzle to be used, how much flow is desired, what size hose is being used to supply the nozzle, and how long the line will be. Match up a couple of columns and the engine pressure is quickly determined. It is then a simple matter of marking the individual gauges for each line with a strip of thin tape (automotive pin striping works well for this) to indicate the proper discharge pressure for each line (fig. 1–8).

Determining pressures for lines other than preconnects laid from bulk or reverse lay beds can become more difficult but will not be overwhelming if some time is spent practicing calculations on the training ground rather than in a classroom. While actually pumping water, the effects of various techniques, equipment, pressures, and flows can be immediately determined. It has been said that experience is the best teacher, and it makes sense that experience in the operation of new tactics and equipment should be gained on the training ground rather than at the fire scene.

We all know that one of the most accurate ways to calculate supply flows and discharge pressures is with the use of hydraulic calculations. Of course, this may not be practical on

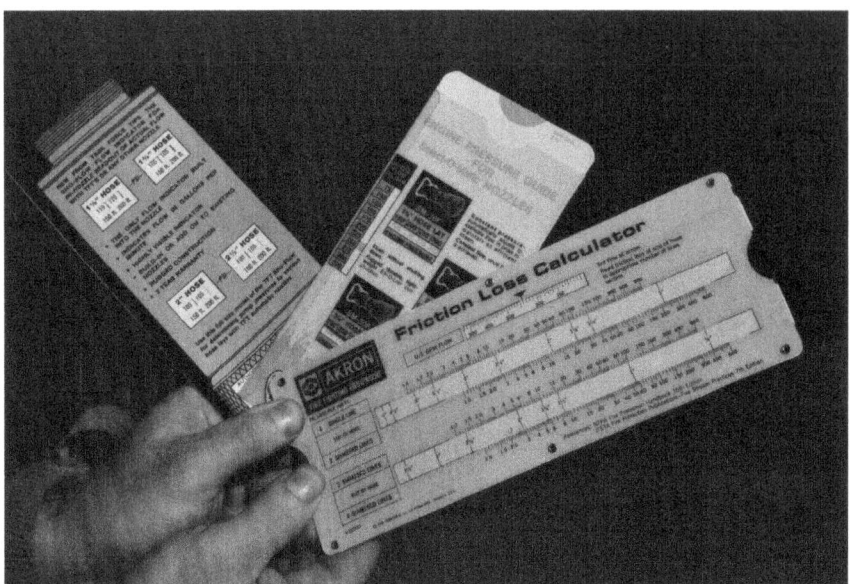

Figure 1–8. Slide-rule-type calculators can provide an easy way of estimating needed pump pressures.

the fire scene, especially when pump operators don't have time to get out a pencil and paper to compute mathematical problems or are not familiar with the actual methods of solving hydraulic problems.

Another method of determining pressures is to use handheld or cloud-based electronic calculators. Their operation is relatively simple, and their answers are adequate for almost all pumping operations. It will take a little practice to learn how to use the calculator rapidly, but it can certainly be mastered after one or two training sessions.

Keep in mind that calculations using charts or calculators can only provide a starting point. Differences in a pump's internal plumbing, variations in actual internal hose diameters, and conditions such as hose construction can all affect the exact amount of water flowing out a hoseline. The most accurate method of calculating needed pump pressures is the use of a calibrated flow meter and this is discussed in chapter 13.

Another consideration to assist in determining accurate fireground water flow calculations is the installation of flow meters, also called flow minders, in place of pressure gauges on the discharges of new pumpers (fig. 1–9).

With these meters, flow calculation is simple: just open the discharge valve, and with the water flowing, adjust the throttle or valve until the desired flow is achieved. The flow reads out directly on the gauge in GPM, making operations much easier since the mathematical conversion of pressure to volume does not have to be computed. Since it is gallons of water, not pressure, that puts out fires, it makes sense for the pump operator and nozzle crews to be thinking in terms of rate of water flow rather than pressure. Flow minders also eliminate problems with different hose sizes on the same line and problems of elevation. What is being indicated at the pumper is what is being delivered at the nozzle. The only drawback is that the pump operator must make sure that line is actually flowing before making the throttle adjustments. Gauges that show both the flow and pressure are now available. If the line is not flowing, the operator then throttles the discharge to the predetermined pressure.

Figure 1–9. Flow meters provide an accurate and simple device for indicating flow amounts and should be considered when specifying any new apparatus.

Factors Affecting Water Movement

If it is desired to move additional water through a single hoseline over a certain distance, it can be accomplished in two basic ways:

- Increase the pump pressure on the supply end of the line
- Increase the size of the line

If the hoseline is already laid and more water is needed, it is a relatively simple matter to increase the flow by increasing the pump pressure, although this method will have its limits.

For example, let's assume a 200', 2½" line is flowing water on a fire in a lumberyard. The line has a 1⅛" smoothbore nozzle operating at 45 PSI, so the flow is 252 GPM. If the officer decides more water is needed, they can radio the pump operator to increase the pressure. If the pressure is increased 20 PSI, the nozzle will be flowing around 290 GPM. Now, because the pressure on the line has increased, not all the extra pressure will be received by the nozzle. Some pressure, approximately 10 PSI in this case, will be lost due to increased friction loss caused by the increased amount of water movement through the hose. Within certain limits, and depending on the hose size and the existing amount of flow, increasing the water available at the nozzle by increasing the engine pressure is the simplest and easiest way of increasing the flow. It must always be kept in mind, however, that some nozzle types cannot safely or efficiently utilize the increase in pressure and flow. For example, handline smoothbore stream tips begin to degrade at nozzle pressures over 55 PSI and reaction forces increase exponentially.

The nozzle crew must be aware of the flow and pressure changes so they can prepare to handle the resulting increase in nozzle reaction. These items are discussed in detail in later chapters.

As the pressure and flow rise, so does the friction loss. At some point, depending on size and length of the hose, the friction loss will become excessive. This is indicated if an increase in pump pressure has little effect on the stream or flow. If the desire is to reduce pump pressure and reduce friction loss, it is necessary to utilize a larger hose size or multiple lines to provide for a more efficient operation. For example, the chart here shows the friction loss in PSI per 100' for a 200 GPM attack line flow.

Hose Size	Friction Loss Per 100' at 200 GPM and 100 PSI
1½"	100 PSI
1¾"	61 PSI
2"	32 PSI
2½"	8 PSI
3"	3 PSI

As can be seen, the larger the hose diameter, the more efficient water movement can become.

If it is desired to operate a 200 GPM stream at the end of 200' of hose, it is necessary to compute the required pump pressure by adding the friction loss to the nozzle operating pressure. If a combination spray nozzle is being used, and its normal operating pressure is 100 PSI, you will need to add this figure to the amount of friction loss in the line to determine what the pump pressures need to be.

Hose Size	Pump Pressure
1½"	300 PSI
1¾"	222 PSI
2"	164 PSI
2½"	116 PSI
3"	106 PSI

As can be seen, the smaller the hoseline, the higher the pump pressure that will be needed to supply the same flow. Although most modern fire pumps can easily supply 350 PSI, they supply that pressure at less than half of their rated capacity. Extended operation at these high pressures is inefficient and increases wear and tear on the pump and drivetrain components and endangers operating personnel. Pumping at pressures over 300 PSI is not recommended.

In addition, if more lines are stretched from the same pumper, managing multiple streams at high discharge pressures becomes difficult for the pump operator. Unless the relief valve or pump controller is properly set and operating, high discharge pressures can become dangerous if one line is suddenly shut off, sending a water surge to the other lines (fig. 1–10).

While 3" hose is more efficient at moving 200 GPM than the other lines (as can be seen in the previous chart), it becomes extremely heavy and not as easy to maneuver when filled with water, especially inside a building. Common sense indicates that either the 1¾" or 2"

Figure 1–10. While manufacturers rate some handline nozzles as flowing upwards to 300 GPM, a common problem on the fireground is believing 1½" and 1¾" hose are actually flowing these gallonages. Friction losses in preconnected piping, friction loss in hose, reductions in pump pressures, kinks in the line, and debris in the nozzle all work to reduce effective flow rates. These firefighters were driven from their interior positions because their 1¾" handline could not provide a flow rate adequate to overcome the heat being produced. It is best not to expect high flows from smaller lines, but rather to select a larger diameter attack line in the first place and then operate it at realistic pressures.

line offers the best maneuverability while delivering the required flow at a reasonable pump pressure. The 2½" line, while needing a low pump pressure to deliver the flow, is heavier and a little harder to maneuver than the smaller lines and, while it could be used to supply 200 GPM, the 3" hose is much more in its element at flows above 250 GPM.

If the original line has to be extended past 200', it will be more efficient to use 2¼" or 3" at the beginning of the line, close to the pump, rather than add on more 1¾'" or 2" line to the end of an existing line.

As a general rule of thumb, hose sizes up to and including 2½" are most useful for attack lines. The 3" and larger hoselines are most useful for supply lines. The average working flows that can be obtained from individual hose sizes at efficient pump pressures, are as follows:

Hose Size	GPM
1"	40
1½"	150
1¾"	185
2"	250
2½"	350

Some nozzle manufacturers advertise that their nozzles can obtain high flows from small lines. While they are essentially correct, to obtain these flows, it is necessary to overcome excessive friction loss generated within the hoseline by increasing the pump pressure. The design of a nozzle can influence optimal nozzle operating pressure, but it cannot reduce friction loss incurred by moving a certain flow through a certain size hoseline. If flows higher than what is shown on the chart are desired from a certain size line and nozzle combination, higher pump pressures must be used.

Simplifying Water Delivery Calculations

If the height of the fire is above the water supply, additional pressure will have to be supplied to overcome the effects of head loss or back pressure. Conversely, if the fire is located below the water supply, compensation must be allowed for the increase in pressure caused by gravity. The rule of thumb is to increase the engine pressure 5 PSI per story or per 10' of height to compensate for elevation. The same figure can be used to reduce pressure on downhill lays, but it is extremely difficult to estimate the amount of depression. Communication between the supply and delivery ends of the downhill lay is the best practical way to determine working pressures.

To further simplify fireground hydraulic computations, many departments have established estimated guidelines for initial pump pressures in various situations. While these estimations will not be as accurate as actual before-scene calculations, they can serve as quick rules of thumb when charging lines. After the normal attack line flows are determined, pressures can be adjusted upward or downward as needed. When using any method, the effects of pump operation must be constantly evaluated and adjustments immediately made to correct deficiencies. If too much pressure is being supplied to a nozzle, the stream will appear ragged-looking and high nozzle reaction force will make the line hard to control.

If too little pressure is being supplied to a nozzle, the stream will appear weak, have little reach, and the flow may be far below what is required to extinguish the fire. The pump operator, as well as firefighters operating the line, must be alert to these visual indicators and make adjustments in pump pressures when needed to provide safe and efficient operation. If streams are weak and the compound or inlet gauge shows a reading of around 10 PSI, steps must be taken to increase the amount of water coming into the pumper. If this is not possible, you will have to maintain or decrease the amount of water flowing out of the pumper.

Some commonly used rules of thumb are as follows:

Handline Nozzles: Smoothbore nozzles should operate at 50 PSI nozzle pressure and combination nozzles at their marked rating from 50 PSI to 100 PSI.

Heavy Stream Devices: Deluge guns equipped with smoothbore tips should operate at 80 PSI, and those equipped with combination nozzles should operate at 100 PSI if rated maximum flow is the goal. About 35 PSI should be allowed for friction loss within the device if older, split waterway guns are used. If modern single waterway, bent loop devices are used, allow 10 PSI for friction loss.

Hoselines: A good way of estimating the starting point flow from any size handline up to 2½" is to allow about 30 PSI per 100' length for friction loss. The nozzle operating pressure must be added to this figure to obtain proper engine pressure. This is just an average estimate, but it will be close enough for initial evaluations. However, the figure will not be as efficient as figuring out exact friction loss using a calibrated flow meter.

Standpipes: When pumping into a dry or non-building fire pumped supplied standpipe, 150 PSI can be used as a starting pressure. If the location of the fire is known, 5 PSI per floor should be added. Since there is a strong possibility that someone is using a house hoseline off the system on the fire, the standpipe Siamese should be charged immediately. Water pressure conditions on the fire floor should be reported as soon as possible so supply pressure can be increased or decreased as needed. Wet standpipes with building fire pumped supplied systems need to be supplied at designed input pressure by pumping apparatus in cases of a building fire pump failure or if the fire department desires to take over the system.

Sprinklers: Lines should be connected to the fire department Siamese whenever it is suspected that a sprinkler system has activated. Most departments do not charge the lines unless smoke or fire is present. The charge pressure should be no less than 150 PSI and no greater than 200 PSI, unless noted on the fire department connection.

Relaying to Another Pumper: When relaying water to another pumper, begin with pump pressures to supply 40 to 50 PSI at the receiving engine when first charging the lines. Voice communication between relaying units is essential to avoid over or under-pressurizing lines. If there is not enough pressure provided on the receiving end, the delivery volume will suffer. If too much pressure is received, energy will be wasted by the supply pumper and the danger of burst hoselines or destructive water hammer increases. If 4" or 5" large-diameter lines are being used, the initial charge pressure should not exceed 100 PSI. If a relay hydrant valve is used, about 30 PSI should be figured for friction loss within the device.

Reverse Lays: Reverse lays are usually supplied with a certain amount of working line and a nozzle connected to the end of a supply line. The amount of pressure needed to properly operate the attack hose should be predetermined. Then, depending on the flow, it is then necessary only to add about 25 PSI to 35 PSI per 100' of supply hose to determine the needed engine pressure.

Keep in mind that these are only rule-of-thumb suggestions to get started. Actual working pressures need to be determined by taking into consideration flow rate, hose size, and amount of hose in the stretch.

Supplying the proper pressures using some general rules of thumb is not as intimidating as attempting to calculate figures from a formula can be. Understanding the basic principles is the first step to efficient and safe fire stream operation. The second step is to continually evaluate pumping performance and make adjustments as necessary. A fire is considered a dynamic force because it is always changing. Likewise, pumping should be considered a dynamic operation because it is always changing, just like the fire it is trying to extinguish.

Review

Most problems with poor streams on the fireground can be traced to improper management of the water supply. Once the officers and pump operators understand the four basic methods of increasing supply flow, the effectiveness of their fire streams will increase. In this chapter, the main discussion has been centered on solving supply problems. In the next chapters, operational requirements to increase the effectiveness of nozzle operation and water application will help in solving problems on the discharge side of the pumper.

When operating a pumper, nothing beats using common sense. If there isn't enough water coming in, there is no way to make more come out unless the supply is increased. Pump operation can be as difficult or as simple as the operator wants to make it. In all cases, efficient pump operation is simply a matter of remembering first to get water into the pump and second to get water out of the pump.

Chapter 1 Review Questions

1. How does one create stream reach?
2. Name two factors that affect friction loss in a hose line.
3. Name two ways that an engine can obtain water for operation.
4. Describe a reverse lay.
5. How many 4" lines are needed to equal the flow of a single 5" line?
6. Name the three basic steps that need to be completed for a fire pump to operate.

Strategic Considerations

How many times have you stretched a preconnected attack line into a structure only to find out it is not long enough to reach the fire? How many times have you started to attack a fire inside a building only to have the crew driven outside because they were not flowing enough water to overcome the heat being produced? How many times have precious minutes been lost as engine companies tried to figure out how they were going to stretch and connect heavy caliber lines to cover threatened exposures? How many times have attack plans worked well yet been abruptly stopped because a tactical water supply operation could not sustain the needed rate of flow?

Any attempt to operate streams on a fire should begin with a strategic water management plan, developed to quickly and effectively utilize water on the fireground. This plan can help a department determine flow rates, size and lengths of lines, and type of stream shaping devices, based solidly on actual evaluations of life hazard, building construction, building contents, and water supply within the fire department's own response area. By developing a strategic plan beforehand, a department can help ensure that supply and attack lines of the proper length and flow will be quickly available for tactical deployment by the incident commander.

Many fire officers feel that tactics and water flow can only be determined by experience. That is not necessarily true. Actual fireground water management experience certainly contributes helpful information when formulating attack plans, but it should not be considered the best or only yardstick when measuring a department's capabilities. There are many smaller departments that experience few fires yet have implemented excellent strategic water flow and supply plans.

By keeping in mind some clear-cut goals, performing an accurate evaluation of hazards, and then forming a set of plans based on facts, not feelings, any fire protection organization can develop effective and safe attack and supply strategies.

It may be that other factors, such as improved apparatus, hose, and hardware, may be needed for a department to more efficiently apply water to a fire, including the way all that hardware is laid out on the apparatus. A strategic evaluation and planning process can determine what equipment is required and where it should be placed. Examples could include having a pre-attached single inlet master stream device to some 3" static hose, having preconnected lines of various lengths in the same size, and considering bulk small handline beds in jurisdictions that have the potential for lengthy and complex handline stretches. Often significant fireground improvements can be accomplished for little to no additional funds, just the required knowledge, understanding, training, and department effort.

For example, I have seen many instances where a department purchased a new tanker or tender without evaluating its effect on operations. In most cases, the largest tank possible is purchased, and then it is found that the 3,000-gallon monster cannot be easily maneuvered on the roads and driveways of the district. The end result is that 3,000 gallons of water arrive on the scene, but because the tanker cannot get close enough to the building to deliver its water directly, some type of time-consuming supply pumping plan has to be put to use.

Other departments have large tankers but utilize portable holding tanks of less capacity than the tanker volume. At the fire scene, the large tanker dumps only a portion of its load and has to wait until the holding tank is almost empty before completing its off-loading.

The simple answer is to purchase a larger holding tank or a second tank which, in combination with the first, is capable of holding at least all the water the tanker carries. By planning first and purchasing second, not only will the budget dollar go farther, but fireground efficiency will be improved.

Setting Goals

The first step in developing a strategic water flow plan is to determine your goals. What do you actually want to accomplish with the water? Answering the following questions will assist in determining your department's firefighting goals.

- What does your department expect to accomplish in the following situations?
  - One-room dwelling fire
  - Multiple room dwelling fire
  - Heavily involved apartment fire
  - Heavily involved commercial occupancy fire
  - Heavily involved warehouse or barn
  - Heavily involved, high life risk occupancy, such as a nursing home or hospital
- What equipment and staffing do you have available to fight each of these fires?
- How much water is available to fight these fires and how will you get it to the scene?
- How many personnel are available to fight each of these fires?

To answer the first question with "put the fire out" is not acceptable from an evaluation standpoint. You must be realistic. For example, in a multiple room dwelling fire, consideration must be given to average room size and type of occupancy, as well as average building height within the department's response district. A bedroom in Beverly Hills would be quite a bit different in size and contents from one in the South Bronx.

When answering the questions, you should list desired goals, such as when and how you plan to stop the fire's spread, perform searches, perform ventilation, and effect final extinguishment.

The first question in the list is directed at evaluation of your district, the last three at determining available firefighting resources. The answers to the first question could possibly dictate changes in the others.

In helping a number of departments write specifications for fire apparatus, I find committee members more often than not arriving at the first meeting with long lists of desired accessories, trim, and paint colors. I've been immediately bombarded with questions about makes and models and estimated delivery times. All of these questions are unimportant until the department answers the first and most basic evaluation question: What do you want the apparatus to do?

"Well, we want a rescue-pumper," someone answers. All that can be assumed from that response is that they want a truck with compartments and pump installed. The unit's exact mission must still be determined before you can begin writing the specifications.

Will the engine always be responding as first due? Will it ever be a water supply company? What extra tools and supplies—medical equipment, hazmat tools, rapid intervention team equipment, and extrication devices—will be carried? What about heavy stream devices? What about foam capabilities? What does its response district look like? Are there steep grades, narrow streets, closely spaced frame apartments, confining alleys, widely spaced hydrants, industrial occupancies, chemical storage, hospitals, high-rise buildings, waterways, interstate highways, or rutted roads? Have your members measured and weighed the equipment you now carry and sized the new equipment you intend to install?

In short, before you purchase equipment or plan tactics, you must thoroughly define and understand their intended use.

For example, purchasing an in-stock pumper is relatively easy. Designing the proper pumper to fit an individual department's exact and specific mission is a bit more difficult. When it comes to stream management, pulling off a preconnect is easy. Making sure it will do the job depends on how much time and effort the department has invested in research, equipment, and training.

As much as we hate to think about it, the mission to be accomplished must be balanced with budget availability. You must be realistic when setting firefighting goals. Some of the nation's largest fire departments, with hundreds of firefighters on duty each shift and extensive numbers of apparatus and other resources, occasionally face fires in which they have no choice but to let it burn itself down into a size that the fire department can manage. Officials in those cities realize there will always be the potential, within their jurisdiction, for a fire they cannot readily extinguish.

A department has to take into consideration what an expenditure of money and time will give the community in return. For example, if a large, well-run company has located a large plant or warehouse in your district, and the contributions to local employment and tax revenues are high, it may make sense to upgrade municipal fire protection resources to help mitigate a potential economic disaster for the community if the company suffers a large-loss fire.

In an opposite example, suppose a chemical blender moves into an old warehouse on the outskirts of town. The building was not designed for chemical machinery and is located at the end of the water supply system. The company is careless with its mixing operation, causing a high fire potential, and is not cooperative with inspectors. In short, it is operating on the cheap.

The potential for fire in the second example is higher than in the first example, but the economic impact to the community in case of a fire loss is much less. Spending large amounts of community funds to extend the water system and equip the fire department for a fire in this specific property may not be cost-effective to citizens paying the taxes. It may be advisable to have the second company provide increased protection for itself in the way of automatic extinguishing systems or sprinklers. If the company refuses the department's demands, it may be necessary to pursue the issue through legal channels.

Legal Challenges

Our society has become more aware of accountability in the last 20 years. The consumer movement is demanding that the public get its money's worth, and if not, is pursuing remedies through the court system. For example, many thousands of automobiles are being recalled each year as manufacturers are being held accountable for defective parts and designs. Since Watergate, more and more public officials are being taken to court and held legally accountable for deeds ranging from misuse of tax revenues to hiring practices.

Insurance companies, after paying out large sums of money to cover fire losses, are attempting to recoup their loss by suing others. The New Orleans Fire Department was forced to battle a $75 million lawsuit in which insurance companies charged that the department used improper tactics to fight a fire in a large manufacturing plant in the early 1980s. The 10-week trial ended in the department being found not guilty; however, the cost to the city and the department to defend themselves was calculated at over $1 million. The New Orleans Fire Department was fortunate: over the years they have had an excellent fire suppression loss record, keeping meticulous records and conducting thorough investigations. But how many other departments could cover the expenses and time required to fight 10 years' worth of litigation?

It might not seem fair, but consumers, expecting the most for their money, are now demanding the best fire protection at a time when the general trend in the fire service is to reduce spending and personnel. Excuses such as, "we needed another pumper," or, "we didn't have enough water" are no longer valid. Sharp courtroom attorneys are going to ask why. Why didn't you make a case to purchase a new pumper before the city council? Why didn't the department plan ahead to provide for an adequate water supply? The million-and-a-half-gallon pond is only a quarter mile from the barn. How could you not have enough water?

A small number of fire chiefs and other experts who are paid to testify at trials are cashing in on this trend. Imagine facing the chief of a large department who describes their well-financed operations to a jury of civilians and then paints a picture of incompetence when describing the tactics used by the department being sued. There are probably more lawyers in this country than there are firefighters. They run law businesses. In marketing their services, the legal profession is leaving no stone unturned to help assure their financial and professional success. This quest for financial and professional rewards means that none will stop at suing a fire department, fire chief, or individual officers or firefighters if they think they can score a huge reward.

Fire departments have a legal and moral responsibility to provide the best possible life safety and fire protection. Public consumer awareness is beginning to force legal action to enforce this responsibility. Looking for a way to spread the costs of fire loss payouts, insurance companies are finding departments without proper operational planning, apathetic officials, and substandard tactics as ripe targets for a lawsuit. Documented strategic planning is one way of reducing the department's risk of being accused of irresponsibility.

Insurance companies have been rating fire departments for years, telling departments how to protect the insurance companies' investments at the fire departments' (taxpayers') expense. Maybe fire departments should be telling businesses how to protect themselves at the businesses' own expense or expect a large loss if an uncontrolled fire overwhelms the responding fire forces. Fire suppression operations must be run as business, and just as ways must be found to limit fire losses, methods to limit liability in case of loss must be instituted as well. This is where proper planning is needed.

Extinguishing More Fire with Fewer Personnel

In modern fire stream management theory, the horse-and-buggy ideas of "1½" on dwellings, 2½" on everything else" are being replaced by equipment derived from modern technology. One of the reasons for the rapid advance in the use of high-tech hardware is that the fire service is slowly losing personnel each year, either through renegotiated labor agreements or through loss of volunteer personnel bases within communities. While most career departments are now having difficulty finding qualified firefighter candidates, the other problem is not having enough employment slots in which to put them.

The fire service overall is having to put out more fires with fewer people than ever before, and the prospect of relieving that problem in these days of budget cuts and lack of community participation seems near impossible. While officials are wrestling with the extremely difficult task of personnel replacement, new hardware, new technology, and updated training must fill the gap by allowing each firefighter to become more productive on the fireground.

By using a strategic fire stream management plan as a starting point, a department can more easily evaluate the advanced hardware now available and can integrate new technology into its operation more quickly.

A word of caution here. Throughout this book, I discuss operations of hoselines with a minimum of personnel. I deplore the operation of engine companies using a minimum of three people. I think it is a shame that firefighting companies throughout the country are forced to operate with fewer personnel than the same communities use on each trash collection or street patching vehicle. Minimum staffing to safely accomplish basic firefighting goals in attack operations requires no less than five persons on engine companies and six or more on truck and rescue companies.

I am also enough of a realist to know that most fire apparatus in this country respond to alarms with fewer personnel than they should. Many of the tactics discussed in these pages describe interior work with a single firefighter operating a hoseline along with a supervisor. In no case should a single firefighter work alone inside a fire building. It is assumed that when one firefighter is operating the line, an officer is alongside to evaluate fire conditions, perform limited truck work as needed, report conditions to the incident commander, and coordinate the work of additional personnel.

As more firefighters are used to advance a line and attack a fire, they will make the effort less stressful and much safer. Describing two-person attack line operation should not be taken out of context and be treated as the norm rather than the exception. The following descriptions are given in the interests of firefighter survival and are certainly not intended to advocate any type of staffing reduction on fire attack units.

Response District Evaluation

To obtain an accurate picture of hazards when evaluating a department's response district, care should be taken when gathering data. A good foundation for formulating a strategic water management plan must contain a comprehensive tactical survey of the department's district, taking into account

- Life safety and evacuation protection considerations
- Building construction prevalent within the response district
- Evaluation of typical building contents and amount of fire loading from contents
- Distance of hazard from apparatus position, including side and rear entries and elevation from ground or apparatus level
- Exposure hazards and where heavy streams should be placed to protect them
- Available water supply, including hydrant or drafting locations
- Sprinklers, standpipes, and other built-in extinguishing systems
- Review of International Organization for Standardization community survey

A fairly comprehensive picture of the potential fire problems within a department's protection area can be developed by using these criteria to evaluate a number of locations and then averaging the collected data (fig. 2–1).

Hazard base information can then be used to determine how, where, and how much water needs to be delivered to a target hazard (figs. 2–2a through 2–5b).

A strategic water management plan is usually based on an average of target hazards with a given response district, whereas a standard preplan is usually used to develop tactics for only a single hazard. The strategic plan can then be used to validate a standard operating procedure for water application. Where a strategic plan is in effect, commanding officers on the fireground should only have to concern themselves with tactically applying the water, not whether the line will reach the fire or whether the flow will be sufficient. The fireground, with its usual noise and controlled confusion, is not the best place to be calculating needed flow rates or attempting to determine what resources are available. Of course, there will be those cases on the fireground in which the strategic plan might have to be modified to cover such unforeseen circumstances as blocked entrances, fire volume dictating the need for increased water supply, or extending lines to reach a distant hazard.

Figure 2–1. Individual apartments in this four-story frame apartment building are rather small in cubic-foot volume; however, planning must take into consideration the total amount of involvement if a door is left open and the fire extends into the common hallway. Certainly, life hazard will be of prime concern and overhead wires will limit deployment of aerial devices. Aggressive attack with high volume handlines will help limit fire spread from its original location, and additional lines must be placed on each floor to limit upward extension through interior walls and openings.

Figure 2–2a. A fire in this garage should present little problem and waterflow requirements are well within the capabilities of being supplied by a single company, using 150 gallons per minute (GPM) hand lines, 250 GPM handlines, or master streams, depending on amount of involvement.

The strategic water flow plan should not be considered inflexible or all-encompassing, however. The idea is to plan a baseline fire flow for effectively handling what you consider normal target hazards but designed to be quickly and easily multiplied. The plan must have built-in expansion for handling potential fires which require flows greater than the baseline flow. The commander on the fire scene can then easily add or subtract from this plan to fit the situation at hand.

Figure 2–2b. This view of the garage from the rear, however, indicates planning is needed for extra companies to protect fuel oil storage tanks in case of fire in the garage. If the garage is well involved, large flows will be needed for exposure protection, far more than what will be needed for simply extinguishing a fire in the structure.

Figure 2–3a. Originally designed as a duplex, this Victorian is now divided into at least eight apartments located on three floors. Many times, modifications are of shabby workmanship and turn the inside of the building into a maze, making it difficult to move lines in for attack and perform primary searches.

Figure 2–3b. Balloon-frame construction practically guarantees any fire on the lower floors will rapidly reach the attic. Measuring indicated at least 300' of line will be needed to reach the upper floor via the rear stairs.

Figure 2–4a. One eastern fire department has this hazard they have nicknamed the "barracks." Officers of the company, normally first-due, have wisely increased the baseline flow rate over that required for other dwellings within their district. If on arrival fire is showing from more than two windows, tactics call for the stretching of a 2" handline to handle increased potential fire volume.

Figure 2–4b. The view from the front makes it seem that the building offers no particular problems. The view from the rear presents problems in reaching basement apartments from the street and the need for ladders in the 40' to 50' range to remove trapped occupants from the third floors.

Figure 2–5a. The size of the structure as well as the setback from the street dictates the need for long hose lengths to reach far areas. While fires are rare in well-maintained houses of this type, it is important to include these occupancies when gathering strategic plan data.

Figure 2–5b. This large frame home appears to offer few problems to attack crews. Its separation from other structures precludes the possibility of communication and a hydrant is located at the corner of its lot. The natural tendency would be for the attack pumper to park near the hydrant for ease in obtaining a secure water supply. From that position, it is 90' to the front door and 150' to the rear door. The house measures 50' × 70'. It can be seen that 150' preconnects would be of little use, especially if the fire was on the second floor or attic. It was found that the basement entrance was located 180' from the pumper at the hydrant, so if the fire was located in the front corner of the cellar, a 300' line would be needed to reach the blaze. Size of interior rooms and distances dictate a 300' line with a 150 GPM flow be stretched for initial attack operations.

After a determination is made on the mission of a company or department, and once the district is evaluated, you should compile a pattern of baseline flows and application tactics for the hazards. To help evaluate these flows and application tactics, the following questions can serve as a guide:

- How much flow do you want to apply?
- Where do you want to apply it?
- How will the flow be distributed?
- How many personnel will be needed for application?
- How will you sustain the supply?

How Much Flow Do You Apply?

The desired flow a department wants to be capable of applying is determined by establishing a baseline flow. *Baseline flow* is a calculation of the water needed to handle an average fire within the department's response area or within certain groups of structures. Equipment should be available to not only flow the baseline rate but should also allow for the rate to be easily multiplied for larger fires.

The late Larry Davis, an expert in rural and commercial fire protection and an author of several books on rural firefighting, developed a time/severity chart which illustrates the effects of flow rates. It can be seen that once a fire starts, it will generally become more severe, reach a peak of severity, and then decline and eventually extinguish itself when no fuel is left to burn. How long this takes depends on quantity and type of fuel involved.

If an attempt is made to extinguish the fire, a flow rate must be used to apply water in sufficient quantity to overcome the heat being produced. If the applied flow rate is less than what is required, the fire will continue to burn until it consumes enough fuel to reduce the heat being produced. At a certain point, the flow rate will overcome the reduced heat output and the fire will be extinguished. In the meantime, all the water being applied to the fire will be wasted. Some of the applied water may have been useful in preventing the fire's spread, but it will have had little effect on total extinguishment.

The ideal situation is, of course, to apply the water to the fire in sufficient quantities to overcome heat production. It has been recognized in recent years that fire will be extinguished more rapidly if water is applied in quantities much higher than what is needed to overcome heat production. When planning, remember that it is certainly much more efficient to have excess capacity available and not have to use it than to not have enough water and watch as fire defies all efforts and destroys the building. There is nothing wrong with providing too much flow. Belting a fire with a higher flow than is required causes the fire to be extinguished much faster. Just be sure to shut the line down as soon as the fire blacks out to prevent excessive water damage.

Personnel Requirements

One of the first objectives to consider when planning flow requirements is determining the maximum flow that can be applied by the personnel normally available on the scene. With today's modern hose and hardware, this flow can be easily increased over what was common years ago. For example, if three firefighters arrive on the scene of a fire, they have a number of water application options.

With one person operating the pump, the traditional option would be to have the other two personnel stretch a 30 GPM booster line or 95 GPM 1½" line to the fire. Unless the fire is extremely small, it is doubtful the booster line will have enough flow for extinguishment. If it is stretched and proves ineffective, every bit of energy and water used up to that point will have been wasted. If the 1½" line is stretched, it may have a better-than-average chance of providing the necessary flow, depending on the occupancy. The flow may be adequate if the fire is in a bedroom of a private dwelling, but if the fire is in the rear rooms of a food store, it most certainly will not be. Figures 2-6a and 2-6b show the life cycle of a fire during different suppression situations.

By changing the attack line size to 1¾" or 2" and by providing a nozzle that can effectively utilize the higher flow rate, much more fire-killing power will be available while needing nominal additional effort to stretch than does the smaller lines. Stretching a high-flow line also allows for excess capacity to be instantly available for unforeseen circumstances, such as discovery of additional fire in concealed spaces or rupture of flammable liquid storage containers. The idea is not to have to run outside and stretch another line if a higher flow is needed (fig. 2–7).

For instance, let's assume it is necessary to crawl down a long, hot hallway to reach a fire in a back room of a restaurant. It is extremely important that the line the firefighters have with them provides enough flow to handle the fire they will be facing. Once they make it to the doorway and prepare to attack the fire, they will have only a few seconds to evaluate the situation, open the nozzle, and apply the water. If their actual waterflow rate is equal to the

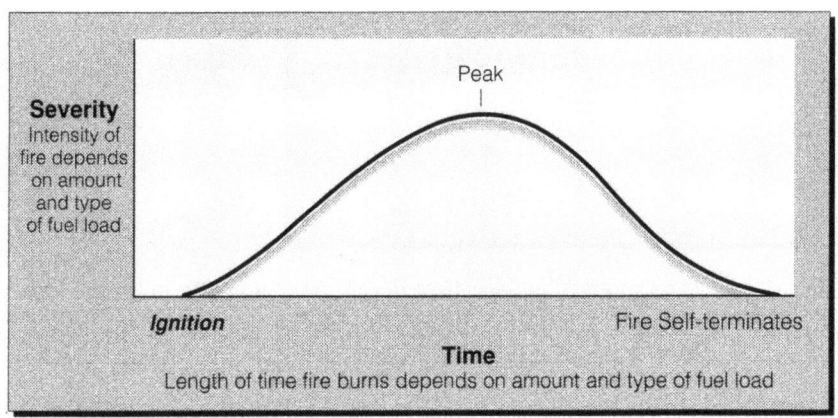

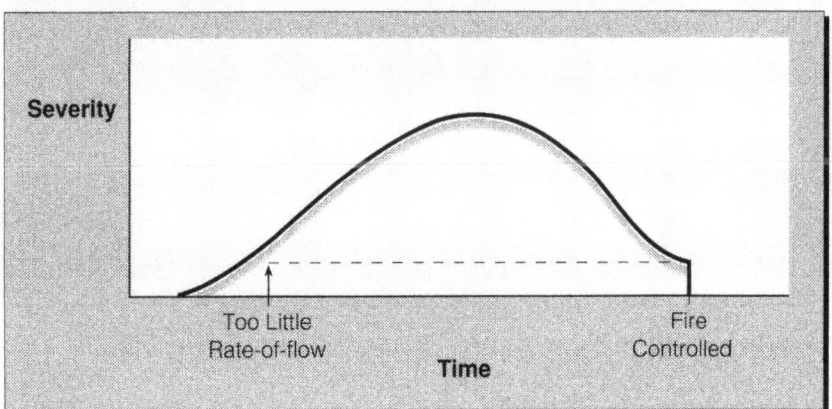

Figure 2–6a. Illustration #1 shows the life cycle of a fire when no suppression steps are taken. All fires will self-terminate when they run out of fuel or oxygen. Illustration #2 shows what happens when a department attempts to control a fire by using too small of a line. The fire will continue to burn until it reduces its own fuel load. At a certain point, the department will be applying enough flow to overcome the heat being produced by the reduced fuel load. Other than confinement, all suppression operations up to this point have been wasted effort. (Images courtesy of Larry Davis)

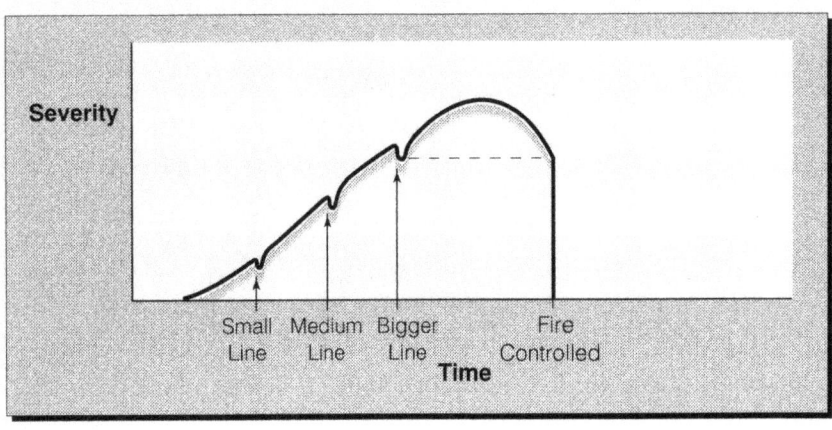

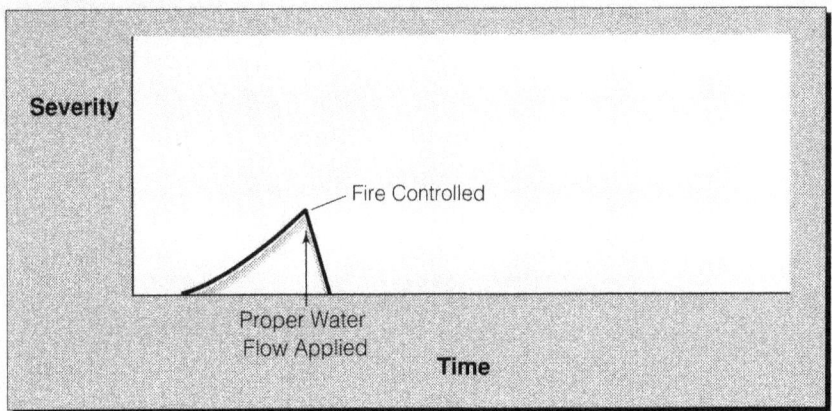

Figure 2–6b. Illustration #3 shows the results of applying too low a flow, then attempting to increase the flow by pulling the next-sized line in an attempt to control an increasing amount of heat output. This usually results in the fire continuing to escalate in size as the increased amount of flow arrives too late to overcome the increased heat production. At some point, the fire will reduce its fuel load by combustion, allowing the fire to be controlled during its decline. Illustration #4 shows proper application of fire flow. Ideally, the department will apply enough flow to stop the burning process before the damage curve peaks. The more flow that is applied sooner, the quicker the burning process stops, reducing the fire's loss. By rapid intervention, quick suppression of the burning process increases the chances for reducing injury and death rates for both occupants and firefighters. (Images courtesy of Larry Davis)

required rate, the fire will be extinguished. If their actual waterflow rate is greater than the required rate, the fire will be extinguished more quickly.

On the other hand, if their actual waterflow rate is less than required, the fire will not be extinguished, and it will become a matter of how long the crew can take punishment from heat and combustion products before they are driven out of position. It is important that the crew carry as much flow rate with them as possible.

It's impractical, based on wanting a high flow rate alone, to decide to carry a 2½" line inside for every interior fire. Balanced against the flow rate are factors of maneuverability, number of personnel needed to move the line, and amount of nozzle reaction the personnel can safely handle. It is wise to evaluate potential fire situations beforehand to determine the probable flows that will be required for extinguishment. Remember, there is no all-purpose

Figure 2–7. The relative sizes of 1½", 1¾", and 2" attack line hose. When equipped with 1½" thread couplings, the larger sizes are often easy to deploy dry as 1½" but offer much more firepower through increased flow rates. All utilize 1½" thread couplings.

firefighting handline. Some are easier to use than others, but good attack practice dictates that a number of lines of various flows and lengths should be available. The selection of which line to deploy should be made on the scene based on fire volume, handling ability, expected fuel package, and size of compartment.

Establishing Flow Rates

To review, a baseline flow is that which can be deployed by the apparatus and personnel available on initial response balanced against the rate of flow needed to combat anticipated fires within the department's response district.

If a department determines that 150 GPM is the baseline rate of flow within its district, based on size and type of occupancies, their construction, and the amount of personnel available to operate the line, this flow can be easily supplied by one 1¾" line preconnected to a single pumper. That pumper should also have additional 150 GPM lines to quickly deploy to areas of possible fire spread, and at least one 250 GPM line for heavy flow on a large fire.

Large supply hose and heavy stream devices should be available so that the single pumper can quickly flow at least 750 GPM on an exceptionally large fire. In addition, the outgoing flow must be backed up by a tactical water supply plan.

A good starting point for flow determination is a review of fire reports for the past 3 or 4 years. Information on amount of attack lines in use, number of personnel, use of hand tools and ladders, amount and size of supply hose or tankers, and amount of fire loss can help determine how much water was needed at each fire along with the average number of firefighters available for its application. This information is composed of facts. Since the fire has already occurred and the amount of water used, how it was applied, and its success or failure is known, the information can provide an invaluable starting point.

After compiling facts, consider probabilities next. A number of different buildings within the department's response district should be evaluated. The more buildings and the greater their diversity, the more accurate the results. The structures should consist of a number of dwellings, both single-family and multiple-family, as well as a number of commercial and industrial occupancies. By measuring or estimating each building's size and calculating flow using the following formulas, a plan of flow probability will develop indicating the estimated amount of water flow needed to stop a fire successfully in each type of occupancy.

Fire Flow Formulas

The first attempt at determining how much water is needed to extinguish a given amount of fire was published in 1959 after much research (including almost 10 years of controlled experiments and large-scale fire testing). Keith Royer and Bill Nelson of Iowa State University developed and published a theory called "the ideal rate of flow." The *Iowa formula* for computing the flow called for in the theory is expressed as

$$V \div 100 = GPM$$

V is the volume of the largest room or area within a structure expressed in cubic feet. The volume of a room can be easily figured by multiplying length by width by height. 100 is a constant factor that divides the cubic footage to give you the minimum GPM needed for extinguishment.

For example, a common bedroom measures 14' by 12' with an 8' ceiling. The room's dimensions multiplied together (14' × 12' × 8') come to 1,344 cu. ft. This cubic footage figure divided by 100 gives a fire flow for that room of 13.44 GPM.

When calculating the formula, researchers found the amount of heat released by burning solid materials was limited by the amount of oxygen in the air. Because of this limiting effect, the fire itself will not obtain high-heat release until it becomes large in size and creates a great draft, increasing the amount of oxygen available. If the fire is venting out the roof, it can be assumed that the added draft caused by the upward movement of combustion products will cause more rapid burning. This increase in heat release must be taken into account if the formula is used at the fire scene. Additional air movement caused by improper use of fog streams and positive pressure ventilation fans can also accelerate burning, releasing more heat than if the fire were left to itself.

The formula is accurate to compute the water necessary for extinguishment of fires within enclosed buildings when natural solid materials are involved. When computed for use in small areas, the indicated flow seems extremely low. As the area increases, however, the flow seems to appear more reasonable.

In the late 1950s, the common fire load on which the Iowa formula is based, consisted of wood and natural material furniture and limited amount of fire loading. These days, taking into consideration interior finishes and furniture being almost 100% synthetic, and the heavy amount of fire loading, the National Fire Academy (NFA), developed a new fire flow formula expressed as follows:

$$L \times W \div 3 = GPM$$

For example, a commercial occupancy is 50' by 75' and is one story tall. Multiplying length (L) by width (W) (50' × 75') gives us the area of the structure at 3,750 sq. ft. Using this number and the NFA formula, the required fire flow for this structure would be 1,250 GPM (3,750 ÷ 3 = 1,250).

Either method will work for estimating fire flow; however, the Iowa formula seems to calculate lower flows than what common sense would determine necessary. On the actual fireground, it is best to go in with the highest flow the crew can safely handle.

The low flow rates for smaller areas were at one time correct, although contributing factors (such as connecting rooms) can make them unrealistic yardsticks.

For the computed fire flow to work properly, water must be applied at or above the critical computed rate, and it must be distributed properly, penetrating the fire environment to reach the burning material (fig. 2–8). The theories behind distribution tactics are explained in chapter 3.

Remember, too, that the cubic footage of adjoining areas that may be connected to the fire area through open windows or doors, and the possibility of spread to the floor above must also be added to the total area when computing the flow.

The formulas' consideration of the amount of fire heat is based on the amount of fire intensity reached 10 minutes after ignition. Remember, when using the formula for planning, if delays in receiving alarms can be anticipated—for example, from a factory district which has little traffic on nights and weekends—the alarm delay will result in longer pre-burn times and the flow called for by the formula will have to be adjusted upward to compensate for handling the extra heat.

Figure 2–8. Flow is sufficient here, but it will never extinguish the fire because it is not being applied directly to the involved area. For the fire flow formula to be meaningful, it is assumed that the required rate of flow be applied directly to the base of the fire.

To extinguish fires, the critical flow rate must be sufficient to handle both the heat generated by the fire and residual heat present in the building when attack is attempted. Take, for example, fires originating on the lower floors of garden apartment buildings which are constructed of poured concrete or concrete blocks. If fire has been burning for a considerable period of time before attack is attempted, the walls act as enormous heat sinks, absorbing a considerable amount of the fire's heat, only to release it when hit by hose streams. The amount of heat that has to be dealt with in this situation can in some cases be greater than the heat being produced by the fire itself (fig. 2–9).

Applying the required flow to the fire must be considered. Computing the required flow and having the hardware to deliver the flow is a waste of effort if the flow cannot reach the fire. If, for example, the fire is in an interior space that cannot be reached through an outside opening, and the department does not make entry to apply the flow to the seat of the fire, all flow calculations and equipment preparation are rendered useless. The formula, adjusted as needed to compensate for additional water flow required in certain high-heat situations (such as fires confined within concrete or block walls, or for fires of greater intensity when synthetics are involved), can give planners a useful yardstick to help compute baseline fire flows.

Individual Line Flows

After a baseline total flow is established, it will be necessary to determine a method of delivering this flow. If only three firefighters are responding on the first-arriving piece of equipment, they will normally be capable of putting only a single line into service. This line, by necessity, should be easily stretched and should flow as much water as possible, taking into consideration how much nozzle reaction can be handled by one person with the other shoving the hose.

Planning also must consider applying flows for extinguishing average fires as well as additional flows that might be needed to head off a fire's potential spread. For example, a large couch ignited by a carelessly discarded cigarette is a simple fire to extinguish if the couch is located in the middle of a vacant lot. The amount of fire present is easily handled with a water extinguisher or a low flow from a booster or 1½" line using just a few gallons of water. If that same blazing couch is located in the living room of a single-family dwelling, however, the same amount of computed flow will be required for extinguishment, but the fire's heat being contained by the building's structure and the potential for fire spread has to be anticipated when stretching a line. In addition to the spread potential, visibility may be restricted by thick smoke from the burning cushions and the heat containment may make it difficult for firefighters to get into position to make an attack. Since the fire is within a structure, the attack crew must also be concerned with life safety and a primary search should be started as soon as possible (fig. 2–10).

Locate the same blazing couch in the basement storage area of an apartment building, and the potential for fire spread will be dramatically increased because extremely poor visibility below ground will make the fire more difficult and time-consuming to locate, giving the fire more time to produce more heat. Ventilation cannot be accomplished as quickly as on the upper floors because of the size, number, and location of windows.

Figure 2–9. To extinguish fire, the flow rate must not only be enough to stop heat generation, but it also must be applied to the fire. The flow rate here is sufficient to begin knockdown. Protecting an exposure can be good practice; however, there was more than one exposure threatened by this fire and the line may have been put to better use by knocking down the main body of fire, thus simultaneously removing the fire threat to all the exposures.

Figure 2–10. Flow rates must be provided to not only extinguish the fire at hand, but to also allow for unexpected developments. While 30 GPM may eventually handle this fire, the lack of protective clothing and low flow rate would provide the setting for disaster if the fire reached a propane tank or the engine compartment where the fuel pump was still operating.

So, even if the flow requirements for the basic amount of generated heat are the same, the fire, when located in different surroundings, calls for lines with increased flow rates to handle additional potential hazards as well as additional personnel to handle search and evacuation operations.

For example, one medium size New England city's fire stream management plan calls for the first-arriving engine to take a position near the fire and press an attack using the tank water. The second-arriving company's initial responsibility is to supply water to the attack engine.

To implement this policy, one of the available engine companies has four preconnected lines available. On a one- or two-room dwelling fire, company water management policy dictates stretching at least one 1¾" line having a 150 GPM flow. That is part of the company's strategic plan. The line officer has a choice of 150', 200', or 300' lines with either smoothbore or fog nozzles. The line officer's decision of what line to deploy is a tactical one. If the situation dictates, and if the fire situation demands increased water flows, the line officer can also tactically decide to expand the strategic plan and lay a hydrant supply line, upgrade to two or three 1¾" lines, or utilize 2½" handlines or master streams.

By comparison, the city's strategic water management plan dictates first-arriving engine placement. The company's strategic water management plan mandates one 1¾" line with a specified flow on a one- or two-room fire; however, the company plan also allows for tactical deployment of line length, additional or larger diameter lines.

Planning need not be this complex. Depending on the situation, planning can be effective even though it seems deceptively simple.

A small rural department in Arkansas had only one truck with a portable pump and a 300-gallon tank. After evaluation of their hazards and available resources, the department decided to use their booster line only on grass and trash fires and to purchase a 125 GPM nozzle and 200' of 1½" hose for use on all other fires. They realized their equipment and water capacity was limited, but they found by testing that the pump would supply 125 GPM at 150 pounds per square inch (PSI), and 200' of line would reach and operate at any structure in their district. The department formulated a simple strategic plan based on their apparatus, available water, and pump capacity. They further maximized their efficiency by concentrating rescue and attack training utilizing the limited equipment available.

Unwritten Planning

I watched as a fire in a large southern city, involving a number of one-story stores, developed quickly into multiple alarms. As each engine company arrived on the scene, they laid a single 2½" supply line from a hydrant and then pulled off two 1½" attack lines, much as if they were preparing to attack a fire in a single-family dwelling. Since most of the working fires in this city involved single-family dwellings, the department had geared its operations, through repetition, to the most common alarm. While this store fire clearly called for master streams, the officers functioned as if programmed in a single mode. To the department in this city, any fire meant a single forward lay supply line and two 1½" lines, each flowing 95 GPM. While this system usually worked for a confined room-and-contents fire, it was certainly not up to the task of stopping a blaze in a row of stores.

Although they did not realize it, this department had a strategic water management plan. It was developed informally, was unwritten, and was based on actual fire experience.

Because of the shift system the department used, it was doubtful that an average engine company officer would respond to one multiple-line fire in a 2-year period. In order to improve this city's fire defenses, formal planning was needed to help ensure deployment of higher volume streams and increased fire scene water supply to help prevent the department from allowing further large losses.

Many departments that do have an understanding of the principles of providing adequate flow rates for both supply and attack lines fall into the same tactical rut as departments having unwritten planning. After researching, they decided to pack their engines with large-diameter supply line and switch to 1¾" attack lines in an attempt to increase waterflow efficiency. Because they figured that the 1¾" line will kill a larger amount of fire compared to the 1½" lines previously used, the department fell into the habit of pulling the all-purpose 1¾" lines on all fires. This is a common result of firefighters becoming overconfident in the line's abilities.

While the flow rate delivered by the average 1¾" line can handle many fires, a surprising number have probably been lost because the line could not deliver a fire-killing flow punch to overcome fires in large fuel loads.

Actions must be taken to prevent unwritten planning from creeping into attack operations. Training must be given on operating all sizes of attack lines and officers must be cautioned to carefully evaluate present and potential fire volume when estimating tactical flow rates from medium-diameter attack lines.

I was surprised to experience that one large Midwestern state did not teach the use of 2½" lines in their Firefighter I and II classes. Instructors need to make sure add this very necessary skill to ensure a firefighter can operate safely and effectively on the fireground.

Evaluating Rates of Flow and Areas of Application

If an organization wants to reduce its loss rate in types of fires that are not normally encountered, their strategic planning should allow for built-in expansion, designed to be easily understood and quickly deployed when conditions warrant. This organization should also plan procedures and apparatus to handle larger fires than those normally encountered by the department.

If larger fires are planned for, then routine fires can certainly be efficiently handled. For example, a new chief was appointed as the first full-time commander of an organization consisting of combination paid and volunteer personnel. Village officials realized the need to revitalize and update the department and felt that someone with extensive experience outside the department was needed to accomplish the turnaround. The new chief was charged by village officials with implementing a strategic firefighting plan in order to reduce fire losses and improve life safety.

To help evaluate the department's tactical operations, first-due engines were sent to buildings in their district and, with long measuring tapes borrowed from the public works department, measured the distances between the street and various building entrances. It was discovered the 150' preconnects carried on the engines would only reach 10% of the doorways, and if the door was reached, would only allow an average of less than 30' of working line within the building. Reviewing 2 years' worth of past fire reports, the data indicated that if a dwelling fire was in possession of more than one room upon arrival of the first engine, the building usually ended up as a total loss. Further investigation considered street layout, building construction, personnel response, water supply, and operations connected to an automatic mutual aid program in which the department participated.

The informal attack plan in effect at this department called for the first-arriving engine, with two firefighters and 300 gal. of tank water, to position at the front of the building. If an attack line was needed, a booster line was used because the personnel considered that line the easiest to stretch and replace. If a supply line was needed, the second-arriving engine either laid a forward lay of one 3" line from a hydrant or helped hand lay a feeder line from the attack engine to a hydrant.

Analysis showed an average initial daytime structure fire response of four firefighters operating two engines, each from a different station. With the system then in effect, one firefighter would size up and attack the fire while the other three busied themselves with laying a supply line. With this system, it took four firefighters to flow 30 GPM. Because of the life hazard potential and the loss record, it was decided to drastically alter the department's tactics.

Using the field data and facts gathered from fire reports, a new strategic fire stream management attack plan was formulated which included the following:

- The main hose beds were changed from forward to reverse lays. This allowed for the hose in the length required to be dropped at the fire scene and for only one firefighter to get hydrant water into the pumper and charge the lines.

- Booster lines were banned from use inside a building because of their low flows and difficulty of stretching by the available crew.

- A 250' 1½" preconnect was added and equipped with a 125 GPM nozzle. This line was used if the company officer felt he could control or extinguish the fire using only the tank water.

- Additional skid load beds were added so that two personnel on the first-arriving engine could lay two 1½" and two 2½" lines in the street and pump them with hydrant supplied water. Portable master stream equipment was relocated to lower compartments so it could be quickly deployed by the engine companies using a minimum of personnel.

- A plan to bolster daytime personnel using village public works employees as trained firefighters was put into effect.

An intensive daily training program stressing interior direct attack with the new water flow plan was instituted to assure personnel became proficient with the changes, and a written standard operating procedure was developed and distributed to all members.

The chief wanted a strategic plan to ensure that a minimum of 250 GPM could be flowed by each of the first- and second-due units. Baseline water flow calculations indicated most average fires could be handled with one 1½" line flowing 125 GPM.

The 250 GPM flow requirement was for potential use in apartment buildings to protect evacuating residents or for use in a well-involved commercial occupancy. In addition to the lines used by first-arriving firefighters, the chief desired immediately available additional lines for other personnel to use on arrival to protect exposures and to back up the first line.

The plan allowed almost 2,000 GPM to be flowed by four personnel if the master stream devices were deployed. Except for nozzle purchases, the changes were made without any financial expenditures, indicating budget constraints need not be a limiting factor when upgrading suppression operations.

After implementing the strategic water management plan, it was felt that the department needed to continue training on size-up, entry, and attack techniques. By putting the strategic water flow management plan into use early in the turnaround process, the foundation was laid for additional improvements in the department's firefighting efficiency.

This example demonstrates that a strategic water flow management plan can help officials ensure that water is available at the fire scene in the quantity needed and that enough hose is pulled to enable the water to be efficiently distributed.

The desired amount of flow was determined by evaluating a large number of buildings within the village. The points of application were determined and measurements taken to ensure lines were provided in the proper lengths to reach the hazards. Planning for the distribution of flow was based on available staffing and considerations of height and size of hazard. The sustaining of water supply utilizing limited personnel was addressed with the reverse lay hose loads. Each element was blended together to form an effective strategic water management plan.

Through adequate planning and training, the same line can serve many missions. Determining the line's mission might seem simple enough, but think back to all those times you have seen a first- or second-arriving engine stretch a line and attempt to utilize a stream which was completely inappropriate and had no appreciable effect on the outcome of the fire. The flow rate should equal or exceed the task assigned.

Because of the limited number of personnel normally available to operate attack lines in the initial stages of a fire, planning should allow handlines to be light in weight and easily maneuverable while providing the maximum GPM flow possible. From a tactical standpoint, it is much more desirable to concentrate a three-person crew on a high-flow line and adopt hit and move tactics than to have three low-flow lines with one person on each attempting to surround a fire. Multiple lines may be needed to cover exposures or to back up the first line, but those lines, as well the first, should be as maneuverable as possible, able to flow a maximum amount of water, and crewed by enough firefighters so they can be effectively and safely operated (fig. 2–11).

One Midwest rural fire protection district protects a village of approximately 4,500 residents as well as 110 sq. mi. of rural farming area in north central Illinois. They determined that four preconnected 1½" lines be carried on each attack pumper, two 200' and two 300'. All have 95 GPM fixed gallonage nozzles.

Evaluation of the district indicated a 1½" line flowing 95 GPM could easily supply the calculated critical rate of flow given the average size of dwelling rooms and the type of construction common throughout the response area. The department felt that additional 95 GPM lines were needed to head off fire extension. In this department's case, a higher per-line flow rate was not an issue. Their most important strategic consideration was distribution of flow to different points within multiple-story balloon-frame structures to prevent fire extension. If

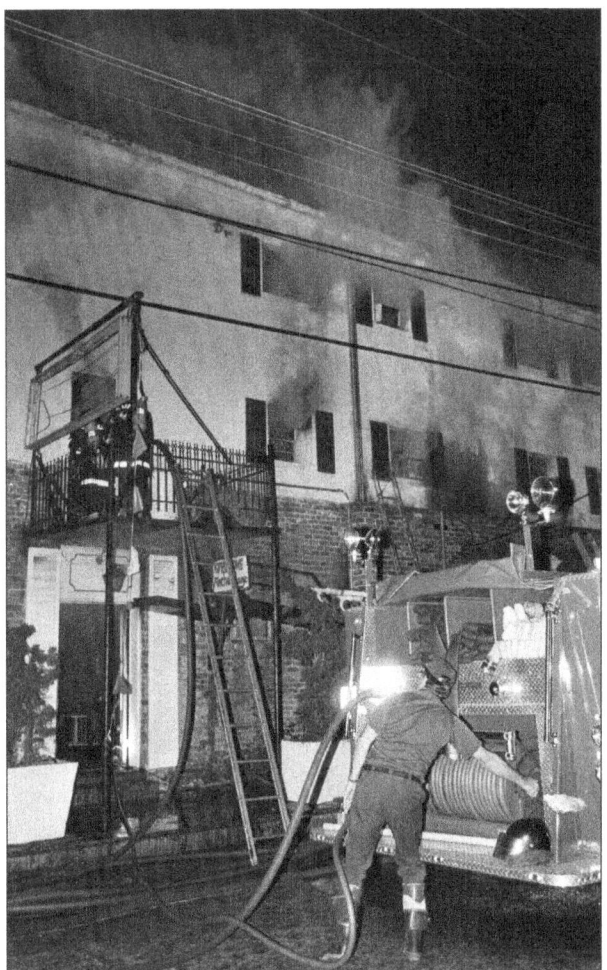

Figure 2–11. A pump operator rolling up a booster line which was used in initial attack on this apartment building fire, which eventually required the services of a third alarm to control. While 30 GPM will handle a great number of fires, at that flow, there is no reserve to handle unanticipated developments.

more flow is needed at a certain point, two lines can be operated together. If a larger line is needed, 2½" hose is available. Not surprisingly, their strategic water management plan calls for a high-volume prepiped deck gun attack on barns and other large buildings if the fire is beyond the capability of the 95 GPM lines on arrival.

Evaluation of Line Sizes

1" Booster Line

A 1" booster line can flow a practical maximum of 40 GPM. At that flow, the small line can handle a surprising amount of fire. However, there is no excess capacity in a 1" line to handle an unanticipated condition. If an attack is made on a vehicle fire, the 40 GPM flow will probably be adequate. If the fuel tank pops open, or if the van is loaded with a hoarder's possesions, the resulting intensity of fire cannot be safely handled by a 40 GPM flow. Booster lines utilizing ¾" and 1" hose are still used in many parts of the country, primarily for use in combating wildland fires. Because of its relatively high friction loss (105 PSI per 100' when flowing 30 GPM compared to 26 PSI for 1" hose at the same flow), the average effective output ranges 12 GPM–15 GPM, which is much too low for structural firefighting. While the booster hose pulled straight from the reel is easy to stretch, going around a corner of a building or up a set of stairs causes friction. This additional drag requires a number of extra personnel to be involved in the booster line's speedy stretch. The booster line is extremely difficult to stretch over 50' from the reel using only one firefighter and is also extremely difficult for one or two firefighters to lead out into a multi-story building. Because the booster line is filled with water, its weight, coupled with its difficulty in stretching and its low flow, makes it an unlikely candidate for an effective and safe structural attack line.

An excuse for deployment of low-flow lines historically given by departments (especially in rural areas) is the belief that use of the low flows will conserve water. We know that in order to extinguish a fire, the fire department must apply enough water to overcome the heat being produced.

A certain amount of water will cool a certain amount of fire. That is a law of physics, and it cannot be changed with plain water alone.

It is generally found that departments that advocate the need to conserve water through use of low-flow attack lines do so as a reaction to a bad experience they had in the past. They may have made attacks with booster lines flowing less than the critical amount of water flow needed to extinguish the fire. The fire continued to burn until, at some point, they ran out of water. Without any waterflow to hamper combustion, the fire accelerated, burning down the building. After being stung with a loss, they began to believe that by conserving water, they may have had just enough left to complete extinguishment.

I've never seen a chief get an award for having the most water left in the tank after the farmhouse collapses. Water is there for only one reason: to put the fire out. The trick is to effectively apply what water is available.

The late Chief Clyde McMillan, inventor of automatic nozzle theory, explained high-flow tactics by telling a story about a man who had two buckets, one filled with 30 pounds of pea gravel and the other containing three 10 lb. bricks. A huge mugger was approaching the man to steal his wallet, so the man had to make a decision.

On one hand, he could throw handfuls of pea gravel at the attacker until his 30 pounds of gravel were gone. But the little pebbles would ineffectively bounce off the attacker because they were too small and too spread out to have any stopping power. On the other hand, if the man threw one of his 10 lb. bricks at the attacker and scored a direct hit on his head, the attacker would have been stopped. Only one quick, heavy hit was needed, and 20 pounds of backup was still left in the bucket. Why throw pebbles ineffectively at a fire with a booster line when you can brick it with a quick hit from a large flow line?

Concentrating the water weight into a heavy hit instead of spreading it out over a period of time is a tactic proven at thousands of live burn demonstrations and countless successful stops nationwide. Less water is used extinguishing identical fires using a high-flow line than by using boosters. You have to be aggressive, find the seat of the fire, and apply the water at the base of the fire to cool the flames. Conserve water only by applying it wisely, not by artificially lowering the rate of flow below the needed fire flow amounts.

1½"–2" Lines

For many years, medium size attack line has proven to be ideal to stretch and operate within a building. A firefighter can take 200' of flat hose from an engine and easily stretch it around corners and up and down stairs with a minimum of effort. When 1½" line came into general use, smoothbore tips were used which flowed from 40 GPM–100 GPM. As combination nozzles became practical for fire service use in the 1930s, normal flows varied between 50 GPM and 95 GPM. Today, the most popular flows for medium-sized handles are between 125 GPM and 160 GPM. These flows have proven extremely effective for trash fires, vehicle fires, and for the majority of interior dwelling fires, forming the basis for a large percentage of firefighting in this country.

Even with the availability of larger sized 1¾" and 2" attack lines, many departments feel comfortable with the flows available from 1½" line. The Detroit and Washington, DC departments still prefer the maneuverability of 1½" line, and its effectiveness is proven daily in both of these busy departments.

While most departments in the United States use 1¾" handlines, few know of its origins and the fact that it was never designed to be used without an expensive chemical injection device.

In the 1960s the city of New York employed the Rand Corporation to help evaluate and plan its fire defenses. This was an unusual move because the Rand Corporation had no in-house firefighting expertise nor personnel with any field firefighting experience. The city wanted to utilize Rand's "think tank" ability to assess the productivity of its fire department and to develop methods to increase that productivity, unbound by traditional thinking. Two major developments emerged from this study.

The standard handline fire stream used by New York at that time was around 260 GPM supplied through 2½" hose. Smaller, 1½" lines were rubber covered and usually used for fire attack and overhaul.

Rand found that flow efficiency of the smaller line could be increased almost 95% by increasing the size of the attack hose from 1½" to 1¾". It was found that the hose could be coupled with standard 1½" threaded couplings without losing much water flow efficiency, and nozzles required only slight modification for increased flow capability. Rand also found that if an additive called rapid water (polyethylene oxide, a straight-chain polymer) was injected into the water stream, the flow increased to the point where the 1¾" hose could flow 250 GPM at normal pressures, as much water as the traditional 2½" line. The rapid water additive, trademarked by Union Carbide as UCAR, tended to align the water's molecules in such a way as to reduce friction loss, actually making the water more slippery.

The rapid water injection system proved to be troublesome and expensive to operate and its use was discontinued after a few years. The 1¾" hose remained, however, and has become an impressive and important medium-sized attack line. Without the rapid water injection, and by using a $^{15}/_{16}$" smoothbore nozzle, flows of 180 GPM were obtained, satisfying the New York City Fire Department's (FDNY) requirements. They also eventually increased the true internal diameter of their 1¾" hose to 1.88". So, once charged, FDNY spec 1¾" hose is closer to 2" than 1¾".

A 1¾" hose is only slightly larger than 1½" hose, but its friction loss is half that of 1½". This means an attack crew can stretch a 1¾" line using only a little more effort than they used stretching 1½", but the flow at the nozzle can, in theory, be almost doubled. The only problem is to make sure that the crew can handle the greatly increased nozzle reaction.

Some officials decided that if 1¾" hose offered higher flows when compared with 1½", water flow could be increased even further by using 2" hose. The 2" is a common hose size for maritime fire protection and artificial snow making, so it was easy for hose manufacturers to oblige and make 2" hose with 1½" thread couplings available to municipal departments.

The 2" hose has about half the friction loss of 1¾" hose and about a quarter of the friction loss of 1½" hose. Its weight when filled with water is slightly greater than 1¾", but its handling characteristics are similar. Many departments have used 2" hose to replace 2½" hose altogether, substituting 3" when supply lines are needed from a pumper to an aerial device or master stream unit.

The Los Angeles Fire Department, evaluating tactics after several large-loss fires in high-rise buildings, experimented with elevated water flow tactics and have standardized on 2" hose with 1½" thread couplings for standpipe work. Coupled to 200 GPM nozzles which operate at 75 PSI nozzle pressure, 2" hose proved considerably lighter and more maneuverable than the 2½" line formerly used yet supplied almost the same flow.

Hose manufacturers state that in the past five years, 1¾" is the most popular size of handline hose shipped to municipal departments. Most privately express wonderment at the reasons why. Many officials feel that 2" hose should be more popular than it presently is. It has the potential to flow almost as much water as a 2½" line but with handling characteristics which rival those of 1¾" line. They see a trend (moving slowly in certain areas, but nevertheless moving) of progressive departments adapting 2" hose as an all-purpose handline, especially in departments where personnel reductions have become critical.

2½" and Larger Lines

In some cases, the fire will be of such a potential or of such an intensity that its control will be beyond the flow capabilities of medium-diameter hoselines. To counter large fire volume threat, the quick deployment of streams flowing 250 to 350 GPM using 2½" or 3" hoselines should be included in every strategic plan. Many fires that could have been easily stopped had a large line been available in time have destroyed buildings because the attack began or continued with a line that could not deliver the flow required.

The practice of abandoning a line not flowing enough water and running to the engine to get a bigger line is wasteful in both time and firefighter stress unless your objective is to burn down the building.

One department in Connecticut has a target hazard, nicknamed the Barracks, in its response district. Built on a hillside, three three-story frame buildings contain almost 100 small, shabby apartments. Plans call for the use of a 2" preconnect if the fire is showing from more than one window on arrival. It is known that the fire could probably be handled with a 1¾" line, but the potential for rapid fire volume increase and the life hazard within the structures dictate the need for the large line to be in place immediately.

If a smaller line is stretched and it is found that the fire involves a hallway with burning apartments on both sides, the 1¾" line may not have enough knockdown flow. If the crew gets into position and finds only one room involved when using the larger line, the 2" line will certainly handle the smaller fire as efficiently as the smaller line.

Some departments have decided to rely only on the all-purpose 1¾" hose for all of their handline operations. Their theory is that with automatic nozzles, the lines can flow as little as a 1½" and, by increasing the pressure, as much as a 2½" or 3" line. This is true up to a point. However, the high pressures needed to flow 200 GPM and above through 1¾" lines makes their operation extremely inefficient and sometimes dangerous.

After extensive initial evaluation, departments have placed 1¾" hose in service for use as a universal attack line, thinking they are obtaining high flows. While on paper, 1¾" hose can deliver an impressive amount of water, on the fireground, departments are surprised to find that flow rates, when measured by flow meters, are considerably less than they expected. The reason is being able to safely handle nozzle reaction forces.

It can be reasonably estimated that most 1¾" handlines in use across the country probably flow less than 125 GPM if nozzles that have an operating pressure of 100 PSI are used. With nozzles that have an operating pressure of 50 PSI–75 PSI, 140 GPM–160 GPM can easily be achieved (fig. 2–12).

Firefighters become reluctant to operate a line that they feel is unsafe to control. In actual firefighting operations, they will make attempts to reduce nozzle reaction to a manageable level by partially closing the shutoff, having the pump operator reduce engine pressure, or, in the case of selectable flow nozzles, moving the selector ring to a lower gallonage. All of these items will have the net effect of reducing nozzle reaction at the expense of flow rate. If

Figure 2–12. The advent of 1¾" handlines and matching high-flow nozzles have done much to increase firefighting efficiency. While they can knock down quite a bit of fire, they should not be stretched if the volume of fire is obviously beyond their capability.

automatic nozzles are used, the problem becomes compounded because the nozzle provides an effective-looking stream at low flow rates.

When planning attack strategy, the effects of nozzle reaction must be taken into consideration. Unless reaction management is planned beforehand, firefighters will take steps on the fireground to make the line easier to handle. All of these steps will result in reduced flow.

Many departments have switched from 2½" supply lines to 3" or large-diameter hoses because the lower friction loss at the same or greater flows makes the movement of large volumes of water much easier. These departments could pressurize 2½" line with pressures over 200 PSI to move more water in relay operations, but they found increasing the hose diameter more practical.

The same is true for handlines. Why pressurize a 1¾" hoseline with 250 PSI engine pressure in order to flow 200 GPM from a combination nozzle when a 2" line will deliver the same water at an engine pressure of 161 PSI, or a 2½" line will flow that amount at only 116 PSI? When flowing water from 200 GPM–350 GPM, the 2" or 2½" lines are much more flow efficient and, because of the lower charge pressure, much easier to maneuver since they are less stiff than the higher-pressure line.

Reaching the Fire

Most departments use preconnected handlines for the bulk of their firefighting. A preconnect can be easily stretched and rapidly put into operation by a minimum of personnel. Its biggest disadvantage is that the length is fixed. It has a nozzle on one end and is connected to the pump at the other. Unless some planning is done before packing the bed, the line can frequently be too short or too long. Excess length is usually not a problem since extra hose is simply flaked around outside the building. If the line is too short, however, much time will be wasted piecing in extra lengths of hose to reach a distant objective. This operation can become deadly if the interior crew runs out of line before the seat of the fire is reached.

I know of one department that has developed a practice of pulling the attack engine into the driveway or up on the front lawn, stopping next to the burning building, because its 150' preconnects are not long enough to reach most structures from the street. The most amazing thing is that this abnormal positioning happens at almost every fire response. No official has ever thought to utilize any of the many lengths of spare hose back at the fire station to lengthen the crosslay preconnects or to provide a long preconnect packed in the rear bed.

Research is very important to determine ideal preconnect length. Different length lines may be provided that can be tactically deployed depending on fire conditions. It is extremely time-consuming and downright dangerous to suddenly terminate an attack operation to splice another length into the hoseline.

Some departments, forced to operate with a minimum of personnel, stretch a combination or *day* line into building fires. A combination line usually contains 100' of 1½" or 1¾" hose with a high-flow nozzle, coupled to the threaded tip of a 1⅛" or 1¼" smoothbore nozzle at the end of 200' or more of 2½" line. This line can be easily handled if the smaller hose is bundled together with straps or hook-and-loop tapes. When stretching the combination line, the smaller hose can be shoulder loaded along with two folds of the 2½" line and advanced to the fire. The 2½" line is placed with the nozzle near a doorway in case the smaller line proves not to have enough flow for the fire. This way, the crew can immediately disconnect the small hose and then go to work with the 2½" flow without having to go back to the engine to stretch another line.

In areas where the distance from the engine to the fire varies greatly, such as in districts with U-shaped apartment buildings, many departments do not use preconnect lines, but use bulk hose beds instead. Bulk hose beds can be located transversely over the pump area or at the rear of the unit. In use, the required amount of hose is selected and advanced toward the fire. When the fire is reached, the pump operator disconnects the next coupling, makes the pump connection, and supplies the line. In actual practice, the use of bulk beds keeps excess hose around a pumper to a minimum and can be used rapidly.

The FDNY's rule of thumb for stretching hose from their bulk beds is to figure one length per floor and then one extra from the point of entry. If the building is over 50' deep, figure one extra length per each 50' in depth.

It is good practice to use larger diameter attack lines for backing up lines operating inside a fire building. Many departments require that any line sent in to back up another must be the next size larger.

High-Volume Handlines

Many departments in rural areas are increasing the effectiveness of preconnected lines by carrying lightweight ground monitors guns attached to a 2½" or 3" line. One person can stretch the line and operate the gun at flows in the 400 GPM–500 GPM range. Many rural departments have made excellent stops of large barn and building fires with these preconnected *bomb* or *blitz* lines. While popular in areas with limited water supplies and large structures, it's ironic that preconnected big gun lines are rarely seen in cities that have water supplies to effectively supply them.

These nozzles and hardware are explained in chapter 8. The decision to place a strategic plan into effect will help evaluate these devices and place them where they will do the most good—in the hands of the firefighters.

High-Rise Buildings

Flows for most buildings can be averaged together to compute baseline flow requirements when planning water flow requirements for general structure fires. However, some hazards, such as high-rise buildings, must be considered separately during evaluation.

Fighting fires in high-rise buildings calls for different water delivery tactics than fires fought from the street. Unless the fire is on lower floors, water for high-rise operations will be available only through the installed standpipe system. Flow tests on all systems in the department's response area must be accomplished in order to establish how much water the department will have available for firefighting. This is a necessary first step when formulating high-rise fire suppression tactics (fig. 2–13).

Figure 2–13. This fire on the roof of a building under construction, if located at ground level, could have been knocked down in minutes with a preconnected line. When planning for high-rise fires, the time required to get lines into operation must be taken into consideration. Height means time. It is not uncommon to not have lines in position for 30 minutes after arrival at a high-rise fire, making planning for large flows mandatory to cope with the fire's increased intensity.

The public should be thankful that high-rise fires are relatively rare when unenlightened departments carry only two doughnut rolls of 1½" hose and a plastic nozzle as their entire first-line high-rise attack equipment. With this equipment and only 65 PSI–90 PSI available on upper floors of many buildings, approximately 60 GPM (at best) will be available to fight a fire. This is less than many departments consider a minimum flow for automobile fires.

The fire spread potential of high-rise fires is frequently underestimated. Large departments such as New York City, Chicago, Boston, Houston, and Los Angeles, which have experienced multi-floor conflagrations, recommend no less than 200 GPM for an initial flow. All have plans for multiplying this flow quickly if demanded by fire volume conditions. All carry 2" or 2½" hose into high-rises in order to provide these high flows.

Experience has shown that many high-rise fires can burn for quite a long period before discovery, especially during overnight hours. It takes time for operating forces to group, travel vertically, perform size up, connect lines, and commence attack. On many occasions, 20 minutes–45 minutes elapse from time of alarm until suppression forces are ready to operate on the fire floor. If entry to the fire floor is blocked by heat and flames, search and rescue operations are delayed until the fire is knocked down.

Because a high-rise building cannot be easily ventilated by conventional tactics, it takes time to remove heat and smoke. A fire can be readily spread downward from the floor of origin by flaming debris falling down pipe and wiring shafts as well as upward, auto-exposing windows above. As the fire spreads upward from the floor of origin, the only way to cut it off is to get above the fire—the suppression forces operating in the hottest, worst possible positions.

When planning water management for high-rise fires, think volume. Large-diameter attack lines along with both low operating pressure smoothbore and combination spray nozzles should be provided. When planning, departments must also make provisions for placing portable deluge guns into operation on upper floors, supplied by lines from two or more standpipe outlets, when the fire volume warrants.

During building evaluations, departments should carefully measure distances from standpipe outlets to all parts of a floor. Partitions, desks, office equipment, and other obstacles increase distances, requiring more hose than was calculated by using a floor plan map alone. Plan on attaching lines to outlets both above and below the fire floor if extra volume is needed. This operation will also require extra hose.

Some departments have found the use of 2" hose ideal for high-rise operations. It is easy to carry and deploy and is capable of supplying needed high flows. A short 6' section of 3" hose, called a *pigtail*, can be connected to a standpipe outlet which will feed two 2" lines attached to a gated wye.

Many departments have problems with frequent failure of building standpipe valves, especially those in older occupancies that have had little or no maintenance over the years. In order to reduce the chance of failure, the use of the pigtail would require the house valve to be operated only once, when the valve is first opened. After the line is charged, flows can be controlled using the gates on the wye, eliminating stress on the building valve.

Review

A strategic plan is a combination of a number of factors that will enable a department to determine its own particular water delivery requirements. The plan includes

- Determination of organizations mission
- Evaluation of individual hazards within a response district
- Determination of flow required for those hazards
- Determination of how the flow will be distributed, either through multiple small lines, large lines, or a combination of both
- Determination of how those flows will be delivered depending on water supply, apparatus, and available personnel
- Evaluation of new equipment and tactics needed to deliver the required flows

Strategic planning will provide the foundation from which to build an effective attack strategy. Properly completed, the plan should provide effective flows through the proper number and length lines that are capable of being safely and effectively handled by available personnel.

Chapter 2 Review Questions

1. What is the most effective handline for use on a commercial building fire?
2. How does a strategic water management plan differ from a preplan?
3. What is the main limiting factor of a 1¾" handline?
4. When determining desired firefighting goals, what are some items to consider when planning operations in a heavily involved commercial occupancy?
5. Using the time/severity chart, define the main factor that will stop a fire.

3

Nozzle Theory

This is probably the most important chapter in the book as the ideas presented here begin to construct the foundation for developing logical tactics with which fire can be sensibly, efficiently, and safely attacked. Before flowing any water, the principles of fire attack and fire behavior and their scientific origins must be understood by firefighting personnel.

When collecting research for this book, a large number of chiefs, line officers, instructors, and firefighters were asked a simple question: On what scientific facts do you base your methods of fire combat? Almost everyone questioned was sure that present firefighting tactics are so soundly based on exacting scientific and practical foundations that they questioned my sanity for asking such a stupid question. When pressed further for details, some had hazy recollections of reading about tactics developed by Chief Lloyd Layman over 73 years ago. Some could recall recent scientific testing by the Underwriters Laboratories Fire Safety Research Institute (UL FSRI) but could not recall if its findings reinforced or disputed Chief Layman's findings. Few who professed to be tactical experts had actually read his books.

It's time the fire service faced the facts. Many firefighters have blindly accepted fire service lore, which has been handed down as sound tactics through generations of officers, instructors, and firefighters, as scientific truth, even though mounting fire losses and firefighter injuries may indicate otherwise.

Basic Methods

Three distinct, basic methods of attack are used for structural firefighting: direct attack, indirect attack, and a combination of the two, sometimes called transitional attack. Each has advantages and disadvantages depending on the fire situation; however, the underlying scientific principles which govern the operation of the tactics are, generally, grossly misunderstood by the majority of the fire service.

Direct Attack Method

The direct method of attack is simply applying water to the base of the burning material, at the flame/fuel interface, where the flammable vapors being distilled by heat from solid material ignite and burn. This application must be in sufficient quantity to overcome the heat being produced, cooling the material below its vaporization temperature, causing the combustion process to cease. The stream is quickly worked around a burning compartment to cool the atmosphere gases and eventually cool the flaming material to a point where it is no longer producing flammable gases.

The burning process creates smoke and combustion gases which are carried upward by the fire's heat. The direct method of attack causes little disruption of these heated combustion products. Reducing heat production by extinguishing the fire at its base stops the burning process at its source, in turn stopping the upward liberation of more heat, smoke, and gases.

The efficiency of this method has been reinforced by documented testing in the late 1980s and early 1990s by the University of Illinois Fire Service Institute as well as in a number of tests by individual departments. What could be simpler? Move in, find the fire, put out the fire. Retain visibility and reduce the chances for steam burns.

For some reason, years ago, the fire service managed to make the simple difficult by the widespread adoption and modification of the second method of firefighting, indirect attack.

Indirect Attack Method

Since the early 1950s, generations of fire officers and instructors have insisted on battling interior fires not by cooling the source of flame production, but by using tactics that attempt to extinguish the fire by cooling an area of heat sometimes quite remote from the fire itself.

"First, for your protection, you throw a fog on the ceiling to cool things down," the instructor says. The new trainee opens the nozzle as instructed, and all of a sudden, the world goes dark. The water they were told to apply has expanded into steam, at least 1,700 times its original volume, displacing smoke, combustion gases, and heat downward and outward. As the superheated ball of steam and smoke swirls around them, their vision is obscured. The trainee loses visual contact with other firefighters and cannot see the fire they were attempting to extinguish. After beating a hasty exit, the trainee immediately questions the idea of trying to cool the overhead.

"You got to do that for your own protection," the instructor says. "If you don't cool down all that heat near the ceiling, the fire could spread over your head and burn you."

The trainee thinks for a moment and says, "Isn't that what we just did—spread the fire? We had to shut down and back out because we couldn't see what we were doing. The others had to leave because they were burned on the neck. We had to wait 3 minutes for the steam and smoke to lift and then, when we went back in, we found the fire still burning in not just one but two rooms. It seems to me that spraying water at the heat in the ceiling caused more problems than it solved." The instructor bulls up, mustering all 20 years of their firefighting experience and grunts, "Kid, don't be a baby. Just because it got hot in there is no reason to gripe. Someday you'll learn to take the steam and heat like a real firefighter."

Sound familiar? If entry is made into a kitchen and fire is found coming from the oven, where is the best place to direct a stream of dry chemical from an extinguisher? In the farthest corner? Over our heads? No, we aim it directly into the oven for a second or two and immediately the fire is out. If a campfire had to be extinguished, is the best practice to throw the contents of a water bucket up in the air in hopes it will come down in the right spot? Of course not. The water would be poured directly on the glowing coals.

Has anyone really questioned why we continue to spray water on smoke over our heads in hopes it will somehow extinguish the fire burning beneath our feet? Where did this practice start? Who's responsible?

Indirect Attack Theory

Spray-it-here-extinguish-it-there tactics had their beginnings when some fairly radical (for the time) theories were explained over 71 years ago in a small, blue book, about the size of an old-time, first grade reading primer. The book, titled *Attacking and Extinguishing Interior Fires*, was published by the National Fire Protection Association in 1952. Its author, the late Chief Lloyd Layman of the Parkersburg, WV, fire department, documented scientific and practical principles for a fire-extinguishing method he called *indirect attack*. He advanced theories of fire development, heat behavior, and extinguishment which have withstood the test of time and still ring true today.

Chief Layman, while commandant of the U.S. Coast Guard firefighting school during World War II, experimented with the use of what he called *finely divided water particles* in an effort to reduce a fire's heat in enclosed spaces. His research attempted to find a more efficient way of extinguishing fuel oil fires, confined within compartments of ships, utilizing the excellent heat absorption qualities of water being converted into steam. In his book, he described how heat was rapidly reduced as a water spray was introduced into the superheated atmosphere of an enclosed shipboard fire environment. Through use of heat measuring devices, researchers found a wide variation of stratified temperatures within a fire area, whereby the hottest part of the fire atmosphere was at the top and the coolest was at the bottom. When water was introduced into the hottest part, the area at the top of the fire, the water was violently turned into steam, expanding to 1,700 times its original volume at 212°F. All the while, this action was absorbing heat, displacing oxygen feeding the fire, and, after a time, if kept enclosed within the fire area, actually extinguishing the fire.

He found that the process of a liquid converting into a gas absorbed great amounts of heat, causing a rapid temperature drop to be indicated on the measuring instruments. The generated steam, liberated throughout the fire area, would first cool the atmosphere, and then, as it expanded, would spread outward and downward, filling the involved compartment and displacing the oxygen the fire needed to sustain combustion. As the source of heat, the fire, was eliminated, the steam cooled and condensed, reducing itself in volume. This created somewhat of a vacuum, drawing fresh, cool air from the outside into the fire area, developing a form of self-ventilation.

After returning home after his wartime service, Chief Layman continued his experiments on structures and attempted to adapt his shipboard fire tactics into indirect attack on a structure. In explaining the scientific principles on which he based his theories, he also became one of the first to describe the various stages of fire development to the fire service, from incipient to flame production to smoldering phases.

Chief Layman stated that while combustion is sustained by a combination of fuel, oxygen, and heat, two of the parts—fuel and oxygen—are present practically everywhere. The only abnormal condition is excessive heat. His fundamental principle of firefighting was that control and extinguishment of interior fires was based on removing this excess heat from the involved area.

He felt the best way to do this was to utilize the heat-absorbing qualities present when water turns into steam.

We should keep in mind that a combination of direct attack and ventilation had been successfully extinguishing fires and removing heat and gases from fire buildings for years. Chief Layman advocated accomplishing the same goal using a radically different approach, keeping the building closed up while attempting to reduce excessive heat contained inside by using the heat absorption ability of steam conversion. The building cannot be ventilated in a normal manner when using these tactics (fig. 3–1).

Tactically, Chief Layman's indirect method of operation was to apply a finely divided stream of water particles (fog) into the hottest part of the fire, the overhead, and then allow time for the resulting heat-absorbing conversion of water into steam to reduce heat, and thus extinguish the fire. During the time the steam was doing its job, the heat needed to remain bottled up in the building to work effectively.

In order to fully comprehend indirect attack, we must understand exactly what he advocated. It must always be kept in mind that the following conditions should be present for his tactics to work properly:

Figure 3–1. This stream's heat-absorbing capability is being carried away by the fire's thermal currents. In addition, the flow rate is too low to overcome heat production and no penetration past the eaves of the building is being achieved. By not understanding the principles involved in its effective application, firefighters have taken Chief Layman's indirect application tactics and used them in ways he never advocated. For indirect attack to accomplish extinguishment, the fire must be confined, allowing the generated steam time to reduce the heat and oxygen content of the fire atmosphere.

- In order to contain the fire's heat so steam can be generated, the building must be reasonably enclosed on arrival and during the attack.

- The fire must be at such a stage to have already generated a great amount of heat before the attack is begun. The hotter the upper atmosphere, the more steam that will be generated and the more effective the attack.

The foundation basis for all his theories was his Coast Guard experience obtained from attacking shipboard flammable liquid fuel fires remotely by dropping hoselines with fog tips down ventilation holes. In most test fires, this action was continued for 20 to 40 minutes in order to give the generated steam enough time to extinguish the fires. Remember, during the tests, the structural material was steel and the primary fuel was flammable liquid.

For structural firefighting, he advocated injecting fog from the exterior of the fire building or from an interior stairway below a fire on an upper floor. He never advocated applying fog for indirect attack while firefighters were within the fire area. He said in his book that this action would run you out of the building (fig. 3–2).

He said that unless a fire had generated a great amount of heat, and unless most of the heat remained confined within the fire building, indirect attack would not be effective. In cases where the fire is free burning, he said direct attack should be used.

An indirect attack through window opening using high-velocity fog cone. Nozzleman should remain below opening to avoid outrush of heated smoke and live steam.

Figure 3–2. In his book, *Attacking and Extinguishing Interior Fires,* Chief Layman stated that you would get burned if caught by rapidly-expanding steam.

Chief Layman himself said his tactics were useful on only a certain number of fires. In fact, he said if the building was vented and the heat had escaped to the outside atmosphere, his tactics would not effectively extinguish the fire. We must keep this fact in mind as we analyze what the fire service did later to twist and bend his theories.

At that point in history, most of the research, training, and direction of the public fire service was controlled by the insurance industry through the National Board of Fire Underwriters and individual state rating boards. They provided much of the leadership and formulated public fire protection policies from the early part of the century until about the late 1960s to early 1970s. These insurance officials rapidly embraced the methods advocated by Chief Layman as a way to reduce fire losses and began a massive campaign to urge the widespread use of water fog by the nation's fire departments.

In 1950, Chief Layman addressed the Fire Department Instructors Conference at Memphis, TN, where he presented his theories in a paper entitled "Little Drops of Water." After some discussion, a group known as the Exploratory Committee on the Application of Water was formed to test Chief Layman's techniques and, ultimately, to spread the gospel of indirect water fog use to the nation's fire departments.

This group, whose members consisted of a large number of representatives from the fire insurance industry, did some testing, but for the most part accomplished their work by conducting water fog application demonstrations around the country. The group became cheerleaders for indirect attack and, before long, according to people who are familiar with the committee and its work, some myths and half-truths began to surface that started muddying up the indirect attack waters.

The tactics described by Chief Layman, first in his paper of 1950 and later in his book, created a fog frenzy within the fire service during the 1950s and 1960s. During that time, according to periodicals of the day, no self-respecting fire official would be caught dead using anything other than fog nozzles. As an example of how far this fog mania was carried, some departments actually modified adjustable fog nozzles, inserting a stop or pin in the pattern adjustment to physically prevent a nozzle operator from flowing a straight stream (fig. 3–3).

Throughout the 1950s, thousands of fire officials applied Chief Layman's tactics to structural fires with varying results. Magazines of the era were full of photos of extensively involved buildings being fought with a single 1½" fog nozzle, the firefighter attempting to brave the heat and flames in order to get close enough to a window to apply the fog. According to these articles, no forward-thinking department would consider using anything but fog and indirect attack. This was taken to the extreme of having the text describe how a fire was stopped using water fog while the accompanying photograph showed the building ringed with straight streams and the roof falling in.

Viewed in retrospect and seasoned with over 70 years of operational experience, Chief Layman's theories and tactics have been proven essentially correct and will work effectively if the building and fire conditions he described are present when employing indirect attack methods. In perspective, however, the principles of indirect attack should have been considered by the fire service as only another set of tools for tactical arsenals instead of being considered the final answer to all structural fire extinguishment.

Applying water from the outside of a fire building made some sense when viewed historically. In the late 1940s and early 1950s, protective clothing, which was basically rubber rail gear, did not insulate as well as it does now, and protective breathing apparatus was not in widespread use for initial attack. The idea that it was correct to attack a fire from the

Figure 3–3. Chief Layman's book, published over 71 years ago, and the tactics he described, were initially misunderstood by many and then twisted further over the years to where many firefighters suffered thermal injuries trying to apply them.

outside—not to have to fight heat and choking smoke—helped to dramatically advance the theory of indirect attack. Many departments had fought fires from the outside with smoothbore stream nozzles anyway. What they thought Chief Layman advocated was that outside attack was the proper approach only if fog nozzles were used.

As departments adapted tactics for indirect attack from outside a fire building, they gradually began to ignore the principles on which those tactics were based. This was the first phase in the bastardization of Chief Layman's theories. Departments didn't read the fine print, or if they had read it, it was quickly forgotten. Exterior attack using water fog on wide open as well as closed up buildings became common practice. With the underlying principles of interior heat retention disregarded, the general misuse of exterior indirect attack more often than not pushed the fire back into the building, causing as much if not more damage and destruction than before fog was widely used.

Years later, while evaluating their loss records, many departments rightly decided if exterior attack with a fog nozzle pushed the fire inward, then bringing the nozzle inside would push the fire outward. That decision, based on actual experience, was the beginning of the second phase of the bastardization of indirect attack. Does it come as a surprise that the fire service began using tactics based on misapplied and misinterpreted scientific principles?

The interior indirect (or combination) attack, as still practiced by a percentage of the fire service today, was invented by the fire service itself to compensate for problems encountered by employing techniques based on earlier self-invented principles. Nowhere in his writing did Chief Layman present scientific arguments that advocated spraying water over the firefighters' heads in a fire situation in order to create steam bath conditions. On the contrary, he said firefighters would be burned if they were unfortunate enough to find themselves enveloped in a hurricane of water converting to steam.

Interior, indirect attack severely punishes the firefighters working in the fire building. Don't forget, Chief Layman's indirect attack principles never advocated placing the firefighters within the area of steam generation. The fire service accomplished that themselves without any help from Chief Layman.

At that time, the popular belief within the fire service was that the success which was being attributed to water fog was a result of the shape of the pattern in which the water was applied. Many incorrectly believed a fog stream possessed magical fire-extinguishing powers far beyond those of solid streams.

Iowa Research

In May of 1954, a battalion chief from Washington, D.C. wrote an article published by *Fireman* magazine in which he said, "This 50 GPM [gallons per minute] of fog has been found as effective in the absorption of heat as a 200 GPM straight stream." Coming from a chief of a large department, this statement sounds impressive. The only problem is that the results he describes in his statement are improbable scientifically because they violate some basic laws of physics.

Wanting to explore the scientific basis for both this new method of fire attack as well as the older, direct method, Keith Royer and Bill Nelson of Iowa State University's fire training institute began a series of comprehensive experiments beginning in 1952 and lasting almost 20 years (fig. 3–4).

Their body of work is considered by many as the most detailed study of structural fire behavior ever conducted. They did not, however, receive the widespread attention from the

Figure 3–4. A photo from a report by the Exploratory Committee on the Application of Water shows the group. The tall gentleman with the black tie in the second row is Chief Lloyd Layman and to his right is Keith Royer of Iowa State University, the father of research that determined that flow rate, rather than application pattern, had the most effect on extinguishing fire.

nation's fire service as Chief Layman's did. This could be because their experiments and testing offered no single, simple fire attack quick fixes to the fire service.

Royer and Nelson had the almost impossible task of distilling this wealth of information, gathered over many years, from hundreds of controlled building fires, into a form the fire service could utilize. Their theories and tactics have been published in limited form by the university and appeared in some national publications. The most widespread distribution of the information they gathered and the tactics they recommended were described in three training films, *The Nozzleman, Coordinated Fire Attack*, and *Where's The Water*, produced in the late 1950s and early 1960s and easily found with some quick internet searching. Iowa State University published a booklet called "Engineering Extension Service Bulletin #18" describing the theories and methods advocated by their research.

The basic purpose of the Iowa experiments was to attempt to answer how to best stop the flow of heat energy being released as a result of uncontrolled burning within a structure. The testing was extremely comprehensive, with each fire fully instrumented to provide temperature, airflow, and visual data in the form of motion pictures. Actually, Royer prefers to refer to the fires as experiments rather than tests. The data gathered at each fire were evaluated not only by firefighters, but also physicists, chemists, and other engineering and scientific professionals at the university. One of the first facts to emerge from their testing was how the amount of heat energy released by burning fuels was controlled by the atmosphere (fig. 3–5).

Figure 3–5. Iowa State published its findings and produced three instructional movies to illustrate their findings in the mid-1950s.

Each fuel has a calorific value, the calculated amount of heat which theoretically could be released by burning. This is usually expressed in British thermal units, or Btus, per pound of fuel. Up until the time of the tests, the amount of water or other agent needed to extinguish a fire was calculated on the theoretical calorific amount of heat the protected material contained. The Iowa testing showed the actual amount of heat released by burning fuels in a structure fire was much less than the calculated value. This lessened value was caused by the limiting factor of the amount of oxygen available in the atmosphere.

For example, an average one-pound piece of wood has a calculated heat release value of approximately 800 Btus. If the piece of wood was actually burned in the presence of pure oxygen, and its heat output measured, the heat release value would drop to about 535 Btus. If the same piece of wood was burned in normal air with a 21% oxygen content, the heat release, as measured by the Iowa scientists, showed a considerable drop in value, averaging around 37 Btus. Their testing proved that normal fuels within a building could not release their entire potential heat value by actual combustion because the oxygen content in air limited the rate of burning. From this research, Royer and Nelson developed the formula for computing fire flow as described in chapter 2.

For the first time, the fire service had a rate of flow formula based on the knowledge of reduced heat release from solid materials and the data amassed in extensive experiments within actual structures.

Royer, in an interview with the author, stressed that the formula was designed to be used as a tool to enable fire departments to determine required flows for hazards within their area. Application of the formula greatly simplified making estimates of water requirements for fire extinguishment. This estimate of water flow requirements forms the basis for one of the Iowa group's basic attack principles: critical rate of flow.

The group found that successful fire extinguishment depended more on applying the water in a quantity sufficient to overcome heat production rather than the form in which the water was applied. If water was applied at a rate lower than that needed to cool the burning material below its vaporization point, the fire continued to burn. The critical flow rate is the amount of applied water which will cool the calculated mass of material below its vaporization point and thus perform extinguishment.

Their research also found that the shape of the stream, be it a spray or straight stream, had less effect in extinguishment efficiency than was previously thought. They discovered that in order to do its job, water must be properly distributed to the burning material. This distribution can be accomplished with either combination fog or smoothbore nozzles, as long as the water properly cooled the burning material. In light of the thinking at the time, however, most of their demonstrations used water in spray form (fig. 3–6).

In viewing the films and reading the results of their research, it must be noted that their tactics advocated application of water from outside the fire building. Though they did discuss interior application, the first priority in the Iowa methods was to knock down visible fire before making entry. In an interview with the author, Royer says their testing did not address the problem of fire spread caused by applying streams from the outside of the building.

The subject of life safety or the effects of steam on trapped victims was never addressed in the three films.

Encouraged by the successful firefighting results from the theories of rate of flow and water distribution, the Iowa group conducted more experiments in buildings of all sizes and collected and assessed additional data.

Figure 3–6. In this demonstration using a propane "Christmas tree," a straight stream moves through the flame mass, moves very little air, and causes little disruption to the fire's path flow.

Royer continued that in 1959, a large-scale experiment was conducted by the Iowa group in an old warehouse located in Des Moines. The basic goal was to extinguish a large fire in the building using high-flow fog streams applied from aerial ladders. The fires were lit and, when conditions dictated, a number of fog streams positioned around the building were opened up. The fire was not extinguished and actually succeeded in burning through the roof before it was brought under control.

After evaluating the data from the fire, the poor performance by the fog streams was attributed to lack of penetration. Even though the fog was carried inside by the fire's draft, the fire continued to boil violently until it vented itself through the roof. Royer later stated that had they simply twisted the nozzles to the straight stream position, penetration would have been obtained and the fire extinguished. This started the group thinking that flow coupled with distribution, not form, was the key to efficient extinguishment.

At this time, the Iowa researchers suspected a number of things taught in early years of widespread fog use were not right. After analyzing and distilling the results of the Iowa State testing, Royer and Nelson advanced a number of fire suppression theories that had solid scientific backing. These theories were refined into firefighting tactics and have been taught at the university to thousands of firefighters.

The Iowa studies proved a number of principles which are basic to interior attack. The rate of flow and water distribution are the two most important factors affecting fire extinguishment. The type of nozzle had little effect on fire extinguishment performance, provided the critical rate of flow was met and the stream penetrated the area of involvement. However, the application of their tactics favored violent movement of a combination nozzle applied through the windows of a burning building, possibly in deference to Layman's tactics being demonstrated in the same time period.

A combination nozzle more easily allowed the stream pattern to be adjusted to fit the environment. In some cases, the variable pattern allowed more effective distribution. The distribution of water in the area of involvement was important. They felt the stream should be swept in the form of a circle, square, or rectangle, hitting the floor, wall, ceiling, wall, and floor in a single motion. This bounced water off all sides of an area making distribution more uniform. They called this a combination attack of both direct and indirect principles.

It was believed at the time that if the rectangle was formed by moving the nozzle in a clockwise manner, the fire would be driven away from the nozzle. If a counterclockwise rotation was used, the fire tended to pull toward the nozzle and the atmosphere was greatly disrupted. Scientists tried to explain that this phenomenon was caused by a relationship between electrical charges in the water and negative, positive, and neutral ions present in combustion. This theory was never proven scientifically, and later testing has shown that moving the nozzle in either direction has little effect on atmospheric conditions.

The critical rate of flow needed for extinguishment was the amount of water necessary to overcome both the heat being produced by the fire as well as the heat already present in the fire atmosphere or stored by solid material in the fire area. Once the critical flow rate was applied, excess application of water disrupted the atmosphere and made ventilation difficult because high steam saturation made air difficult to move. The atmosphere created by over-application of water was determined to be too hot for firefighters to comfortably penetrate but not hot enough to convert water to steam.

According to Royer, to prevent over-application of water, the nozzle must be shut down immediately after the fire blacks out and flame production is no longer visible.

Some of these principles discovered by the Iowa research have found their way into the mainstream of fire service teaching, while others went languishing in relative obscurity outside of the state of Iowa. By applying these principles to the use of direct attack, and by understanding the limitations of indirect attack when used inside a fire building, a more sensible approach to interior operations began to emerge.

Evaluating Indirect Attack Tactics

When contemplating using indirect attack, its relation to life safety must be seriously considered. The life-safety issue was not addressed either by Chief Layman or in the extinguishing tactics advocated by the Iowa group. Remember, Chief Layman's tactics were based on having the steam take time to find its way around an enclosed space and extinguish a fire. If a fire building has most of its windows broken or missing and doors removed, the principles of indirect attack will not work effectively, as the steam needed to accomplish extinguishment will travel where the fire's heat takes it, usually to the outside atmosphere, not reaching the burning material.

Viewed from another perspective, in order for indirect attack to work properly, the building has to be closed up. This means intact windows and doors. Now, if the building does have intact windows and doors, we must then assume the possibility of its being occupied.

If indirect attack is used from the outside of a closed up, occupied building, the superheated steam, smoke, and combustion gases will be driven into the building and possibly down on any trapped occupants. While indirect attack might fill the building with steam, it does little to help the comfort of trapped victims. In fact, it may seriously injure or kill them.

In some departments, especially in the geographical areas of the south and west, initial attack is still commonly being made from the exterior of a fire building with the spray from a combination nozzle directed through a window into the overhead. Talking to some officials and firefighters in these departments gives one the feeling of being caught in a 1950s time warp.

If a fire is within a structure, is of any great volume, and has ventilated itself to the outside atmosphere through windows, applying a fog stream toward the inside of the building from the area of exiting flames will push the fire inward, spreading it to other parts of the building. The steam created by this action loses its desired extinguishing effect as it is rapidly carried back to the outside atmosphere by the fire's natural thermal currents, unless continued operation of the stream keeps combustion products confined in the structure. Many fires have been successfully extinguished by using exterior tactics. It remains a viable consideration if the building cannot be entered. However, if the safety of any trapped victims or the limiting of unwanted interior fire spread are considerations, fog or indirect exterior attack is certainly not the best tactic to choose.

Pushing a self-ventilated fire back into the structure will not only spread the fire internally but will certainly make survival difficult for any victims still trapped inside. While it may be much easier on the firefighters to make an external, indirect, or combination attack, expanding steam will drive smoke and combustion gases to the floor throughout the building, making quick entry by firefighters for primary search extremely difficult and punishing.

Wielding a narrow-cone fog pattern within a building produces air movement similar to that of a large fan. Tests have shown that a 95 GPM spray nozzle operating at 100 pounds per square inch (PSI) nozzle pressure can move from between 10,000 to 20,000 cu. ft. of air per minute, more than most smoke ejectors (fig. 3–7).

The common technique of ventilating an area by directing a fog stream out a window demonstrates the great amount of air-pulling power possessed by a combination nozzle (fig. 3–8).

Within a fire building, this air movement becomes a dynamic force which can push or pull fire, smoke, and combustion gases into both wanted and unwanted areas. If not handled with care, this rush of air movement can bring unwanted steam and heat down on the nozzle crew and any trapped victims.

The design of combination nozzles contributes to this increase in airflow. As water moves through a combination nozzle, it hits the center baffle, turns 90°, hits the nozzle sidewall, and turns another 90° before exiting. This action causes the water molecules to fracture and pick up additional air. In the area outside the nozzle baffle, a vacuum is created and air is pulled from the outside of the spray stream to the inside. This rush of air mixes with the water as it is propelled forward. Water viewed in glass is clear. Water exiting a combination nozzle is silver, not clear, because it is mixed with air.

The higher the nozzle exit pressure and the higher the flow rate, the more air is moved. The wider the fog pattern, the more air is moved. To move less unwanted air, a straight stream is the safest to be used inside buildings.

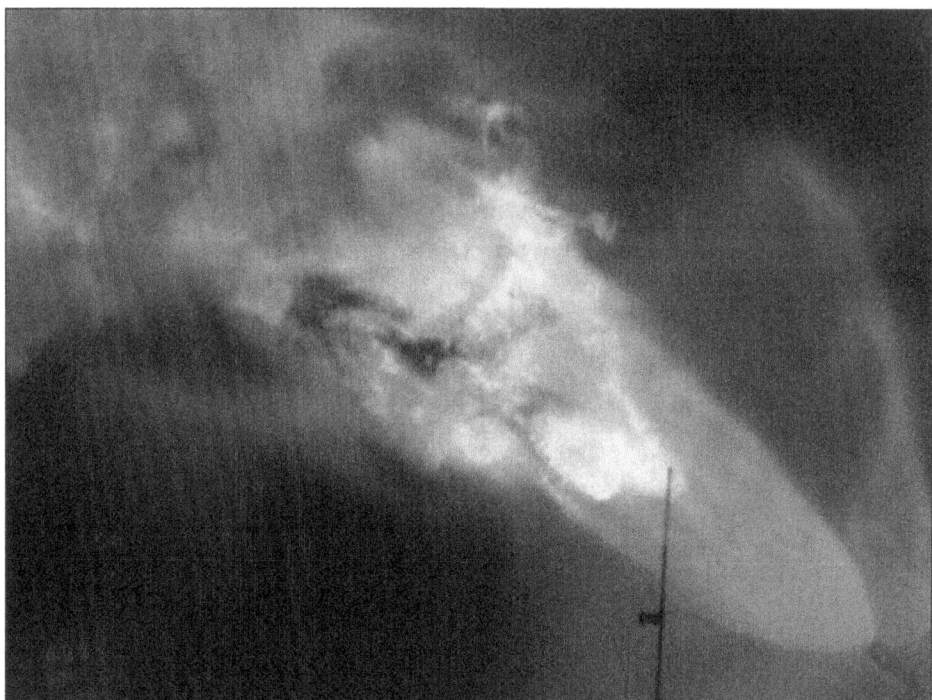

Figure 3–7. A narrow fog stream inserted into the flame mass causes its increased air movement to pick up the flames and torch them far ahead of the stream. Imagine this happening in a number of connected compartments within the building.

Figure 3–8. When a combination nozzle is placed in the wide fog position, the area of lowered atmospheric pressure common just ahead of the baffle actually sucks the flames toward the nozzle operator, a good reason to never use this pattern during fire attack.

Let's assume an engine company is advancing down a hallway in a well-involved, dark, smoke-filled apartment building. The nozzle crew, their visibility reduced, cannot see a fire in a side room and pass it by as they work their spray stream toward a fire down the hallway. The same suction effect used to ventilate a fire area to the outside can now pull the fire out of the bypassed side room and down the hallway behind the nozzle, trapping the crew in the hallway with flames on both sides.

Another example of the misuse of a combination nozzle's airflow characteristics would occur as a crew works its way down the same hallway. The officer decides to use inside indirect tactics, directing a spray stream into the overhead to generate steam for extinguishment and for protection. The hallway is long, and fire is lapping outward from two or three apartments ahead. Beyond the involved apartments are more apartments yet to be searched. The spray stream will create steam in the overhead as it passes by the fire involvement and the stream's air-moving abilities will push the created steam, heat, and combustion gases further down the hallway. If doors are open, this steam balloon will fill the far rooms and adversely affect any victims or searching firefighters occupying those rooms. If the doors are closed, the steam balloon will pack up in the hallway and when it builds up enough pressure to overcome the airflow effects of the stream, will explosively blow back on the firefighters, picking up flames from the fire rooms along the way.

This rollover or blowback effect, discussed by worried firefighters for untold hours in drill sessions and on the training ground, can be easily caused by the firefighters themselves through misapplication of spray streams.

Sometimes, fire movement caused by spray streams can have an adverse effect on operating crews. Extensive research in the late 1980s by the Fairfax County, VA, fire department and the University of Illinois Fire Service Institute has proven that a fire can be pushed by a spray stream into an area which has enough volume to accept the fire. An open window venting into the outside atmosphere is a good example. In these same series of tests, when attempts were made to push the fire back into an unventilated room, the fire created pressure which rapidly overcame nozzle velocity and violently forced the flames back on the nozzle operator.

Recent research on fire movement and behavior has proven the following:

- Because of air movement caused by the combination nozzle's exit velocity, fire can be easily pushed by a spray pattern into other interior rooms, to the outside atmosphere, or into other open or void spaces within the building, such as walls and attics, during normal attack.

- The fire cannot be pushed easily into an unventilated area. An example would be an attempt to push a fire in a high-rise building back down a hallway into its unvented room of origin. When tests simulating this were attempted, the fire violently reacted and repeatedly rolled over and around the nozzle operator. The amount of fire an unvented area could receive depended on the amount of atmosphere cubic footage contained in the area. A fire burning in a hallway could be easily pushed into an uninvolved apartment beyond the fire until the apartment became packed up and atmospheric pressure in that area was greater than the nozzle velocity. Then the fire would pop back on the nozzle crew. This pushing effect can violently be multiplied by rapid steam expansion if the spray pattern was directed into the heat of the overhead.

- If a wide spray pattern is used, this pushing movement is less rapid, but the stream's entrained air accelerates burning. Even worse, if a narrow spray pattern (power cone fog) is directed at the fire, the stream's air movement is concentrated, and the resulting velocity picks up the flames and blowtorches the fire into other areas.

Put simply, fire and combustion gases can be pushed by a spray stream into areas beyond, over, and to the flanks of the fire if the atmosphere has enough capacity to accept them. Fire and combustion gases cannot be pushed into an unventilated area or an area of small volumetric capacity and be expected to remain there, such as in the confines of an unventilated basement area.

Underwriters Laboratories Fire Safety Research Institute Research

Probably the most comprehensive and validated testing of fire service attack methods has been that conducted by UL FSRI in the last few years.

Their initial 3-year study, published in 2018, centered around research on fire service ventilation and suppression tactics, for example exterior/interior suppression called *transitional attack*, to provide a comprehensive assessment of firefighting tactics (fig. 3–9).

One of the more controversial conclusions reached by the studies was that transitional attack, knocking down a vented fire from the outside and then rapidly changing to an

Figure 3–9. The UL FSRI began their studies by building actual buildings inside their mammoth burn cell, then burning them, gathering massive data using extensive instrumentation and audio-visual recording. (Photo courtesy of UL FSRI)

aggressive interior attack, materially affected the efficiency and safety of the attacking firefighters as well as increasing the survivability of any victim trapped inside.

The study developed creditable data which shows that directing a straight stream from a typical 150 GPM or so handline into a window venting fire accomplishes materially reducing the interior heat and provides cooling water to all sides of an involved compartment. The tactic is to direct the stream at the ceiling with gentle side-to-side movement. Care must be taken to not whip the stream which tends to block the flow path of exiting heat and gases. Once the fire is knocked down from the outside, a rapid advance on the fire from the interior is made much safer.

Many instructors, including the author, taught this tactic as *breaking the back* of the fire, reducing its intensity and making it safer to gain entrance to complete extinguishment (fig. 3–10).

Transitional attack greatly differs from the earlier late-1950s method in that a fog stream is not used for the purpose of creating steam and the higher flow rate will overcome a larger volume of fire than the typical 60 to 95 GPM streams used in the 1950s.

Even better, all the UL FSRI testing, methodology, tactics, and conclusions are available on its website, https://fsri.org/research. Any serious student of fire attack should consider it a must to go there and take part in their documentation and accompanying online classes.

Figure 3–10. A number of test fires were conducted on Governor's Island with the cooperation of the New York City Fire Department (FDNY). (Photo courtesy of UL FSRI)

Stream Selection for Penetration

Using transitional and direct attack calls for applying water directly to the flame/fuel interface to stop combustion reaction by cooling the burning material and to cool down the compartment. To accomplish this task, the water stream must be capable of penetrating through heat and flames to reach the burning material. The best stream shape, of course, is one that holds the water together in a tight mass so as to be better able to fight the effects of high heat and thermal drafts. The first choice should be a straight stream from a low-pressure combination nozzle or a smoothbore nozzle which will provide more water mass than a stream from a high nozzle operating pressure combination spray nozzle.

While the tiny water droplets created by air entrainment and spray stream patterns are considered more efficient for indirect attack, the tendency of these droplets to vaporize in the atmosphere before cooling the burning material makes spray streams undesirable for direct attack operations. Remember, direct attack calls for directly cooling the compartment and burning material. Simply put, a straight stream will more effectively accomplish extinguishment by transitional or direct attack.

We must consider a stream's solid material-cooling ability when discussing direct attack. About 30' from the nozzle, a straight stream provides about 2 sq. ft. of coverage. If, for example, a $^{15}/_{16}$" tip is being used at 50 PSI nozzle pressure, it will flow 185 GPM, providing 92.5 GPM per square foot of coverage. If a combination nozzle is flowing the same 185 GPM but the pattern is set at about 30°, the area of coverage will be approximately 100 sq. ft. This provides approximately 1.9 GPM per square foot if all the water is able to reach the burning material. It must be remembered that the finely divided water droplets are easily pulled upward into the fire's thermal column, and when this happens neither they nor the steam they generate can cool the burning material. The concentrated water mass of a straight stream provides the superior penetration needed to directly halt the combustion process. The distribution of the high concentration of water per square foot depends on effective movement of the nozzle by suppression crews.

Indirect, spray nozzle attack on the fire's heat requires that the generated steam must eventually settle on the burning material to stop combustion, the source of heat production. If the fire is properly ventilated, which it should be to provide for the attack crews' and trapped victims' safety, the indirect attack theory goes out the window along with the upward movement of the steam blanket indirect attack's means of extinguishment.

Life Safety During Firefighting Operations

Safety for both the firefighters and trapped victims must be the most important consideration when operating streams within a fire building. Testing has now proven that a direct attack on the burning material with a straight stream, coupled with coordinated ventilation, is still the best way to protect firefighters and victims from heat and smoke present in most interior fires.

The direct straight stream attack generates a minimum amount of unwanted steam and does not cause the dangerous driving effect that could threaten trapped occupants or spread the fire. Direct attack can be made with either smoothbore nozzles or combination nozzles set on the straight stream pattern position. To safely extinguish the fire without burning the nozzle crew, the idea is to keep vaporizing water out of the superheated overhead when firefighters are operating under or around it.

One of the hottest, most dangerous areas within a fire building are the areas behind and over the fire. In many departments, ladder or rescue companies will enter the fire area from the side opposite the attack crew to perform search operations before or while water is being applied. There is no doubt this area is tough to operate in, but prompt rescue is a priority if a victim is trapped in the behind-fire area. Taking someone out of a building from behind the nozzle is merely a removal. A rescue is when someone is removed from the area beyond the fire. This beyond or behind area can be three-dimensional if an open stairway or other vertical shaft is allowing the fire to extend upward. Then the behind-fire area can also include a number of floors above the original fire.

To enable rescue crews to operate with some level of safety in the behind-fire area, it is extremely important that the nozzle crew cause as little unwanted fire movement and steam production as possible. Pushing a fire on top of a primary search crew could just wreck their whole day.

If the base of the fire is located in the overhead, then direct attack calls for water to be applied upward to reach the burning material. This should not be considered an indirect attack because the stream is cooling the base of the flames, not attempting to create steam to extinguish a fire located elsewhere (fig. 3–11).

Other Testing

In 1980, the fire department of Osaka, Japan, embarked on a program that attempted to determine the most efficient water droplet size for reducing heat in a fire atmosphere. Their theory was that if a firefighting crew had a hard time making entry into a fire area, especially in high-rise construction, this ideal fog could be injected into the heat of the fire, reducing its temperature without causing excessive water damage. Apparently, tactics in use by the department at that time utilized quite a quantity of water which caused excess damage to the floors below.

Research told them that a water droplet about 200 to 300 microns in size exhibited the most efficient cooling and extinguishing effects. This discharge appears similar to that exiting a paint spray gun. The nozzle they developed has a tip that flows about 48 GPM at about 142 PSI and a fog pattern reach of about 26' and a forward stream reach of about 75'. The device looks like nozzles used on gas pumps.

In test fires, they reported rapid temperature declines after the fog was injected into the atmosphere, allowing entry and normal firefighting to continue using the straight stream pattern. Their conclusion was that since the test fires were knocked down rapidly and little water seeped to the floors below, the system proved efficient and was recommended for service.

Around 1985, the Fairfax County and Montgomery County, VA, fire departments, along with the National Institute of Standards and Technology, conducted a series of tests using this device as well as other nozzles from around the world, which attempted to confirm the optimum size for a fire extinguishing water droplet. The firefighters were instructed to extinguish fires in large piles of oak pallets using their own department's standard operating procedures. Because the testing was designed to determine the ideal firefighting droplet size, the tests were conducted using direct attack with spray patterns. The area from the ceiling to floor was wired with a number of thermocouplers to record temperatures.

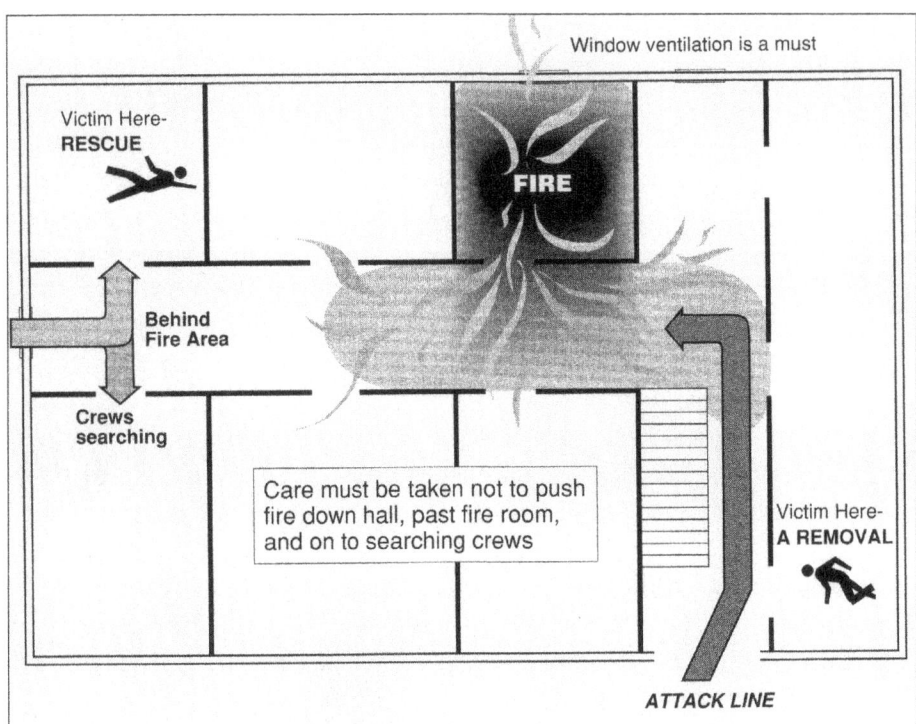

Figure 3–11. It is sometimes necessary to assign crews to perform primary searches beyond or over a fire before attack lines are in full operation. It is extremely important that the attack line be positioned and operated in a manner so as not to push fire and combustion products toward the search crews. The use of straight streams, coupled with proper ventilation that controls the flow path, will lessen the danger to the search crews. Operating in the behind-fire area can be dangerous and most certainly will be punishing. Many big cities classify a victim found behind the nozzle as removal and only victims found by crews searching behind or over the fire as actual rescues.

According to firefighters I interviewed later, during the tests using fog streams, firefighters received first- and second-degree burns on the face, neck, and wrists even though they were wearing full protective clothing, including hoods. The tests were exhausting, but the crews could not be relieved, since it was believed that changing personnel could skew the results.

In some tests, rapid steam generation during attack caused the temperature sensing devices to record a great amount of inversion or reversal of the fire atmosphere's temperature, whereby the overhead was cooled to the original floor temperature and the floor registered temperatures around the 1,000°F mark. The testing concluded that while water fog with a droplet size of between 250 to 350 microns proved to be the most efficient form of water for fire extinguishment, the injuries, reduction in visibility, possibility of unwanted fire extension, and handling difficulties caused by operating the nozzles at their standard 100 PSI exit pressure made the small-droplet water fog the least practical form of water for manual application by humans from within the fire environment. Remember, the testing did not consider life safety for either the suppression crew or civilians who might be trapped in the fire area.

One interesting issue comes to mind when reviewing the data supplied by the Osaka Fire Department. Because of the short reach of the fog stream, the nozzle crew must be no more than 26' away from the fire to effectively apply the water, which placed them in the vortex of the created heat, steam, and smoke.

Osaka department also reported the following:

- Once the temperature of the test fires had been lowered enough to allow entry, they found that further application of water had little effect on reducing internal temperature and caused water damage because no evaporation was taking place.

- Using the fog stream only was likely to cause the fire to rekindle, so the use of a straight stream to continue the attack was recommended after gaining entry.

- If the fire required the application of large amounts of water, the new nozzle system, with its short range and limited flow rate, had limited usefulness.

What actually happened was that the Osaka Fire Department testing reinvented the theory of reducing a fire's temperature with water fog, using the same arguments of efficiency and reduction of water damage that were advanced in this country 73 years ago. They did prove that an atomized droplet of about 300 microns is more efficient in cooling the fire's heat than larger or smaller drops. It is not clear, however, how they intended to manually apply this atomized water at close range without causing injury to firefighters. It is interesting to note that, using their theory of cooling the heated atmosphere, they still needed a straight stream for extinguishment after the temperature of an area was reduced.

Indirect Approach Applications

There are certain fires for which the indirect approach is ideally suited. For example, a fully involved, unfinished attic can be easily handled by poking a small hole in the ceiling below the fire or opening up the attic access scuttle and inserting a fog nozzle set on a pattern narrow enough to deliver water to the underside of the roof. The generated steam will be reasonably contained within the attic space and will extinguish the fire with a minimum of discomfort to the nozzle crew. As the steam builds up, it will blow back downward, indicating the application should be stopped. From the point of view of life safety, we can assume it would be unlikely that a victim would be trapped in an unfinished attic. If there were such a victim, and the attic were fully involved, it would be unlikely the victim would survive. In any event, the fire would have to be knocked down for the rescuers to enter, and the indirect method would be the quickest and safest method for the firefighters to use.

Development Of Sensible Tactics

By examining each theory of fire extinguishment and combining test results with a great amount of practical application experience, a more sensible approach to safe and efficient fire attack becomes clear:

- The rate of flow and proper distribution are the most important factors governing water application for fire extinguishment.

- Steam generation for the purposes of extinguishment, while helpful in very few, specific situations, is not desirable when any firefighters or victims are exposed to its effects or when the fire is not totally confined.

- Straight streams from either low-pressure combination or smoothbore nozzles are approximately equal to each other in effectiveness for most interior firefighting.

- Tests have proven that atomized spray streams with an ideal droplet size of about 300 microns have proven to be the most efficient for fighting fire, if human survival within, above, or behind the fire situation is not a consideration. The mist is so fine, however, that it is difficult to successfully apply by manual firefighting operations without exposing nozzle crew members to unnecessary thermal injury.

- When the rate of flow accomplishes extinguishment, water application should be stopped immediately to prevent disruption of the fire atmosphere. Continued application of water can cause imbalance in atmosphere steam saturation, which is discomforting to firefighters attempting to work within the environment. It also can cause excessive water damage.

- Applying water fog streams to the overhead for protection causes superheated steam to be generated rapidly and to descend violently on the nozzle crew. The steam can penetrate turnout clothing and woven hoods, causing thermal injuries. Fire experience has proven that keeping water out of the overhead to create steam will reduce injuries, increase visibility, and reduce fire spread. If you don't like being burned, don't spray water into the heat at the ceiling.

- Most instructors these days, teach to keep the straight stream moving up, down, and all around. This provides cooling to all surfaces of the compartment by direct water application and reduces the generation of unwanted excess steam.

- The fire should be attacked as soon as the nozzle crew is in a position to hit the fire with a straight stream; it is neither necessary nor desirable from a safety standpoint to get on top of the fire before opening the nozzle. Good practice is to cool the surfaces on your way in and then hit what's burning when you get there.

- Never "pencil" flow on a fire. Shutting the stream on and off before the fire is completely extinguished is a sure way of getting seriously hurt or killed. This technique caused the death of a firefighter in Homewood, IL, a few years back and seriously burned his partner. Never shut the stream down until the fire is extinguished. Always keep in mind that a closed nozzle extinguishes no fire.

- Ideally, ventilation should be accomplished through carefully selected channels by suppression personnel specifically assigned to the task. The UL FSRI studies specifically call for controlling the ventilation flow path to ensure that the fire is not unwittingly fed excessive amounts of outside air, and in some cases, controlling the door through which the attack team enters. Ventilation must not be considered as merely a byproduct of nozzle operation or stream pattern selection. There are some who advocate blowing a fire ahead of the nozzle crew through the use of a spray stream. This is a dangerous practice that can push fire, combustion products and steam into unwanted or unknown areas, spreading the fire and seriously threatening trapped victims or personnel searching these areas. If it is desired to utilize the air-moving capabilities of fire streams to assist in ventilating an area, their operation should not begin until all factors of victim and firefighter safety as well as unwanted movement of fire, steam, and combustion products have been carefully considered.

- Straight streams perform best when operated aggressively. Crews should plan to advance quickly on the base of the fire. While the most efficient tactic is to prevent generation of flammable gases by directing the stream directly on the burning material, sometimes obstacles such as furniture, partitions, office files, or machinery prevent direct application. In these cases, water should be distributed by deflecting the stream off of the ceiling, walls, and floor by quickly sweeping the water column as far ahead as possible, bouncing large, cooling water droplets into the fuel and fire area.

- Almost any fire department using any types of tactics can handle a room-and-contents fire, some better than others, but nonetheless successfully. Too often, these successes are the basis for all firefighting done by a department. Be careful about falling into the *room-and-contents* or *dwelling fire syndrome*. I've seen departments that could handle a bedroom fire quite well and then proceed to burn down every store, gas station, multiple dwelling and business occupancy to which they were called. They suffered from the room-and-contents syndrome. They did not plan for flows over 150 GPM, using large caliber streams or provide adequate water supplies for fires in areas larger than 200 sq. ft. When was the last time your crew practiced stretching and operating a 300' 2½" line?

Preventing and Mitigating the Effects of Flashover

Retired Deputy Chief Vincent Dunn of the FDNY described flashover as the sudden full-room involvement in flame caused by thermal radiation feedback in his book *Safety and Survival on the Fireground*.

As a fire develops, its heat is absorbed into ceiling and upper wall panels and reradiated back into the space, gradually heating the atmosphere, which contains combustible gases and the solid contents of the involved space. When the combustibles are heated to their ignition temperature, both the space and any combustibles it contains, simultaneously ignite.

One myth which should be put immediately and completely to rest is the belief that a spray stream directed into the overhead will provide protection from the effects of flashover. Flashover happens quickly, many times taking the crew unawares, and immediately exposes them to the effects of more than 1,000°, enveloping them from all angles. Chief Dunn describes a point of no return from which a fully protected firefighter will probably escape a flashover space as no more than 5', allowing the firefighter about 2 seconds to retreat or be seriously injured.

It takes more than 2 seconds for a nozzle operator to sense the situation, reach up and change a nozzle's pattern to wide spray. In the time it takes to fool with the nozzle, the crew will have already been seriously burned or killed.

When a fire involves an enclosed space and the fire's heat builds up to a point where reradiation is occurring, flammable gases are being distilled from the heated material, filling the space's atmosphere. When the gases ignite, sporadically in the case of rollover or completely in the case of flashover, the entire space is filled with fire fueled from burning gases.

This situation can be likened to a propane cylinder leaking from a partially-opened valve. If ignition occurs, certainly the effects of flame travel and exposure of surrounding areas are a consideration. But the main consideration in handling this type of fire is to shut off the gas supply. If the fire is extinguished without closing the valve, the fire is almost sure to reignite. In this case, the fire is only a symptom; the fuel supply is the disease.

In any space containing heated gases which are likely to flashover, cooling the compartment immediately, as in using a straight stream to sweep the ceiling, walls, and floor, is a life-saving measure. After much research, I have yet to find a situation where a flashover occurred when a stream of sufficient volume was operating in the fire compartment.

If excessive heat is causing the crews to hug the floor, it can be reasonably assumed that conditions are building for a potential flashover. It certainly makes sense to begin cooling the compartment to help prevent further flammable gas distillation. This is not the same as applying a spray to the upper atmosphere of the area to make steam. One experienced large-city officer described the two different applications tactics as, "Solid cools, steam burns."

Recent research, testing, and evaluation of practical firefighting experience has helped prove that some relatively unscientific methods of fire attack used by some firefighters can in fact, cause injury and spread the fire to unwanted areas.

By combining the principles of transitional and direct attack with indirect attack, and by realizing no one form of water application can efficiently extinguish all fires, a sensible tactics plan can be developed which can kill a great amount of fire without killing the persons performing the application.

Chapter 3 Review Questions

1. What is the most effective and safest method of attack?
2. Describe transitional attack.
3. What are the most important factors in water application for extinguishment?
4. What is the primary cause of flashover?
5. Why is it important to control the interior ventilation flowpath?

Nozzle Basics

It can be intelligently argued that the most critical piece of equipment used by firefighters is the nozzle—that lowly and often mistreated piece of aluminum, rubber, and brass that firefighters always seem to take for granted, but the device on which firefighting success and survivability are solidly based. Without the nozzle, all other firefighting devices are rendered useless. Hose, pumpers, aerial devices, breathing equipment, and turnout clothing cannot be effectively utilized for fire combat unless the nozzle does its job of shaping water into a useful form for firefighting.

To completely understand firefighting, a firefighter must understand nozzle operation. Suppression personnel should never be allowed to flow water in combat without knowing the hows and whys of fire stream production. Imagine an army going into battle without their troops having practiced marksmanship and learning to repair their weapons—the result would be disaster and death. Likewise, a firefighter must understand the primary weapon in the firefighting arsenal, how it should be fired, and, if broken, how it should be fixed so it can effectively fire again.

Some nozzles are unique in design and need special instruction to comprehend their proper operation. These nozzles are described in later chapters. The most widely used nozzle in fire service use is the *fog* or *combination* nozzle. It is best referred to as a combination nozzle because it will flow not only a fog or spray pattern but a straight stream as well. This chapter will explain basic nozzle theory and how it applies to conventional combination nozzles.

Basic Nozzle Design

All nozzles are designed to help put out a fire by getting the right amount of water in the right form and in the right place. To accomplish these tasks, the nozzle performs four main functions:

- Controls flow: The size of the exit orifice in the outlet of the nozzle controls how much water is passing through the nozzle at a given pressure. Each nozzle type has a standard operating pressure for rating purposes: 50 pounds per square inch (PSI) for smoothbore tips, 100 PSI for standard combination nozzles, and 50 PSI–75 PSI for low-pressure combination nozzles. Combination nozzles are available in different style configurations that provide different means for controlling flow. These nozzle configurations will be discussed later.

- Provides reach: Any nozzle creates a restriction at the end of the waterway which changes pressure into speed or velocity. Velocity provides the reach necessary to distribute the water to where it is needed. It may not be practical to approach a fire at close range on the fireground due to excessive heat or a weakened structure. In these cases, the effective reach of a fire stream becomes important both from firefighting and safety standpoints.

- Creates shape: Selection of nozzle type determines the shape of the discharged water. For example, a smoothbore nozzle will produce a long cylinder of water and a combination nozzle will provide a spray pattern. Different firefighting situations may require water to be applied in different forms.

- Determines water direction and form: A *distributor* or *cellar* nozzle will form its discharge into a different style of stream than that formed by a foam playpipe. A low-pressure combination nozzle provides a spray stream that exits a nozzle differently from that of a standard-pressure combination nozzle. Because the stream is denser, the solid-stream discharge from a smoothbore nozzle has greater penetration power in high-heat situations than that of a straight stream provided by a standard-pressure combination nozzle. The effectiveness of the firefighting action provided by a particular nozzle is in many cases a subjective determination by firefighters based on experience in actual use.

Nozzle Components

A nozzle usually consists of two basic parts. The first part is the shutoff. The shutoff device allows the nozzle operator to control the timing of a nozzle's discharge by allowing the flow to be started or stopped. In the case of combination nozzles, the shutoff may be built into the nozzle pattern control where the stream is turned on or off by twisting the nozzle. A ball shutoff, operated by a lever or bail and provided between the hose and pattern adjustment control, is common in most nozzles nowadays. Some break-apart nozzles have both types of shutoffs.

The second part is the stream shaping device. Smoothbore nozzles use a tapered tip to shape the stream. Combination spray nozzles utilize a variety of methods to provide various stream patterns. Teeth are provided around the outer edge of the exit orifice in combination nozzles to help deflect a portion of the nozzle's discharge into the center of the pattern as well as helping break up water into smaller droplets.

Nozzle Types

Structural firefighting nozzle designs fall into three basic categories:

- Smoothbore nozzles
- Combination spray and straight stream nozzles, also called fog nozzles
- Specialty nozzles, such as distributing nozzles, foam nozzles, or piercing nozzles

Smoothbore, combination nozzles with automatic nozzle pressure control devices, and low-pressure nozzles will be described in later chapters. This chapter will deal mainly with the most popular nozzles used in the fire service: the combination (spray and straight stream) type.

Nozzle Basics

Combination nozzles were made generally available to the American fire service in the 1930s, first by a nozzle design imported from Denmark by American La France, followed closely by the Mystery nozzle design, developed in Germany and manufactured in this country by Elkhart Brass Manufacturing. In reviewing published information available from the period before the war, it seemed the combination nozzles were generally considered specialty devices used more for flammable liquid, coal bin, and other difficult-to-extinguish fires rather than for general structural firefighting.

The development of nozzles that provide broken or spray streams can be traced back to 1863 when a patent was issued to Charles Oyston of Little Falls, New York for a device called Improvement in Nozzles. His nozzle, which used a standard smoothbore nozzle as its center, had hooks which could be moved into the stream by sliding a sleeve. These hooks caused the stream to be divided into a crude spray pattern. It is not known if his device was ever put into widespread production, but in a letter to the Scientific American on January 27, 1877, he said, "If water in the form of spray be a good extinguisher, as it undoubtedly is, as numbers of proof exist in our factories and picker rooms, why do not our fire departments use it in that form in all cases where they can?"

He goes on to say,

> I am well aware that this statement may seem extremely absurd to firefighters who have never experimented with this line; but before they condemn it, let them take out a couple of engines and try the experiment. The barbarous system now in use, that so frequently desolates portions of our cities, fills our houses with mourning and our cemeteries with new-made graves, must give way to the dictates of science. Humanity demands it, and I call on the scientists and chemists throughout the land to aid in introducing this needed reform.

Although Oyston comes across as quite a salesman from his writing, it appears that attempting to introduce change into the fire service late in the late 19th century was as difficult as it is now.

Though the theory of combination nozzle use is detailed in other chapters, a brief overview of the spray nozzle's development and movement through fire service history is helpful in understanding its position in the fire service's arsenal (fig. 4–1).

World War II

World War II was the single most important event in the 20th century that initiated sweeping changes in the state of firefighting. Besides the introduction of foaming liquids and improved breathing apparatus, the most dramatic development in structural firefighting was the introduction of fog streams into general firefighting use.

The Navy was instrumental in developing water fog techniques for fighting fires in flammable liquids and taught the methods to hundreds of thousands of sailors during the war years. Their tactics were based on a simple principle: cooling the burning liquid below its vaporization point, thereby causing the halt of flame production.

The Navy developed a set of nozzles specifically for this use. The Navy all-purpose nozzle had twin discharge ports, one of which was a smoothbore nozzle orifice located above

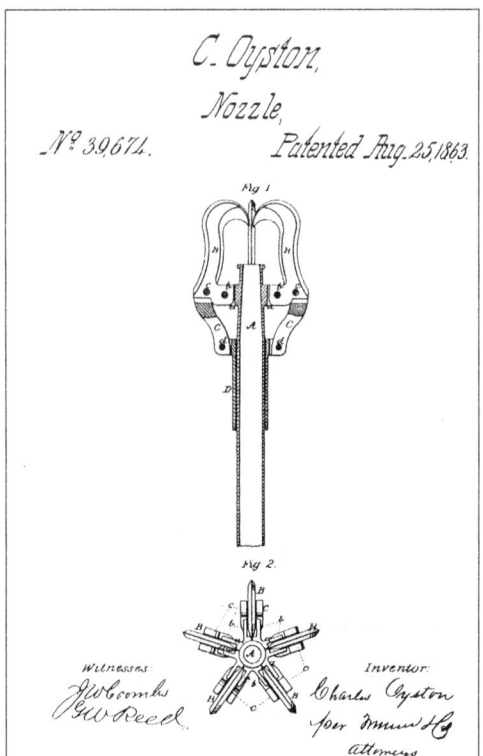

Figure 4–1. The first spray nozzle for fire use was patented in 1863 by Charles Oyston. While attempting to promote his device, many claims were made advocating the effectiveness of a broken water stream over a solid stream for firefighting. It is an argument which continues over 160 years later.

a fixed pattern, impinging a stream fog head. The Navy's tactics utilized the fog head for close-in cooling purposes and the smoothbore discharge for longer-range cooling (fig. 4–2).

Although the nozzle was not designed specifically for structural firefighting, the Navy and other branches of service used it for that purpose, and it was placed in service by thousands of fire departments during and after the war. The nozzle can still be seen deployed by a few departments in New England. The fog head is removable, so an extension applicator can be affixed and used to reach fires in remote areas or to apply a cooling spray of water over the top of a flammable liquid fire.

One of the drawbacks to using the Navy-style nozzle for structural firefighting is its low flows. The 1½" version flows 52 gallons per minute (GPM) through the fog head and 82 GPM through the ⅝" smoothbore opening, and the 2½" size flows 132 GPM through the fog head and 200 GPM through the 1" smoothbore opening. These flows are considered dangerously low for structural firefighting, especially when higher flows are available through contemporary nozzles.

While the Navy-style nozzle is still used in some commercial shipboard firefighting, the Navy as well as the Coast Guard have themselves standardized on the peripheral deflection nozzle, similar to those used in the civilian fire service.

The Early Days of Fog for Structural Use

After Chief Lloyd Layman's book was published in 1952, and the demonstrations of fog nozzle fire attack spread across the country, the fire service interest in spray nozzles begin to develop.

Keith Royer and Bill Nelson of Iowa State University performed extensive testing in the late 1950s and early 1960s which did much to help further the research into the use and design of combination nozzles. Their research led directly to the development of the constant flow nozzle, in which the flow remains the same no matter what stream pattern is selected. Previous fog nozzles discharged varying flow depending on stream position.

Figure 4–2. The Navy-style nozzle creates a fixed-pattern spray stream by directing small streams of water against each other. Called an *impinging stream nozzle*, the droplets produced by this method are extremely uniform in size.

During the late 1960s, in Gary, Indiana, the late Chief Clyde McMillan pioneered research into the development of the automatic pressure adjusting nozzle. Also in the 1960s, manually adjustable flow controls began to be installed on nozzles and, in 1971, Elkhart Brass introduced the first complete, automatic pressure adjusting nozzle.

Other manufacturers quickly followed, and the automatic nozzle became a popular type of nozzle purchased by municipal fire departments. Basically, they always were able to flow a stream with adequate reach.

Within the past 20 years, a reassessment of firefighting tactics by various departments and universities has started a trend toward the increased use of smoothbore and combination nozzles that operate at lower pressures.

Stream Pattern Control

One of the primary advantages attributed to the combination nozzle is its ability to provide a number of stream shapes to fit various firefighting situations. While the straight stream position is the most effective and safest setting to be used while working on a fire inside a building, some fires (flammable liquids fires, for example) are more effectively controlled by the use of a spray stream (fig. 4–3).

There are three basic ways for a nozzle to create a spray stream:

- Peripheral deflection: Having the stream exit the nozzle in a hollow circle and creating different stream shapes by moving a barrel up and down. Most combination nozzles on the market are of this type.

Figure 4–3. Most nozzles used in the fire service are of the peripheral deflection type. Stream shape is determined by the size and position of the gap between the nozzle's center baffle stem and the outer barrel.

- Impinging stream: Holes drilled in the tip allow tiny, solid streams to strike one another, creating a spray pattern. The old Navy nozzle and the bayonet piercing nozzle are examples.
- Swirling deflection: Although widely used for spray devices in industry, it is rarely used in fire service nozzles. The old Coast Guard nozzle tip and high-pressure fog guns are examples.

Controlling Gallonage

Because of its ease of manufacturing and design flexibility, the peripheral deflection style of nozzle is the most common type of combination nozzle found in the fire service. It is available in four basic design styles. The style determines how the nozzle handles its water flow:

- Single gallonage, variable flow: This nozzle's gallonage rating is determined by the flow from only a single pattern position. As the pattern is changed, the flow changes. This style was the most widely used by the fire service until the early 1960s when it was superseded by the constant flow type. The variable flow design is still found in lower-priced industrial nozzles (fig. 4–4).

Figure 4–4. Examples of industrial-type nozzles. The nozzle at the left is a fixed-gallonage, variable flow style, designed for use on interior hose racks. The nozzle at the right is also designed for industrial use; however, it incorporates a floating stem in its design which maintains constant flow at all pattern positions. While both nozzles appear similar, they will exhibit quite different extinguishing effectiveness when used on structural fires.

- Single gallonage, constant flow: This nozzle is similar to the variable gallonage type, except for the addition of a second interior collar (or floating stem), which keeps the gallonage constant no matter what stream pattern is selected (fig. 4–5a). For the nozzle to flow its rated gallonage, the supplied nozzle pressure must be the same as the nozzle's recommended operating pressure.

Figure 4–5a. Older combination nozzles, which were practically all manufactured before 1965, and more current industrial and wildland nozzles are single gallonage, variable flow. As the pattern is changed, the size of the gap between the center baffle and outer control ring varies, which allows more water to flow in the wide spray positions and less in the forward stream position.

- Adjustable gallonage, constant flow: This nozzle has a flow control ring that allows the operator to manually select the desired gallonage. A few adjustable gallonage designs are dual gallonage and used mostly in fighting wildfires. On these nozzles, the gallonage is selected by rotating the pattern selector. All adjustable gallonage nozzles are also constant flow. Once the gallonage is selected by the operator, the nozzle will flow that gallonage in all stream positions. Falling out of favor for structural firefighting, they are still widely used in wildfire firefighting.

- Constant pressure, variable flow: These nozzles are commonly called *automatics*. The size of the discharge orifice is controlled by a spring that causes the exit baffle to open and close depending on the amount of water pressure pushing against it. Automatic nozzles theoretically maintain approximately 75 PSI or 100 PSI nozzle pressure by changing the flow to keep the nozzle pressure constant (fig. 4–5b). The amount of flow change depends on the nozzle design. Once the tip sets itself for the proper flow and pressure, the nozzle will flow a constant gallonage in all pattern positions (fig 4–5c).

Figure 4–5b. An older Mystery nozzle demonstrates how much the gallonage varies as the pattern is changed. Although the nozzle is rated at 95 GPM, its actual flow changes from 140 GPM in the wide spray position to 50 GPM in the forward stream position. All flows were obtained at 100 PSI nozzle pressure.

Figure 4–5c. Reduced gallonage in the straight stream position is a good reason hose rack or industrial use nozzles should not be used for municipal fire department operations. Many departments using these nozzles on standpipe hose packs do not realize that flow is reduced in certain pattern positions.

Shutoff Devices

Although the twist-type shutoff nozzle is common in wildfire applications, most shutoff devices used by the fire service for structural firefighting in industry or rural areas of the country are ball type (controlled by a handle or bail) due to its low cost. The ball has a hole bored into its center and it rotates against a seat to seal and prevent leakage. When the handle is pulled backward (toward the hose end), the ball rotates, allowing the nozzle waterway to align with the hose waterway, and water to pass through the shutoff area. If the shutoff is not aligned properly during manufacturing or if the nozzle is operated with the shutoff partially open, turbulence is created. This turbulence causes the water to tumble through the nozzle, degrading the stream and reducing the nozzle's flow.

Task Force's shutoff model uses a slide valve, or a hollow cylinder through which the water passes, which mates up against a circular seat to stop the flow. The advantage of this type of shutoff is that it creates little turbulence in the water flow if operated in positions other than fully open. It is normally supplied built into a nozzle with an automatic tip.

Combination nozzles can either have the shutoff built into the nozzle body or the shutoff can be a separate unit with threads on the discharge end on which a selection of tips, more hose, or accessories (such as piercing nozzles) can be mounted. The main advantages of a nozzle with an integral, non-removable shutoff are economy and compactness. While nozzles with separate shutoff units (called *break-apart* or *advanceable nozzles*) cost a little more and are a few inches longer in length than nozzles with integral shutoffs, they are much more versatile (fig. 4–6).

The break-apart feature can be used if more line is needed. The tip is removed, and additional line can be added directly to the threaded end of the shutoff without having to trace the line back to the pumper. If a different tip is desired (a smoothbore tip, piercing nozzle, or

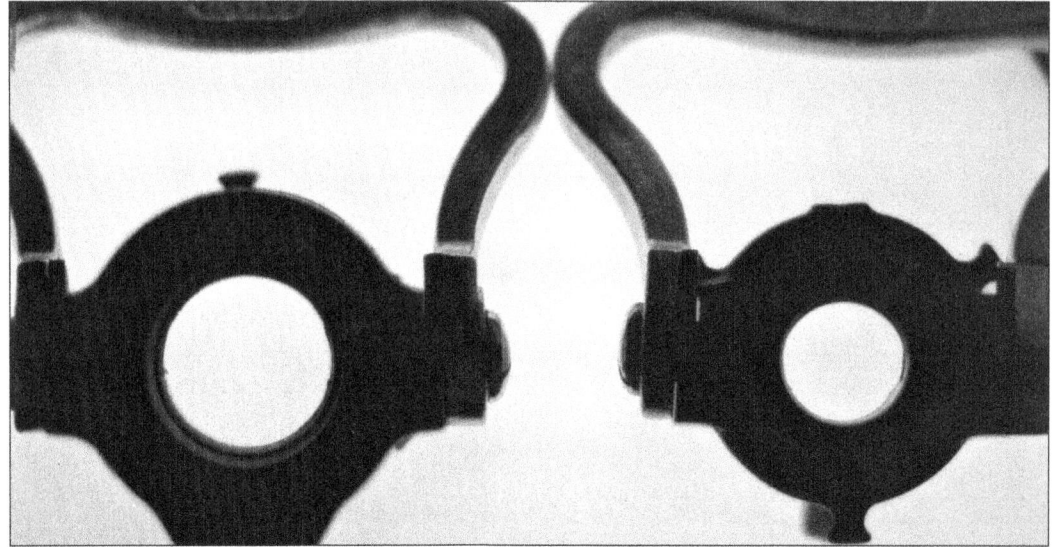

Figure 4–6. A view through the waterways shows the difference between shutoffs designed for large-volume handlines (left) and the older standard shutoff for lower flows (right). Although both have 1½" threads on both ends, it is wiser to purchase the shutoff with the largest waterway because it can be used on any size handline without restricting flow.

foam nozzle, for example), it is little trouble to replace the combination tip with another. If a nozzle with an integral shutoff is attached to the line, the line must be shut down at the engine before the nozzle is removed. This operation can become time-consuming if someone has to trace the hoseline back to the pumper to ensure that the proper discharge is being closed.

Combination nozzle tips used on break-apart nozzles can also contain their own shutoffs which are usually operated by twisting the pattern control barrel counterclockwise, past the forward stream position. This allows for the tip to be removed from the ball shutoff, the extra hose attached, and the tip used on the end of the additional line as a complete nozzle. Not all tips have a shutoff, so it is wise to investigate before they are purchased or placed in service.

A department wanting to standardize equipment for increased efficiency in maintenance and stocking repair parts would be wise to select a shutoff that can be used on all sizes of handlines. Besides the inlet thread size, which can be easily changed with adapters, the basic difference between different sized shutoffs is the diameter of the waterway. A shutoff designed for use on 1½" line or supplied with nozzles flowing up to 125 GPM will generally have an internal waterway diameter of 1". Shutoffs designed for use on lines of up to 2½" will have a waterway measuring 1⅜" in diameter, even if the inlet and outlet threads are 2½". When making purchases, it is wise to specify the larger shutoff even if it is purchased for use on smaller lines. The friction loss inside the nozzle is minimized and the shutoff can be used on larger lines if tactics change or if a larger nozzle is out of service for maintenance.

Also, by utilizing shutoffs with the same internal size waterway, the seals, handles, body, and other parts remain the same size. If only the larger size waterway shutoff is purchased for use on all size hoselines, it will become easier to keep a full stock of repair parts on hand, since they will be common to all hose sizes.

Flush Features

All newer nozzle designs feature a flush provision which enables the nozzle operator to clear the nozzle of debris and stones.

Turning the pattern selector completely counterclockwise activates a cam that opens the exit orifice past its normal water flow position.

On some nozzles, the flush position is activated by turning a gallonage or flush position selector located between the pattern control and shutoff. Likewise, this activates a cam that opens up the exit orifice to allow debris to pass.

When flushing nozzles, the nozzle need not be shut down. Any nozzle, when operated in the flush position, exhibits reduced nozzle reaction. Because of the large amount of water discharged during flushing operations, it is wise to direct the stream (if possible) into areas where the water will do little damage. If something becomes lodged in the nozzle while advancing on an interior fire, the flushing procedure usually cannot wait until the line is taken outside. During attack, immediate flushing may be necessary for crew protection and should not be delayed.

Sometimes the item lodged in the nozzle is too large to pass through the flush opening. If the nozzle has a separate shutoff, it is usually an easy matter to remove the tip, poke the debris out the back of the nozzle, and replace the tip on the shutoff so firefighting can continue. If the nozzle tip and shutoff is one piece, the line will have to be shut down at the pumper so that the entire nozzle can be removed for clearing.

Some older automatic nozzles have a screen installed in the inlet swivel to catch debris before it enters the nozzle barrel. Debris lodged against the inlet screen cannot be detected as easily as debris caught in the exit opening, which show their presence by deforming the stream. The debris caught farther back can easily cause restriction of the water flow, and since the automatic tip mechanism will adjust itself to provide a good-looking stream with the reduced water supply, the suppression crew could be flowing much less water than they anticipated or require for the fire at hand. When using nozzles equipped with inlet screens, streams should be carefully and continuously evaluated during use to ensure flow requirements are being met. If it is suspected that the inlet screen is clogged, the line should be shut down at the engine and the nozzle removed for inspection. Nozzles equipped with inlet screens must be removed after each use and inspected for debris which may have been caught during operation.

Teeth

The fog patterns formed by the early peripheral deflection nozzles had a center empty of water. This area contained air at a lower pressure than the area surrounding the outside of the stream and could cause unwanted flame turbulence when used on flammable liquid fires.

In 1942, Elkhart Brass introduced teeth on their Mystery nozzle line in an attempt to reduce this turbulence. The teeth, machined of brass and closely spaced around the outside edge of the exit opening, diverted a percentage of the spray pattern discharge into the center of the cone. This made the nozzle more efficient on flammable liquid fires since more water was broken up into fine drops that could more easily be vaporized by the fire's heat. The center, being filled with water, did not exhibit the effect of sucking flames into the nozzle core as the early, hollow center nozzles did. This full fog caused less flame agitation when attempting to cool flammable liquid fires.

Teeth have become standard on most all nozzles and are available in different forms:

- Fixed rigid: Machined from hard metal and permanently attached to the nozzle.
- Fixed flexible: Molded into the rubber bumper of the stream pattern control. These are the most durable along with fixed replaceable.
- Fixed replaceable: Molded from plastic or rubber and interchangeable with spinning teeth on some nozzles. They offer the advantage of durability and can be easily replaced if damaged.
- Spinning replaceable: In the past, molded from plastic. Now, generally made of metal. All spinning teeth are durable and replaceable. After each use, the nozzle should be checked to make sure it is free of debris and that the teeth spin freely.

The principle of having a spray stream cone filled with water was considered desirable as combination nozzles began to be used for structural firefighting because the action of the teeth increased the amount of available water droplets. When converted to steam by fire's heat, these droplets tended to increase the efficiency of indirect attack tactics.

As water passes out the nozzle and a portion is deflected inward by fixed teeth, the area of the stream affected by the teeth opens, appearing as small stripes or fingers in the stream. Since the fixed teeth are fairly efficient in filling up the interior of the fog cone with water, this fingering appearance is of little consequence when fighting structural fires.

Most spinning teeth and some replaceable fixed teeth will not have any effect until the pattern is set wider than 30°. When set on a narrow spray pattern, the teeth divert little to no water into the center of the stream.

Protection Myth

Some educators advocate approaching liquefied petroleum gas fires from behind wide-angle spray streams. They also advocate the use of spinning teeth because they feel it gives the best pattern for what they term *protection*. Many departments have purchased nozzles with spinning teeth especially for their perceived protection qualities.

Spinning teeth deflect little water into the center of the pattern when the nozzle is adjusted at its widest angle. When approaching fires, especially those in liquefied petroleum gases, the fire will actually intensify because of the increased airflow from the wide-angle flow pattern, and the reduced pressure in the center of the nozzle will vacuum the fire right up next to the nozzle. Viewed from the side, it can be seen that while all this fire movement is being accomplished, little or no actual cooling or fire extinguishment is taking place.

Approaching a flammable liquid or liquefied petroleum gas fire in this way simply exposes the firefighters to more danger and is more an experience in firefighting stamina and personal courage than an exercise in practical firefighting. When exposing personnel to danger by decreasing the distance between the firefighter and the fire, some mitigating firefighting action must take place, or the likelihood of injury will drastically increase.

If the object is to advance on some plumbing involved in fire in order to turn off a valve, the easiest way is to direct two narrow fog patterns on the plumbing and to move forward at a normal pace. As the crews get closer, the streams can be widened somewhat to about 30° patterns, and someone can reach in and turn the valve off behind this protection. While this movement is taking place, water is cooling the fire, tanks, and piping, helping to reduce the chances of a supply container boiling liquid expanding vapor explosion or metal failure from excessive heat, which could render the valve inoperative.

If a number of spinning teeth are broken and a fire is approached behind a wide spray pattern without attempting extinguishment, the low pressure present in the hollow center will pull the fire right into the nozzle and the broken teeth will spit it out in a circular pattern around the firefighters, similar to a firework pinwheel. This effect is not a problem with the nozzle or its teeth. It is a problem with tactics. No fire should be approached behind a fog pattern without some cooling or extinguishing action taking place. All the wide pattern accomplishes is to give the firefighters a false sense of security. This can easily be broken with the tip of a helmet brim or coat collar, and all the while brings them closer to danger without reducing that danger.

When selecting nozzles for interior fire attack, the effect of teeth on the stream is not as important as flow characteristics, pattern control, and volume control, and it would be senseless to base a final purchasing decision on style of teeth alone.

Combination Nozzle Types

Single Gallonage

The single gallonage nozzle flows its rated capacity at a specified nozzle pressure, usually 50 PSI–100 PSI. While the nozzle is rated at only a single gallonage at its rated operating pressure, it will flow less or more water than its rating depending on the actual nozzle pressure. If less than its operating pressure is supplied, the rate of flow will be decreased and (depending on how low the pressure is) the stream will lack reach and firefighting action. If the single gallonage nozzle is supplied with more than its operating pressure, the volume will increase, but so will the nozzle reaction. Depending on how high the pressure goes, the stream will begin to break up and effective firefighting reach will be decreased. If the nozzle pressure varies a considerable amount above or below the rated operating pressure, the single gallonage nozzle may or may not produce streams that can fight fire effectively, depending on the amount of increase or decrease in pressure and flow. Despite these problems, the single gallonage nozzle is rugged in construction, simple to use, and less expensive than other types. If departments were not advised as often as they are by salesmen promoting their latest products, the fixed-gallonage nozzle would accomplish the bulk of the firefighting in this country.

Adjustable Gallonage

The adjustable gallonage nozzle is similar in design to the fixed flow nozzle except for a selector ring that allows the nozzle operator to select various orifice openings with which the flow can be varied. The selector ring is marked in various GPM flows, usually in ranges of 10 GPM–30 GPM for booster lines, 30 GPM–125 GPM for 1½" lines, 95 GPM–200 GPM for mid-range lines, and 125 GPM–250 GPM for 2½" lines. When operating these nozzles, it must be remembered that each marked flow is calibrated to move that amount of water at the nozzle's rated nozzle pressure. For example, a department determines the proper engine pressure necessary for supplying a 150 GPM flow at 100 PSI nozzle pressure while using 150' of 1¾" line by previous testing. If the gallonage selector is changed to the 95 GPM position without changing the pump pressure, the gallonage will decrease and the nozzle pressure will increase.

Gallonage	Nozzle Pressure	Nozzle Reaction Force
150 GPM	100 PSI	76 lb.

If engine pressure is not changed, the following happens when the gallonage selector is moved to 95 GPM:

Gallonage	Nozzle Pressure	Nozzle Reaction Force
107 GPM	127 PSI	57 lb.

It can be seen that while the nozzle pressure is increased, the gallonage is decreased and the resulting combination produced a decrease in nozzle reaction.

Conversely, if the selector on the same nozzle calibrated to flow 150 GPM at 100 PSI is changed to the 200 GPM position without changing the engine pressure, the gallonage will increase, but the nozzle pressure will decrease.

Gallonage	Nozzle Pressure	Nozzle Reaction Force
150 GPM	100 PSI	76 lb.

If the engine pressure is not changed and the gallonage selector is moved to 200 GPM (fig. 4–7), the following will happen:

Gallonage	Nozzle Pressure	Nozzle Reaction Force
177 GPM	81 PSI	80 lb.

If the gallonage selector is set to a higher setting, the nozzle will flow more water and the nozzle pressure will be lower; however, the combination will slightly increase the nozzle reaction force.

If the engine pressure is not adjusted to maintain 100 PSI nozzle pressure each time the gallonage selector is changed, the nozzle will flow more than the marked gallonage if adjusted below a calibrated setting and will flow less than the marked gallonage if adjusted above the calibrated setting.

It sounds confusing, but if a little thought is applied, the operation is really fairly simple. To repeat, most departments using selectable gallonage nozzles think they are flowing what is marked on the nozzle. This is only true if the engine pressure is changed each time the selector ring is moved. In actual practice, the department should determine the engine pressure needed to flow a gallonage in the middle of the nozzle's range. Then, if more or less water is wanted, the nozzle operator selects a lower or higher marked setting knowing that the actual water flow will differ from what's marked. What is important is that the operator will know they are getting less or more water by looking at the nozzle settings. This operation is greatly simplified if flow meters are used on the pumper discharges to indicate accurate flow rates.

Figure 4–7. The gallonage selector on a selectable gallonage nozzle. This selector may also incorporate the flush activation feature.

Nationwide Nozzle Survey

In a nationwide survey conducted by Elkhart Brass, a clear preference was shown for smooth bore nozzles in both residential and commercial attack applications (figures 4-8a and 4-8b).

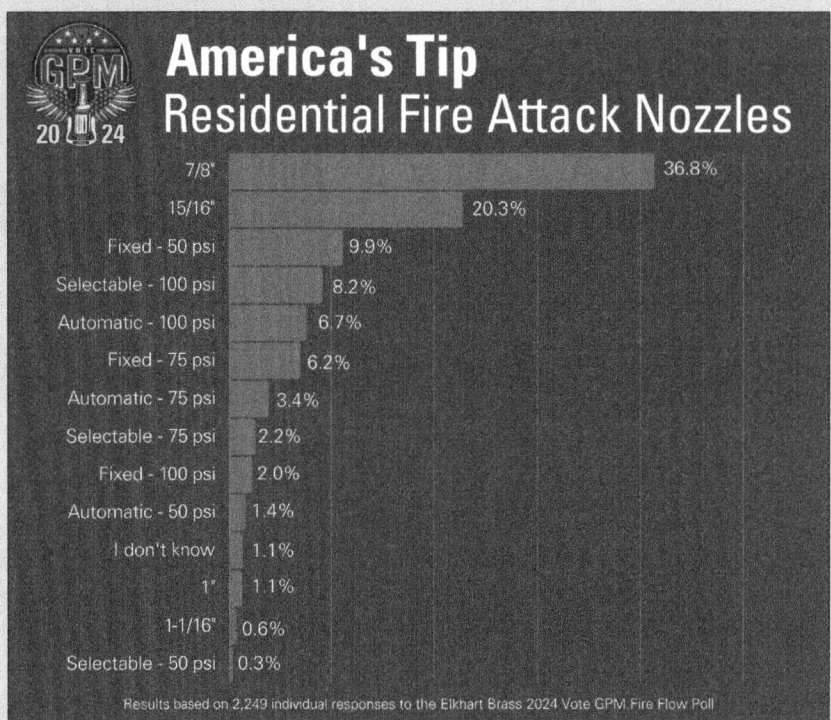

Figure 4–8a. Survey results ranking preferred fire attack nozzles for residential fires.
(Image courtesy of Elkhart Brass)

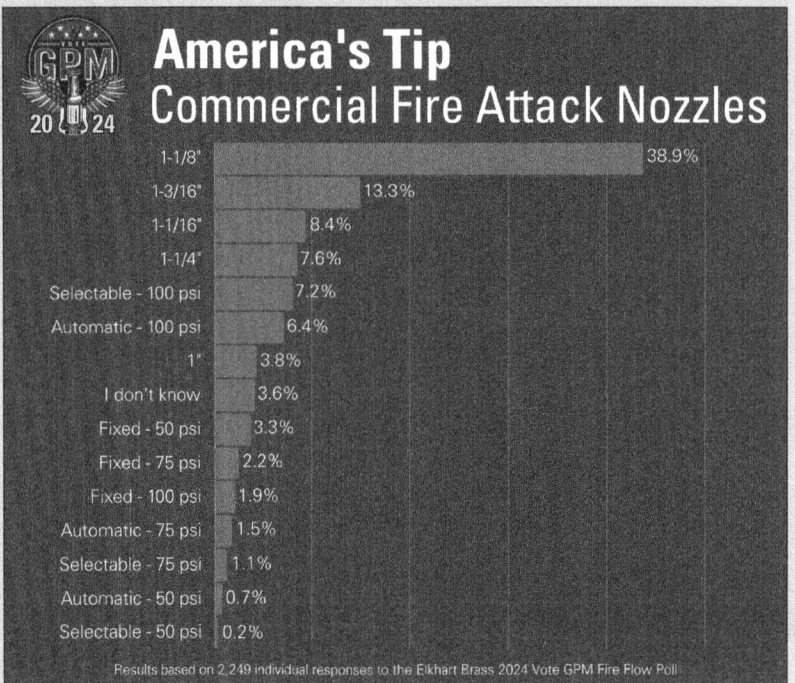

Figure 4–8b. Survey results ranking preferred fire attack nozzles for commercial fires.
(Images courtesy of Elkhart Brass)

The survey also shows a preference for smooth bore nozzles across the country with combination nozzles showing strength in the southern and western areas (figures 4-9a and 4-9b).

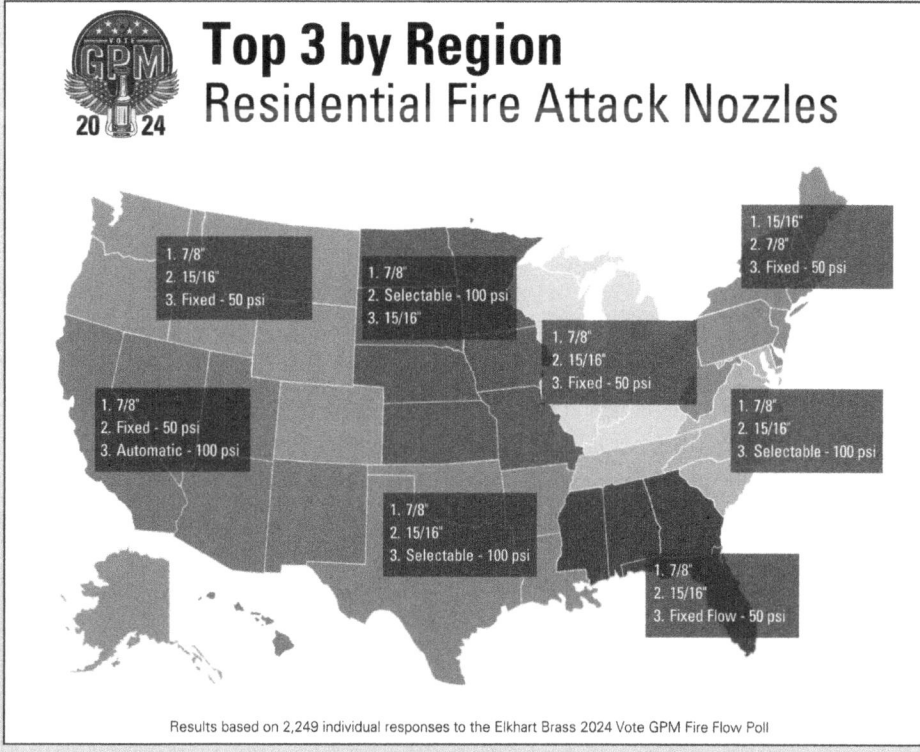

Figure 4–9a. A chart showing residential fire attack nozzle preferences by region. (Image courtesy of Elkhart Brass)

Figure 4–9b. A chart showing commercial fire attack nozzle preferences by region. (Images courtesy of Elkhart Brass)

Using the Adjustable Gallonage Nozzle in Place of the Automatic Nozzle

Some departments purchase adjustable gallonage nozzles thinking they can manually duplicate the operation of automatic nozzles. In theory they can; however, the nozzle operator has no way to sense minute amounts of nozzle pressure variation, which dictates how much to raise or lower the gallonage setting to maintain a constant nozzle pressure. The automatic nozzle reacts more quickly to adjust to different pressure variations than a nozzle operator manually changing the gallonage ring.

An advantage the adjustable flow nozzle has over the automatic in varying pressure situations is that the operator can fairly accurately estimate the flow being applied if the operator changes the selector to adjust for a different incoming supply pressure. If the incoming pressure changes again, rather than have the nozzle adjust the stream (which may provide poor flow performance without the operator sensing the change), the adjustable flow nozzle's baffle remains fixed where it was last set. If the pressure drops, the stream's reach and flow characteristics will deteriorate, serving to alert the nozzle operator that a pressure variation has occurred. The operator can then reselect a different gallonage and, if it is determined that the flow is less than required, steps can be made either to increase flow by calling for more pressure to continue the attack while flowing water at a lower rate or discontinue the attack altogether.

To review, we ask the following question: can an adjustable gallonage nozzle accomplish the same pressure adjustments as an automatic nozzle? The answer is yes, it can, but during an actual fire attack it is doubtful if it would prove effective.

When a large, adjustable gallonage nozzle is used on a master stream device, the operator simply dials in the gallonage setting that gives the best working stream. On a handline, the effects of varying firefighting flow and nozzle reaction must be considered before changing gallonage. On all adjustable gallonage nozzles, the stream's reach and firefighting action must be monitored continuously for effectiveness and the nozzle manually adjusted accordingly.

One operational problem, which seems to forever plague the use of adjustable gallonage nozzles, is attempting to fight a large fire with the nozzle set on its lowest gallonage setting. No matter how many times the nozzles are checked, they almost always seem to be set on the low gallonage setting. I've sometimes suspected the gallonage gremlin, a type of ghost who creeps into unsuspecting firehouses in the dark of night and slips all the selectors to low-flow marks. Nozzle operators must be trained to check the selector before donning face pieces. It should be checked again, if possible, before the operator opens up the line. It can never be checked too many times.

Distributor and Piercing Nozzles

There are two special nozzles that should be carried as standard equipment on all pumpers: the distributor nozzle and the piercing nozzle.

Distributor Nozzles

The *distributor nozzle*, also called a cellar or Bresnan nozzle, has a rotating head designed to spray water laterally and upward. As the name implies, this allows for water to be widely distributed below the operating crew, making the device extremely effective on inaccessible fires in basements, ship holds, or storage areas. Some distributor nozzles have a series of smoothbore discharges while others have up to four small fog nozzles. In actual operation, the type of head makes little difference, although the smoothbore outlets tend to produce heavier, coarser droplets while the fog heads tend to produce a slightly wider area of distribution (fig. 4–10).

It is important to insert a shutoff in the line when using a distributor nozzle. Because it throws water backward down the hoseline when charged, it can become impossible to handle if not already inserted in the access hole. It is good practice to have 50' of line between the shutoff and the nozzle to allow for proper and safe operation.

In use, a hole is cut in the floor directly over the fire (or so it is hoped). The nozzle is placed in the hole and the line charged. The hoseline should be worked so the nozzle first hits the floor and then is pulled to travel up to the ceiling; this is then repeated. It is important that a shutoff be used and a length of hose attached between the shutoff and the nozzle. In this way, the water flow can be easily controlled. When in operation, water should not be sent until the nozzle is the hole. The nozzle will fling water in all directions and firefighters as well as building contents in the vicinity will immediately become soaked if the line is charged before the nozzle is in position. It is also recommended that a second hoseline be stretched to protect the operators of the distributor nozzle from the fire below.

While available in 1½" flow sizes, the distributor nozzle is most effective when flowing over 250 GPM. If it is decided to use a distributor nozzle on and in an inaccessible, unseen

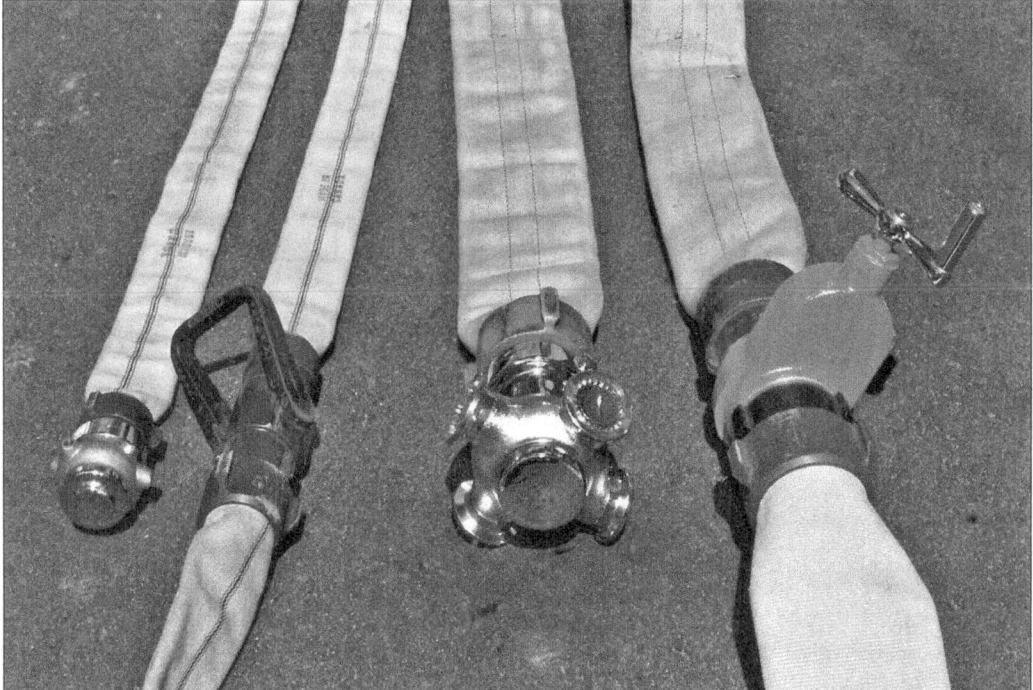

Figure 4–10. Two examples of distributor nozzles. Both are designed to provide effective streams over or under the floor of operation, placing water in areas otherwise inaccessible to firefighters.

fire, it is best to send as much water as possible to more quickly knock down the blaze. If the area is large, it may be necessary to cut additional holes and move the nozzle or use multiple nozzles to provide effective coverage. The best method of attacking below-grade fires is to ventilate and then make hoseline entry from an area opposite the ventilation channel. This action will force smoke and heat away from the nozzle crew. If the intensity of the fire makes entry impossible, the distributor nozzle can be helpful in reducing the intensity so entry can be made more easily (fig. 4–11).

The use of a distributor should be considered as a first step. It will rarely accomplish complete extinguishment. Nozzle crews should have charged lines available and be ready to advance as soon as the fire is knocked down and steam is seen coming from the holes cut for distributor nozzle insertion.

Some departments have found the distributor nozzle effective for attack on confined attic fires. The effects of the spinning nozzle head will drive water upward, making the nozzle effective in applying water on the underside of wood-shingle roofs.

The distributor nozzle can also prove useful in knocking down venting high-rise fires. The line is dropped from the floor above until the water enters the fire compartment. Considering that half the water being discharged will reach the involved area, it should be considered as a plan B in case interior attack is stalled due to fire conditions.

While many fire officers will only have a chance to use a distributor nozzle once in their career, when its use is dictated, nothing else will accomplish the task as effectively.

Figure 4–11. In use, the distributor nozzle rotates, throwing a wide pattern both outward and upward. This makes sure that the underside of the floor above is thoroughly soaked and reduces the heat in the under-floor area.

Piercing Nozzles

Originally designed to apply water to cotton bales and hay storage, the use of the *piercing nozzle*, or bayonet nozzle, can also prove effective on vehicle fires and certain confined structure fires as well (fig 4–12).

Figure 4–12. Examples of a distributor and piercing nozzle

A piercing nozzle punched through a headlight or taillight will quickly allow water to be applied to vehicle fires within those enclosed areas before the locks are attacked. It is certainly much safer for a firefighter to release a hood latch without flames lapping around their hands. Most piercing nozzles flow around 95 GPM and distribute the spray 90° outward. A newer style, 4" piercing nozzle has been designed which has a higher flow rate and projects more water forward, making the nozzle more useful for structural firefighting.

For fires in enclosed spaces such as attics, the newer style piercing nozzle can be punched through the ceiling from below and project its stream upward into the fire. If a stubborn basement fire is encountered, it is easier to drive the piercing nozzle through the floor than to cut holes for a distributor nozzle (fig. 4–13).

The piercing nozzle is also useful for overhaul. For example, it can be pushed up from below to apply water under shingle roofs. It can also be useful when overhauling piles of lumber or other stacked materials, especially when used in conjunction with Class A extinguishing agents.

The only limitation to the use of a piercing nozzle is its low flow rate. Some engine company officers have been known to drill out wider additional holes providing for a more forward water throw and increased flow rate.

Opposed Discharge Nozzle

The New York City Fire Department (FDNY) developed a need for a nozzle that could quickly flow water 90° from the hoseline to stop fast moving fires in cockloft spaces and ducting. The nozzle has two opposed delivery tips which negate nozzle reaction forces. While shown with the FDNY stacked tips having a 50 GPM flow rate, it would make more sense to remove the smaller tips and provide larger sized tips, say ⅞", that would increase the flow to 150 GPM on each side. Again, this is nozzle is not something that will be used every day but will excel when its use is dictated (fig. 4–14).

Figure 4–13. This department in North Carolina carries a piercing nozzle on a crosslay preconnect to be used primarily to knock down closed-up mobile home fires. Using Chief Layman's indirect attack principle, the nozzle easily pierces the thin sidewalls of a mobile home and cools the interior compartment before ventilating and entering the unit.

Figure 4–14. The opposed stream or cockloft head is attached to the end of a 5' pole with a shutoff on the inlet. It can be inserted in the overhead or under a floor to provide knockdown in otherwise inaccessible areas.

Bent Discharge Nozzle

Another nozzle designed by the FDNY for high-rise firefighting is the bent discharge nozzle, commonly called the *floor below nozzle*. The idea is to apply a 250 GPM stream to a venting compartment fire to which interior attack is delayed or hindered by fire floor conditions. It breaks down into two pieces for ease of maneuvering inside elevators and narrow stairwells and can be put together and quickly deployed to stop a fire from spreading to the floor above. Since being introduced to the market, it is proving to be deployed by many departments outside of big cities that face a multi-story high-rise threat (fig. 4–15).

Figure 4–15. The bent discharge or high-rise nozzle is designed to apply approximately 250 GPM through a high-rise window to knock down the intensity of the fire if the interior crew is being delayed or cannot safely advance due to extreme fire conditions.

High-Pressure Delivery

During World War II, experiments with tree sprayers utilizing low volume, high-pressure pumps demonstrated that the finely atomized discharge was effective in cooling certain burning flammable liquids under certain conditions. The war production effort was having a profound effect on availability and allotment of raw materials and the nation's production capacity was being taxed to the limit. Material had to be transported overseas to fight the war, and space on ships, the primary means of transportation, was limited. While protein foam was widely available, it was heavy and bulky to transport. High-pressure fog offered an alternative as an effective extinguishing agent that required no extra space, other than the pumper itself, for overseas deployment.

The Army Air Forces developed simple, easy to construct crash trucks called the Class 125 and Class 135 series, which carried 300 gal. of water and front mounted centrifugal pump rated for 60 GPM at 600 PSI–800 PSI. The Class 125 pumper had a piston pump midship mounted and the Class 135 had either a front or midship mounted centrifugal pump. Otherwise, the units appeared identical. Three ¾" handlines were carried which flowed from 25 GPM–28 GPM depending on the nozzle size. Two of the lines were used to discharge high-pressure water fog and the third was used as a foam line. No more than two lines could be flowed at one time because of the pumps' low capacity. The Navy had similar high-pressure fog units called *fixed fog nozzles* (FFNs).

According to Army wartime firefighting manuals, the high-pressure lines were intended to be used on straight stream to blast the fire away, forming a rescue path. The streams were changed to fog patterns and held parallel to the surface of the burning flammable liquid in an attempt to cool it below its vaporization temperature and accomplish extinguishment. It's interesting to note that the manual never warns that it is practically impossible for water fog to cool low flash point liquids, such as gasoline, below their flash points. Firefighters were cautioned against using fog and foam together because the water would break up the foam blanket. After the war, many of these units found their way into civilian fire departments and were put to use fighting structural fires. With the increasing popularity of fog firefighting after the war, many departments, especially in rural areas, attempted to multiply the perceived firefighting effectiveness of spray streams by applying them with high pressures.

Arguments advocating the use of high-pressure fog stated

- The high-pressure delivery, 400 PSI–750 PSI at the nozzle, atomized the droplets into a fine mist, allowing a more efficient delivery.

- Water was conserved by the use of the low flows. Since the flow was in the range of 35 GPM–40 GPM with two lines operating, it was claimed the tank was not emptied as quickly as with conventional lines.

- Water damage was minimized because (it was claimed) the fog stream was so efficient, all the water applied to the fire was vaporized, so none was left to cause water damage.

- The action of delivering water fog at high pressure greatly multiplies its extinguishing power. If you like booster lines, you will love high-pressure fog.

From the 1950s to early 1970s, high-pressure fog enjoyed great popularity, especially with rural departments. Stories abound about mutual aid arriving on the scene of a barn fire or a block of stores ablaze in the center of town and finding two or three high-pressure lines as the only ones deployed and being operated on the fire by the first arriving department.

A mystique about high-pressure fog's extinguishing action developed in departments using it, a feeling fueled by the manufacturers. One claimed that by throwing a fog between the fire and the unburned portion of the house, a solid, dense, "cold steam barrier" was created to stop the fire. The illustration accompanying the claim showed a firefighter attempting to stop a fire while standing outside a building, using a wide-angle pattern directed over the totally involved roof, between the burning and unburned sections, as flames billowed out of windows below. And this action was claimed to be scientific.

There are a number of actual reasons for high-pressure fog's popularity. The firefighter, after squeezing the trigger of the nozzle, receives a certain amount of nozzle reaction, enough to make the stream feel powerful but not enough to cause the line to become hard to control. This action has the psychological effect of making the nozzle operator believe they are wielding more firefighting action than is actually being delivered.

The line was certainly more lightweight than a charged 1½" and, because of its low volume, proved easy to maneuver while flowing. After the fire, it took little effort and time to reload back on the reel. These physiological handling factors, when coupled with the mistaken belief of the water's effectiveness, the high-pressure delivery, and the concern about running out of water, contributed to the system's use, especially in rural areas.

One advantage of high-pressure fog is that if the flow being delivered is sufficient to extinguish the fire, the driving force of the high-pressure delivery tends to propel the water a longer distance than if the same flow is discharged at normal nozzle pressures. The droplets exiting the nozzle under high pressure tend to be smaller in size and, theoretically, should be more efficient in converting into steam. As was discussed in earlier chapters, small droplets, while more efficient, are difficult for firefighters to apply because of their light weight and lack of penetration force. The kinetic action of high-pressure delivery adds additional energy to propel the water droplets farther into a fire situation than if delivered under normal nozzle pressures.

Even though high-pressure delivery will blow the fine droplets a farther distance than if standard pressures are used (increasing their penetration and, theoretically, their effectiveness), it must be remembered that sufficient flow to overcome the heat being produced is what will efficiently and safely extinguish fires. Small water droplets delivered to the fire, no matter what nozzle pressure is used, will not extinguish a fire if the flow is not sufficient or if the water does not reach the burning material.

Conserving Water Versus Extinguishing the Fire

These days, there are still some chiefs who feel that an attack should be mounted with low-flow fog lines to conserve water, no matter what flow is required to extinguish the fire. It is sad indeed to watch buildings burn to the ground while more than enough water for extinguishment is standing by on the fireground in the tanks of idle apparatus. No matter how you look at the situation, a single 1½" line flowing 125 GPM will put out more than four times the fire of a single high-pressure fog line. If the fire requires 250 GPM for extinguishment, the laws of physics dictate if that flow is not reached, the fire will continue burning. A single attack pumper can easily supply that flow through a single 2" or 2½" line. Of course, that flow could also be applied by eight booster lines, but much more staffing and apparatus would then be required—and not a single booster stream will have the punch of the 2½" line.

There is something to be said, however, for the stream exiting the nozzle at pressures higher than 100 PSI. Nozzle reaction is a combination of nozzle pressure and amount of flow. If the flow is kept between 60 and 80 GPM, water can exit the nozzle at pressures over 100 PSI without causing excessive reaction. A relatively moderate reaction force is present, but it tends to be somewhat absorbed by the hard hose. If a pistol grip is provided behind the nozzle, the line can be more easily controlled.

An interesting case history is the study of the Chicago (IL) Fire Department's use of high-pressure delivery apparatus and how the flow was increased to create a more effective firefighting weapon.

Fog Pressure Apparatus in Chicago

After the war, the Chicago Fire Department entered into a massive apparatus replacement program and, by 1956, practically every frontline engine was new. Since hydrants were readily available, the wisdom at the time decided that to make the pumpers simple and efficient, no on-board water was required. The result was that over 100 pumpers were purchased without booster tanks (fig. 4–16).

During periods of civil unrest beginning in the 1960s, then-Fire Commissioner Robert J. Quinn purchased a number of mini-pumper apparatus on which were mounted high-pressure pumps. These units provided rapid stream deployment on the just-completed expressway system and for structural firefighting where the high-pressure stream could hold the fire in check while conventional engine companies were leading out larger lines to hydrants. The units, called fog-pressures, eventually totaled to 13 and were usually manned by an officer, driver, and a single firefighter. They were equipped with four-stage hydraulic power take-off pumps which had a capacity of 85 GPM at 800 PSI when pumping from the tank.

When originally delivered, each of the two-reel lines were equipped with FMC high-pressure fog, gun-type nozzles. Because of their fragile cast-aluminum construction, low flow rates, and their tendency to freeze during Chicago winters, these nozzles were replaced with Elkhart S pistol grip booster nozzles rated 30 GPM at 100 PSI. When operated at 800 PSI pump pressures, these nozzles delivered in excess of 70 GPM. When the steel-braided ¾" hose wore out, it was replaced with conventional 1" rubber hose which reduced friction loss. In addition to the higher flow, the streams exhibited a great amount of forward driving force which allowed the stream to penetrate deep into fire areas and easily deflect around corners and into far rooms. Firefighters and officers assigned to these units said they could easily handle two and sometimes three totally involved rooms of fire with a single line (fig. 4–17).

As newer major pumper apparatus were purchased, they all were equipped with two booster reels. Operating procedures were changed to allow initial attack by the first arriving

Figure 4–16. An example of a Chicago Fire Department fog pressure unit in service in 1962. The effectiveness of these apparatus was increased by the replacement of low-flow, gun-type nozzles with nozzles flowing around 75 GPM at 800 PSI engine pressure. (Photo courtesy of Bob Freeman)

Figure 4–17. While the Chicago Fog Pressure lines proved effective for interior firefighting, they were also used to help stop a fire from spreading before lines from conventional engines could be charged. In this case, the line provided protection to the engine crew until their 2½" line could be charged.

engine using the tank water rather than by always laying to a hydrant, as was done previously. The standard-pressure, 30 GPM booster line proved unsatisfactory for interior firefighting, and most had been removed from engine apparatus by the beginning of the 1980s and replaced by a bed of preconnected 1¾" line.

In Chicago's experience, the standard-pressure, 30 GPM booster line never matched the overall firefighting success of the 70 GPM high-pressure line. It must also be considered that these lines were operated by experienced firefighters making aggressive interior attacks. Thousands of amazing stops were made with high-pressure units, especially from alleys and other areas inaccessible to larger pumping apparatus. Needless to say, many Chicago firefighters hated to see them taken out of service.

What was realized from the Chicago experience was that when the nozzles and hoses were changed to permit higher flows, the increased flow rate, coupled with the high-pressure delivery, greatly multiplied the units' firefighting effectiveness.

Modern Experience with High-Pressure Delivery

In the early 1970s, New Zealand began to overhaul its fire service, departing from its British origins and looking to other countries for new firefighting hardware. New Zealand has a national fire service with approximately 14,000 members.

Using high-pressure fog fire apparatus left behind after World War II by American troops, the national fire service gained valuable experience with high-pressure delivery.

Prior to the modernization program, British-style first-aid lines were used for initial attack. This system consisted of a reel of ½" inside diameter hose and operated at 150 PSI, providing an extremely low flow. While appreciating the quick-deployment advantages of the reel system, the United Fire Training and Research Center realized the reel line's limitations and looked for ways of increasing its effectiveness by increasing the line's flow (fig. 4–18).

Officials found that by using an American-built three stage pump and matching nozzles, they could flow 60 GPM from 300' of 1" reel line at 500 PSI engine pressure. They consider

Figure 4–18. New Zealand firefighters have found that 60 GPM delivered through 1" hose by a high-pressure pump is ideal for the type of fires they fight.

this flow ideal for fires in private dwellings, in which develop the largest percentage of structure fires. Most private housing in New Zealand is of wood frame or stone construction and averages around 1,200 sq. ft. in area.

This is similar to many British departments who deploy high-pressure, 0.866" reel lines as a first-in attack line that flows around 35 GPM at 500 PSI.

Keep in mind that while the flows seem extremely low, their small compartment size and stone and masonry construction tends to limit heat generation.

U.S. Air Force High-Pressure Operation

While this book concentrates on waterflow for structural firefighting operations, it should be noted that the U.S. Air Force uses ultra-high-pressure pumps for aircraft and flammable liquid incident responses. Their smaller rapid intervention units carry 400 gallons of water, foam, and a pump that supplies 90 GPM at 1,200 PSI. The agent is deployed through a 60 GPM bumper turret and 15 GPM handlines. The larger crash vehicles have pumps rated at 320 GPM at 1,350 PSI.

According to their testing, they feel that the ultra-high-pressure delivery provides greater knockdown while using less agent.

Though a department has many choices to make, understanding the basics of nozzle design provides the foundation for forming the building blocks of effective and safe structural water delivery.

Chapter 4 Review Questions

1. What are the four main functions of a nozzle?
2. What is the primary advantage of a break-apart nozzle?
3. What is the primary advantage of deploying a distributor nozzle?
4. What is the main disadvantage of high-pressure water delivery?
5. What is the best use for an opposed discharge nozzle?

Automatic Nozzles

Constant pressure or variable flow nozzles, more commonly referred to as *automatics*, are probably the most misunderstood and misused pieces of water delivery equipment in the fire service today. Not only are automatics the highest priced nozzles available but they also require a considerable amount of evaluation and training in order to be used efficiently and safely on the fireground. With all these demands required for their proper use, one has to wonder why the fire service has made them one of the most popular nozzles.

The theory behind automatic nozzle operation was developed by the late Chief Clyde McMillan in the late 1960s. By trade, Chief McMillan was an engineer for United States Steel in Gary, Indiana. He commanded the Gary Fire Task Force, an auxiliary fire service that assisted the regular department on extra alarms, helping augment firefighting personnel, providing additional heavy streams, and supplying mutual aid response to surrounding departments. Because efficient development of heavy streams was a specialty of the task force, Chief McMillan often had to cope with problems caused by a feast-or-famine water supply, common with initial heavy stream operations.

At one time or another, most all fire officers have been faced with the problem of ineffective streams as a result of insufficient water availability. In many cases, during the initial stages of placing heavy streams into service, only limited flows are available before additional supply lines can be laid. Unless small, low-flow tips are used when water supply is limited, the streams exiting high-capacity smoothbore or fixed-gallonage fog nozzles will be weak, barely making it over the side of the apparatus and able to hit only the ground, not the fire. If the smaller nozzles or tips used to develop streams with limited water supply are not changed to larger sizes to accommodate the additional supply more efficiently as additional water becomes available, the water stays in the hose because smaller nozzles cannot easily pass through the additional capacity at normal operating pressures.

Chief McMillan reasoned that if a tip with a variable orifice could be designed, its smaller openings would provide a stream with at least some effective reach in the initial stages. If it then automatically enlarged as additional water became available, it would provide the gallonage needed to overcome heat production. Even if the starting flow did not carry enough gallonage to extinguish the fire, it had at least some reach which could serve to slow the fire down or protect exposures.

Since stacked smoothbore tips have always been available, the means to accomplish this flow change was usually at hand. At many fires, however, the smallest tip may not have been small enough, and when additional water supply became available, chiefs rarely ordered the lines shut down so tips could be replaced and the flow increased.

Chief McMillan decided that if some mechanism could be provided to regulate nozzle pressure without the firefighter having to get involved, or without having to shut the device down to change nozzles or tips, the stream from the heavy stream device could utilize available water more effectively. He developed a hydraulically controlled conversion unit designed to be installed in existing fixed-gallonage, high-volume fog nozzles. He attempted to interest other nozzle companies in his invention, and when that proved unfruitful, eventually started production along with his family in the basement of their home. The company, founded in 1969, was named Task Force Tips in honor of his department.

Elkhart Brass was performing research along similar lines and in 1971, introduced the first complete automatic nozzle, the SM-100. Both McMillan's tip and the Elkhart nozzle were master stream devices, designed to regulate nozzle pressure in the 300 to 1,000 gallons per minute (GPM) flow range.

This equipment was followed quickly by handline nozzles, Elkhart's SM-10 in 1971, Task Force's HTFT in 1973, and Akron's Akromatic in 1977. All handline nozzles used essentially similar mechanisms, which each company developed for use in their own master stream devices and modified for use with lower flows.

The first automatic nozzles were designed to more effectively apply master streams where the monitor gun could absorb the varying amounts of nozzle reaction. The primary problems automatic nozzles address are providing effective reach with limited water supply and providing for more effective operation by maintaining a constant nozzle pressure by adjusting water flow upward or downward depending on availability. Manually operated, adjustable-gallonage nozzles were available at this time and performed in almost the same manner as automatic nozzles. One problem with the use of adjustable nozzles was the operator remembering to change gallonages when operating pressures changed. The automatic nozzle attempted to solve that problem by providing adjustments without operator attention.

Automatic Nozzle Design

An automatic nozzle is similar in operation to a fire pump relief valve. When properly set to a predetermined maximum pressure, a fire pump relief valve senses an increase in discharge pressure. The bypass valve opens when that pressure overcomes the resistance of a large spring, discharging the excess pressure and water back into the intake side of the pump. This bypassing action helps prevent excessive pressure rise at the end of the hoselines (fig. 5–1).

The automatic nozzle's pressure control mechanism also utilizes a large spring to balance the pressure. However, on the automatic nozzle, the spring system is preset by the manufacturer rather than being adjustable. As additional pressure (water weight) hits the pressure control device, it stretches the spring, enlarging the exit orifice and allowing more water to flow through the tip, thereby preventing the nozzle pressure from rising.

Some nozzle designs use a modulating system where water pressure acts on a piston that helps balance the pressure exerted on the baffle assembly by the spring. In theory, a

Figure 5–1. This cut-away illustrates the operation of Elkhart's automatic nozzles. As exiting water pushes against the center stem baffle, it compresses a calibrated spring, causing the nozzle to maintain its designed operating pressure. As more water force pushes against the baffle stem, the opening between the stem and waterway is made larger, preventing nozzle pressure from rising but allowing more water to flow.

smoother transition in flow should be obtained as pressures fluctuate, though in actual use, nozzles utilizing this design concept operate in a similar manner to nozzles using a straight spring mechanism. If the nozzles were out of view and only the streams were observed, most firefighters would detect little differences in each design's reaction to changes in pressures.

As the extra water being delivered by the higher pressure is allowed to pass, the nozzle operating pressure is maintained at a certain level even though the flow has increased. Automatic nozzles are designed to maintain approximately 75 or 100 pounds per square inch (PSI) nozzle pressure by varying the flow. Because the orifice automatically adjusts itself to the flow available during operation, it is frequently difficult to determine the actual amount of water flowing.

The idea of automatically regulating the nozzle pressure is the most misunderstood principle of automatic nozzle operation. It must be remembered that in an automatic nozzle, the flow rate varies while the nozzle pressure remains constant. The nozzle does not regulate flow; it regulates nozzle pressure. Variation in flow is a secondary result of the device adjusting to maintain nozzle pressure.

In a February 1984 Fire Engineering article titled "Down to Earth Talk About Nozzles," Deputy Chief Edwin J. Spahn described the operation of automatic nozzles.[1] He said, "Volume causes the fire to become controlled. Pressure is the factor which allows the volume to be projected upon the fire in an orderly and efficient manner." He was absolutely correct.

A few paragraphs later he stated, "The prime feature of this [automatic] nozzle is to have the nozzle pressure remain essentially constant over a wide range of GPM nozzle flow. This ability to maintain relatively constant pressure ensures good stream characteristics without the constant attention of the nozzleman." Correct again, but the statements contradict each other. Each statement standing alone is true. On one hand he says volume is important in putting out a fire, and on the other hand he says the automatic nozzle is effective for firefighting because it controls pressure by varying that volume. Which is it?

---

1. Spahn, Edwin J. "Down to Earth Talk About Nozzles." *Fire Engineering* 137, no. 2 (February 1984).

The automatic nozzle itself can be more efficient in operation than other types of nozzles because it maintains a constant nozzle pressure. Just because the nozzle is operating efficiently, however, does not mean the flow being delivered can extinguish the fire. Automatic nozzle utilization requires close attention by the pump operator in order to maintain adequate flow rates.

One fire chief told a story of a salesperson demonstrating the virtues of an automatic nozzle to the department. They said, "This nozzle will deliver an effective fire stream no matter what the flow." Yikes!

An effective fire stream must provide sufficient water volume to overcome the production of heat and the reach to get that water volume to where the burning is taking place. There is no such thing as an effective fire stream if its flow is not sufficient. From the salesperson's statement, it can be deduced that they had little experience fighting fires. The salesperson's belief that a good fire stream is a good-looking stream emphasizes the amount of misunderstanding in the field that surrounds automatic nozzle operation.

Conversely, nozzle pressure, even though it may be considered secondary in importance to flow, must be properly supplied to any nozzle to efficiently shape and deliver the water flowing through it.

For example, if a fire requires a 150 GPM flow for extinguishment, the flow could be sustained by carrying the water in paper cups given enough people. While the required amount of flow can be provided with these paper cups or, for that matter, from the open butt of a 1¾" line, the application of the water to the fire would not be as effective, convenient, or safe as 150 GPM flowing from a properly supplied nozzle. Proper nozzle pressure is what gives the stream effective reach.

Adequate nozzle pressure without proper flow, or adequate flow without proper nozzle pressure to provide reach, will not effectively extinguish fire. An efficient fire stream is formed by the correct combination of flow and reach. What the automatic nozzle does best is to make the job of providing proper nozzle pressure easier for the pump operator. But making the computations easier cannot by itself ensure that effective streams will hit the fire. In actual practice, many users become complacent and think falsely that the nozzle is regulating flow. The pump operator and nozzle crew still have the responsibility to make sure adequate volume for the fire at hand is being supplied to the nozzle and is being effectively applied by the crew.

High-End Performance

Since pressure is what makes water move through a hose, more pressure applied on the hoseline's supply end will cause more water to move through the hose. We know the nozzle constricts the flow at the delivery end of the hoseline to build up pressure, creating speed or velocity which, in turn, gives the stream reach. When the exit hole in a nozzle is of a fixed size rather than being adjustable, more engine pressure will cause the nozzle pressure to increase, forcing more water out of the tip. If the nozzle pressure increases to a point over the nozzle's normal operating pressure, more water will have to move faster in its effort to escape through the tip since water cannot be compressed (fig. 5–2).

When a fixed-gallonage nozzle is flowing water in this over-pressurized situation (which usually begins about 15 PSI–20 PSI over the nozzle operating pressure) the rapidly moving stream's quality will begin to deteriorate because turbulence created by the increased velocity will cause the stream to tumble and break up. In addition, pressurizing a nozzle past its normal working pressure causes an increase in nozzle reaction; the line becomes stiff and makes the nozzle difficult to control. This is more apparent in smoothbore nozzles rather than combination tips.

If additional engine pressure is forcing additional water down the hoseline to an automatic nozzle, it handles the additional pressure by opening up the tip, allowing the extra water to pass without increasing its velocity. The water volume passing through an automatic nozzle will increase, but the operating pressure will remain relatively constant. The extra water

Figure 5–2. Handline operators must be aware that at a large fire, the activation of master streams tends to rob water from smaller-diameter lines equipped with automatic nozzles. Here, the combination of a partially closed shutoff and reduced water supply forms an ineffective stream which is not performing effective firefighting and is endangering the operator. The habit of not opening the shutoff all the way is a hard one to break, even if circumstances dictate that larger flows are needed for effective firefighting.

volume exiting the nozzle will cause an increased amount of nozzle reaction and will typically be greater than the reaction force caused by the combination of high nozzle pressure and increased flow than is present when a fixed-gallonage nozzle is over pressurized.

The Automatic Nozzle's Ability to Increase Flow

If more water volume is desired, the automatic nozzle offers an operational advantage because of its ability to easily handle higher flows through its elastic hole. For example, if a department has been using single-gallonage nozzles on 1½" lines set at 95 GPM, changing to an automatic will allow that line to flow up to about 130 GPM just by increasing the engine pressure without causing excessive nozzle operating pressure that will degrade the stream. Keep in mind that increasing the flow will cause a rise in nozzle reaction force. If it is desired to utilize the same flow that was obtained with previous fixed-gallonage nozzles, the automatic nozzle will also flow the 95 GPM if the engine pressure is adjusted to supply that flow.

Most single-gallonage nozzles can be converted to flow higher gallonages, but this conversion is done in a shop, not on the fireground. But even if such a conversion is made, the fixed-gallonage nozzle can only effectively and safely flow its rated capacities at normal nozzle pressures. If the nozzle pressure fluctuates, the stream could lose firefighting effectiveness unless some means is provided to compensate for the fluctuation.

After the fire is knocked down, the automatic nozzle's shutoff, as with any nozzle's, can be partially closed to reduce the flow if less water is wanted for overhaul operations. The difference is that the automatic's baffle will adjust to provide an effective, low gallonage stream useful for mop-up.

Automatic Nozzle Operation at Lower Flows

If the pressure and supply situation is reversed and the nozzle pressure decreases, the automatic nozzle will make the hole smaller in an attempt to provide a stream with effective reach. This action will continue until the flow reaches the lower limit of the automatic nozzle. Then the nozzle will act as a fixed-gallonage nozzle. This also occurs at the top flow limit.

As the automatic nozzle adjusts itself to balance nozzle pressure with a lower flow, the result could be a stream with good reach but not necessarily the required volume. An automatic nozzle can easily deliver the volume necessary to overcome heat production during most interior attacks. However, visual assessment of the stream and physical evaluation of the nozzle reaction are not as effective in estimating flow from automatic nozzles as they are with conventional combination or smoothbore nozzles. Flowmeter devices are desirable to properly determine actual fire scene flow with any nozzle, especially with automatics (fig. 5–3 and fig. 5–4).

Figure 5–3. With 45 PSI at the wye, flowing through 100' of 1¾" hose, the flowmeter indicates no measurable gallonage is flowing. There is nothing wrong with the nozzle, but it was designed to regulate the outlet pressure at 100 PSI, far above the nozzle's inlet pressure. Automatics are not the best choice for operation when pressures may fluctuate below its rated operating or regulating pressure. After the disastrous fire in Philadelphia's One Meridian Plaza building in 1991, it was determined that they tried to stretch and operate streams equipped with 100 PSI automatic nozzles when the standpipe pressure was only 40 PSI–45 PSI on the fire floor.

Figure 5–4. Conversely, when the automatic nozzle was swapped out for a ¹⁵⁄₁₆" smoothbore nozzle, and the hose length and 40 PSI inlet pressure at the wye maintained, the nozzle flowed 110 GPM at around 18 PSI nozzle pressure. While not ideal, the stream has some firefighting effectiveness and may prove useful until the standpipe pressure problem can be mitigated.

Many departments not understanding the automatic nozzle's low end operating characteristics have lost buildings and endangered or injured firefighters because the crews assessed the stream they were flowing visually and mistakenly judged it adequate for attack use. Because the reach looked effective, the crews believed nozzles were flowing enough water for the situation. As attacks were advanced, the fires overwhelmed the water being applied with disastrous results.

A department wanting to utilize automatic nozzles efficiently and safely must be sure all operating personnel have an understanding of basic fire stream hydraulics and be able to differentiate between effective nozzle pressure and effective fire flow. The automatic nozzle is an excellent firefighting tool when its operation is understood. Operators must comprehend that the automatic nozzle has a different method of operation and requires more detailed operations in pumping, water supply, and nozzle operation than other styles of nozzles.

Utilizing Automatics

When interviewing chiefs, line officers, and firefighters for the first edition this book, it was found that most departments who at the time were using or considering the purchase of automatic nozzles had followed only salespeople's recommendations and had never fully tested and evaluated the nozzles for themselves. If the automatics replace standard-type nozzles, and if hose sizes or engine pressures are not changed, the automatics will perform, flow-wise, almost exactly as the nozzles that preceded them. Here, the only difference between conventional nozzles and automatics is the increased price of the automatics. If a department is not willing to change operating procedures when changing to automatic nozzles, it may be more cost-effective to save their money and design their operations around the use of conventional nozzles. Too many departments purchase automatic nozzles thinking they are the latest thing in firefighting technology, yet they never utilize that technology to realize the nozzle's potential because they don't understand that automatics must be pumped and operated differently from conventional nozzles.

Automatic Nozzle Master Stream Operation

Since automatic nozzles were first developed for use on master stream devices, a review of the large nozzle's operating characteristics is a good starting point for better understanding automatic nozzle operations.

When used on master stream devices, automatic nozzles can offer certain advantages:

- The nozzle will help develop a stream with effective reach from low flows, provided the nozzle pressure is high enough to operate the automatic pressure control mechanism.

- As more water becomes available, the nozzle will open its orifice without any attention by the nozzle operator, allowing the extra flow to pass through and be applied to the fire.

- If other devices equipped with automatic nozzles are operating on the same supply circuit, the nozzles will share their water, equalizing the flow between them.

- If the water supply fluctuates during the fire, the nozzle will automatically adjust itself to the increased or decreased flow, helping maintain the stream's reach.

Remember, these advantages are obtained from the automatic nozzle when mounted on a master stream device. Handline operation presents a slightly different set of advantages and disadvantages which will be discussed later.

In the early stages of large fire combat, the effective reach of a high-volume stream may be a more desirable consideration than its flow until more or larger lines can be laid to supply more water to the devices. Effective reach can be critical in the early stages of large fire attack when attempting to protect exposures or retard a fire's travel. A lower volume stream actually striking the exposed surfaces is certainly more effective than a stream flowing a higher volume but having no reach. As more water becomes available and attack on the main body of fire is attempted, the device will most probably need to flow a higher volume to effect extinguishment.

Some fire officials advocate delaying attack operations until enough water is available to sustain the operation for any length of time. Others advocate hitting the fire as soon as possible with all the might you can muster, even if it means running out of water before the operation is completed. These officials contend that if the water is hitting the seat of the fire, the attack will be successful 99 times out of 100. When it is not, the building may have been destroyed anyway no matter how much water was made available.

I am among those who feel the attack should be pressed as soon as possible with as much water artillery as available. This does not mean dumping water indiscriminately toward a building. It must be applied wisely in order to do its job of extinguishment. Properly operated, automatic nozzles can assist in this type of attack by helping to provide initial streams with effective reach.

A heavy stream device, whether mounted on a vehicle, aerial ladder, or as a ground monitor, will absorb and mitigate most reaction caused by the flowing water. A handline, on the other hand, is dependent on the nozzle crew holding the line to mitigate the effects of nozzle reaction.

If two or more automatic nozzles are supplied from the same pumper and are supplied at approximately the same pressure, they will equalize their delivery with one another, providing two or more streams with effective reach.

If a non-automatic nozzle is being supplied along with an automatic from the same water supply circuit or same pumper, the non-automatic will not share its water but will attempt to provide its full flow by robbing the automatic stream. The automatic will adjust its orifice to compensate for its reduced flow. The worst situation that could develop when the flows balance is that the non-automatic stream would have no reach because too big a tip is being used and the device is not generating enough pressure for proper operation. Meanwhile, the automatic stream, while reaching the fire, could have insufficient flow for control or extinguishment. To rectify this problem, it helps if all nozzle styles are matched when used on the same supply circuit.

While a good-looking stream can easily be obtained from an automatic nozzle, it is almost impossible to judge how much water is flowing unless calculations were done beforehand or if flowmeters are installed on the pumper. If an ineffective stream is pouring out of a device equipped with a conventional nozzle, an evaluation is quickly and easily made by the looks of the stream alone. The commander can then make the decision to either shut it down or make it more effective by providing the device with more water. Since the lack of effective flow is not as apparent with automatic nozzles, this evaluation might be delayed or not made at all.

In actual fireground operation of heavy streams, conventional nozzles with increased available volume could be providing ineffective streams because the tip size is too small, even though their reach is effective. After the high-volume water supply operation is in service, it may be decided to replace the automatic with smoothbore tips to counteract the stream-bending effects of wind and heat or to provide more stream penetration through high heat areas. But any decision to fine tune the delivery system is usually made at a point after the initial attack has been underway and the facts of building construction, weather, water supply, and tactics can be thoroughly evaluated.

All things considered, given the helter-skelter atmosphere of a fire attack's early stages, utilizing automatic nozzles on large-caliber appliances should give firefighters a better chance of applying master streams more effectively, which is why most of the ladder waterways and platform monitors are equipped with automatic tips nowadays.

Handline Operation Using Automatic Nozzles

While factors such as low gallonage, effective stream reach, automatic flow of higher gallonages, and sharing of water are important characteristics for master streams, they can cause operating problems when using automatic nozzles on handlines. It must be remembered that a firefighter, not a fixed device, will be absorbing the nozzle's reaction forces and the punishing effects of a fire's heat during handline operations.

The more time a department invests in testing, evaluation, and training, the more flexible and efficient the automatic nozzle's operation will become. One of the automatic nozzle's desirable features is that it has the built-in capability for high flows, but keep in mind that high flows mean increased nozzle reaction. Because the automatic reacts almost instantly to pressure variations, allowing the higher or lower flow to pass through the nozzle, the nozzle operator can be surprised with its sudden change in thrust.

Of course, automatic nozzles can just be screwed on the end of the hose and operated in the same manner as conventional nozzles, but it makes little sense to pay $600–$1,000 more for a device if its capabilities will not be fully utilized.

There should be some type of nozzle operating plan and extensive training program in place when automatic nozzles are deployed on attack lines to help increase their firefighting effectiveness. This planned method of pressure and flow supply must be evaluated by each company using the nozzles and thoroughly understood by firefighters and pump operators alike. This will ensure the safe operation of the lines.

Besides operating an automatic as a conventional nozzle, there are three basic operating methods commonly utilized for handline operations with automatic nozzles.

High-Pressure/Maximum Volume Method of Operation

When automatic handline nozzles were first introduced, one of the recommended operational methods was having the pumper supply the automatic lines with pressures ranging 200 PSI–300 PSI. This high-pressure delivery ensures that high water volume is always available at the nozzle. If the reaction proves excessive and hard to handle, the nozzle operator was instructed to reduce the reaction by partially closing the shutoff. In actual practice, the only advantage to this method is that the highest flow possible is always available at the nozzle, although it is rarely used because the reaction generated cannot be safely controlled by the one or two firefighters initially operating the line.

There are many disadvantages to the high-pressure/maximum volume method of operation:

- The hose becomes extremely stiff because of the high pressure and is hard to maneuver, especially inside buildings.

- The operator cannot handle the high flow alone. In actual practice, the shutoff is almost always operated partially closed because of the excessive nozzle reaction, which provides for lower flows than anticipated or planned.

- If high pressure is carried in the line at all times, and if the nozzle shutoff handle is accidentally knocked open when advancing the line, a more dangerous situation could be created as the unwanted flow exits at a higher reaction force than if normal engine pressures were being used.

- When flow is gated at the nozzle, fire flow cannot be accurately determined unless flowmeters are in operation and the pump operator is in constant communication with the crew.

- Balancing pressure between lines flowing at high pressure becomes difficult, if not impossible, for the pump operator to manage.

- The hose is being operated at the top limit of its designed operating pressure. Any defect or worn spot, coupled with the high internal pressure, can quickly cause early and sometimes violent failure.

- When providing high pressures, the pumper is operating at a reduced capacity. For example, a 1,000 GPM pumper can supply 1,000 GPM at 150 PSI but only 500 GPM at 250 PSI. If standard pressures are being supplied to handlines, the total flow of the pump is available. Theoretically, if pressures in the range of 250 PSI–300 PSI are supplied only two handlines can be effectively supplied by a 1,000 GPM pumper.

- While offering impressive flow figures on paper, this method can become dangerous in practice due to high nozzle reaction forces. It is not recommended for use on pre-connected lines.

High-pressure operation can be compared to a new pumper powered by a 500 hp diesel engine. It is ridiculous to consider always operating the engine at full throttle. One of the reasons for specifying a large engine is because it has reserve power to pull steep grades or transport large amounts of water more efficiently. Just because an automatic nozzle has increased flow capabilities available (reserve power, so to speak) does not mean that it should be run wide open at all times. It makes for a much more efficient operation if the use of this reserve power is planned for in advance and called upon only when needed.

This misuse of the automatic nozzle's reserve power capabilities can be illustrated in the following example.

Years ago, a fire department in New England removed all of its new automatic nozzles from service after a bad experience during their first use at a fire. The department's previous procedure called for an initial attack line to be stretched to the seat of the fire and a water supply plan using additional pumpers and tankers to be activated as soon as any working fire was encountered.

When automatic nozzles were purchased, the department decided to use the high-pressure supply approach but did not have a detailed operating plan worked out, nor did its personnel have enough training in automatic nozzle use when the first fire hit.

The nozzle operators advanced the line to the fire and flowed water the way they did when using conventional nozzles, although this time, instead of flowing 95 GPM, the line was flowing much more. Hit-and-run tactics were not used to conserve water, and the additional nozzle reaction did not alert the operators to the fact the flow was increased. The outcome can be easily guessed.

The lack of advancement due to high reaction forces caused the firefighters to pour water on the smoke and run out of water before the fire was extinguished. The crew backed their limp line out of the building and the fire caused considerable damage before water supply operations allowed the reattack to continue.

Afterwards, the chief's reaction to the nozzle's poor performance was to remove the nozzles from service and pronounce to one and all that the automatic nozzle concept was useless. While the automatic nozzle helped contribute to the problem, the operational mistakes were not the nozzle's fault. They were a combination of too much water being delivered in one spot and not enough training given in its application and wise use.

The high-pressure method of operation proves more practical when supplying automatic nozzles from long, reverse lay lines, where the pumper is located at the water source. After the lines are stretched, the operator can charge the lines with, say, 250 PSI as an initial pump pressure. The nozzle crew, while bleeding air from the lines, can test the effect of the pressure before making the attack and, if necessary, can communicate pressure adjustments to the pump operator before moving in.

When using this method during the initial stages of attempting to get water on the fire using a reverse lay, the pump operator can more easily estimate pump pressures and should be able to deliver effective flows more quickly than with other methods that require mathematical calculation.

Fixed-Gallonage Method of Operation

Many departments utilize automatics simply as fixed-gallonage, high flow nozzles. This operation is easily accomplished when using preconnected lines by consulting the pressure and flow chart supplied with the nozzle, determining what flow is required, and matching the indicated engine pressure with the hose length. Provided the engine pressure is maintained, the nozzle operator will have a good idea of the amount of water flowing. If more flow is needed, the nozzle operator can ask the pump operator for more pressure. If flowmeters are not in use, the pump operator can fairly accurately estimate the amount of flow by consulting the manufacturer's pressure chart and can communicate the amount to the nozzle crew.

Since conventional fixed-gallonage nozzles are normally operated in the same manner, except for increasing reserve flow, the only disadvantage to this type of operation with automatics is the additional expense of purchasing automatic nozzles over conventional types.

Predetermined Pressure Method of Operation

An expanded form of the fixed flow method requires more training and coordination, but it will more fully utilize the automatic's built-in flow versatility.

During initial district evaluation, it should be determined what flows will be required for certain operations (i.e., trash and automobile fires, smoke showing, flames and smoke showing). Using the manufacturer's chart, the required engine pressures should be predetermined.

One department uses the following guidelines on its 200' 1¾" preconnects:

Fire Condition	Charge Pressure	Flow
Auto-trash	120 PSI	90 GPM
Investigation, light smoke	150 PSI	150 GPM
Fire showing	200 PSI	200 GPM
Overhaul	110 PSI	60 GPM

By having the pump operator pump to predetermined charge pressures, the nozzle crew has the available flow to handle the situation at hand without having to estimate its volume at the nozzle. This method can efficiently utilize the automatic's flow control characteristics while providing the nozzle crew with a known amount of water. As with any operation, it must be understood and practiced before use to establish the necessary coordination between the pump operator and the nozzle crew.

Nozzle Evaluation

Since high flows are more readily available with automatic nozzles, training in handling the increased nozzle reaction is necessary. The following steps should be performed to effectively determine how much flow is available and how much reaction can be handled when using automatic nozzles:

1. When automatic nozzles are considered for service, everyone involved with their use should understand the relationship between engine pressures and required flow. This is best accomplished by using a flowmeter to calibrate each line and recording engine pressure and flows for later use.

2. Once required flows are determined, each line should be handled by the nozzle crew at that flow to evaluate hose stiffness, nozzle reaction, and other handling characteristics. It is one thing to desire 200 GPM from a 1¾" line and quite another to handle the actual flow. No amount of pad and pencil calculating can equal an hour of moving the high flow line around inside a training building.

3. A baseline flow should be established by determining how much water at what pressure can be safely handled inside a building by a normal size operating crew. This determination should be considered as the practical limit of flow and should not be exceeded unless specifically ordered otherwise by the officer handling the line on the fireground.

4. If additional flow is desired, the pressures needed to deliver that flow should be determined and recorded. It is then easier for the pump operator to increase the flows to predetermined levels when requested.

5. With the flowmeter in place, partially close the nozzle shutoff, noting the effect on flow. Nozzle operators must understand that partially shutting off the nozzle can have profound effects on the flow and could confuse the pump operator. It is not good practice to use the shutoff as a flow control when attacking fires because the lessened flow may not be enough to extinguish the fire.

Handline Nozzle Reaction

Since automatic nozzles can flow higher gallonages more quickly and easily than other styles of combination nozzles, particular attention should be directed toward training firefighters in the safe handling of higher nozzle reactions. Most operators close the shutoff until a high nozzle reaction becomes manageable, knowing the automatic flow control mechanism will adjust the stream. This is not a good practice because, depending on the weight and physical condition of the operator, the flow being delivered can be throttled down to a point where it becomes ineffective in overcoming heat production.

Depending on the amount of the nozzle crew's firefighting experience, they may not immediately notice that their stream has become ineffective, and they could be placed in danger as the fire rapidly gains headway. A safer practice is to have the crews train with different flows and determine how much can be safely handled with the shutoff completely open, then set standard pump operating pressures accordingly (figs. 5–5a, b, c, and d).

Some instructors use a rule of thumb that determines a nozzle operator can safely handle their body weight in GPM at 100 PSI nozzle pressure. If a firefighter weighs 150 lb., that trained person should be capable of handling a flow of 150 GPM. This rule of thumb considers only initial handling, not advancing or working the line for long periods during interior operations. How long the firefighter can handle that flow depends on how much energy

Figures 5–5a and b. Departments must understand the operating characteristics of automatic nozzles to be able to utilize them safely and efficiently for fire attack. Visual assessment of the stream to determine if the calculated flow is being delivered will not work as well with automatic nozzles as it does with other types. These photos show variations in flow that can occur with the operator sensing little difference in stream reach. Appearance of the stream provides little indication of flow rate.

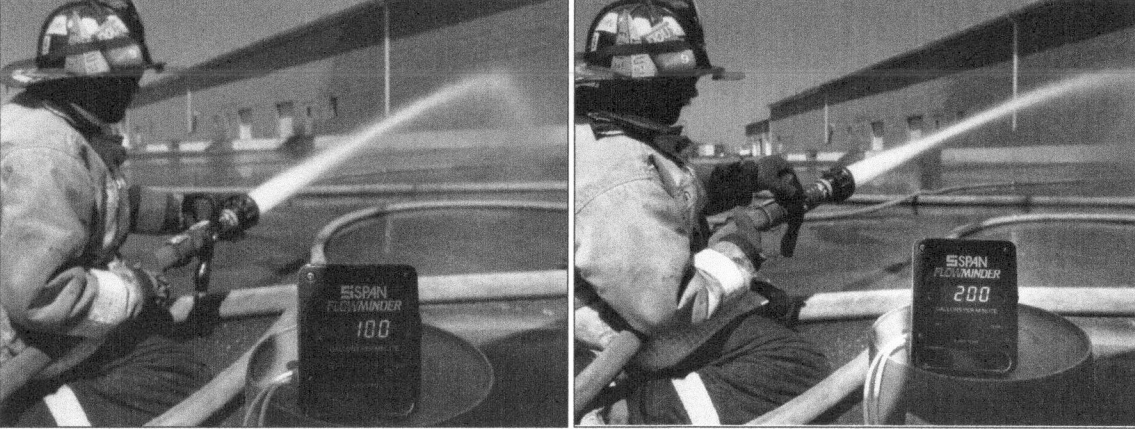

Figures 5–5c and d. At 200 GPM, the operator experienced problems handling the nozzle reaction and attempted to reduce the reaction by partially closing the shutoff. At that point, all calculations for estimated flow were negated by the attempt to control output at the nozzle.

was expended stretching the line and the firefighter's physical condition. One 150 lb. firefighter may be able to hold the line flowing 150 GPM longer than another because of differences in muscle tone, lung capacity, and weight distribution.

If the nozzle reaction formula given in chapter 7 is used to compute nozzle reaction, then the rule of thumb changes. If the nozzle reaction is expressed in pounds of force, the nozzle operator should be able to handle about one-half of their weight in nozzle reaction force. A 150 lb. firefighter should be able to handle 65 lb. of force in reaction. When using the formula or consulting the reaction charts in chapter 7, it should be remembered that reaction is caused by a combination of flow and pressure, although flow will contribute more to reaction force than will nozzle pressure (fig. 5–6).

If more water is needed at the nozzle, more pressure can be requested from the pump operator. The nozzle crew, knowing additional water is on the way, can prepare for the additional reaction force. After the fire is knocked down, it is then certainly acceptable to use the shutoff as a flow control device.

Figure 5–6. Whenever a nozzle operator feels uncomfortable with safely handling the nozzle's reaction force, the first thing that should be done is to partially close the shutoff until the reaction is reduced to a perceived safe level. This action reduces the reaction force by restricting the flow. An automatic nozzle will start closing the outlet until a stream is reached that may have some reach but possibly not an effective flow rate. The stream here has neither. (Photo courtesy of Jo L. Keener)

Automatic Nozzle Flow Ranges

Depending on their manufacturer, automatic nozzles are offered with differing flow rates. The department's needs should dictate which flow range will be most effective for their operation. Because of their size, units offering a more limited flow range have a pressure control mechanism tailored to flow within a certain range and are usually less expensive than larger units with wider flow ranges.

As with any nozzle equipment, when purchasing automatic nozzles, a department should determine which flow range is most practical for their particular operation and how much money they want to spend per unit. If the nozzle will be continually used on the same size line, such as a nozzle expressly purchased for use on a preconnect, it may make more sense to specify a nozzle with a flow range to suit only that size line. If the department wants to take advantage of the automatic nozzle's flow flexibility and utilize the nozzle with inlet adapters on different hose sizes, a nozzle having a wider flow range might be a more cost-effective purchase (fig. 5–7).

Figure 5–7. An automatic nozzle's flow range and operating pressure should be permanently marked on the nozzle.

Pump Operation for Automatic Nozzles

Retired District Chief Andrew O'Donnell, former director of training for the Chicago Fire Department, said about automatic nozzles, "They're just like a car. You have to learn how to safely drive them or you're going to get hurt."

The first key to effective and safe automatic nozzle operation lies in the hands of the pump operator. If the pump operator does not thoroughly understand how automatic nozzles operate and does not supply proper operating pressures, all bets for safe operation are off.

For example, two 200' long lines of 1¾" hose are operating from the same pumper. The engine pressure is 150 PSI and each line is flowing approximately 150 GPM. The operator sets the relief valve or pressure controller on the pumper at 150 PSI and, when one line is shut down, the excess pressure on the other line should be relieved at the pump.

If a 300' line is connected and the officer wants a flow of 180 GPM to cover an exposure, an engine pressure of 220 PSI will be needed. To supply the increased pressure, the operator must reset the pumper relief valve or pressure controller to the highest supplied pressure and the other lines requiring a lower operating pressure are then gated down. If one of the 150 PSI lines is shut down, its pressure will travel to the other 150 PSI line. It will not be relieved at the pump until the pump pressure exceeds 220 PSI.

Because the other automatic nozzle senses this increased pressure, its baffle will immediately open up wider, allowing the excess water to pass and greatly increasing the nozzle's reaction force. In effect, the automatic nozzle acts as a relief valve but with a much weaker spring than the one on the pumper, making the automatic nozzle's activation much faster.

In many cases, the nozzle operator is not prepared to handle this rapid increase in reaction force. As described earlier, automatic nozzles that are operating together on the same supply circuit will attempt to share their water and equalize flow. If operating from a wyed line or from a pumper without a set relief valve or pressure controller, they will always give their water to one another when one line is shut off. This sharing of water can immediately cause a great amount of nozzle reaction on the opposite line or lines.

While this surge also occurs with conventional nozzles, the amount of nozzle reaction increase is not as great as with automatic nozzles, as the flow rate is fixed rather than variable. The automatic nozzle's rapid discharge of increased flows during pressure surges could immediately cause critical problems for the nozzle crew in handling the line, especially if operating on snow or ice or from a ladder.

Depending on the design of the pumper's relief valve or pressure controller, the device may not operate fast enough to prevent a dangerous pressure surge when using automatics. Since the nozzle's spring is more sensitive than that installed in the pumper's relief valve, the automatic nozzles quickly absorb all the extra pressure from shut-down lines possibly without the pumper relief valve or governor activating.

Operational Hints

If a line becomes kinked or caught under a door as it is being advanced, or if debris is trapped within the nozzle, its flow will be reduced. With a conventional nozzle, this reduction of water supply can usually be detected immediately by the degradation of stream quality. If an automatic nozzle is being used on the line, the nozzle will adjust itself for the reduced flow, but the crew may not notice the resulting reduction in water delivery, which could create a problem in safely controlling the fire. The pump operator should monitor line flows with flowmeters, but fluctuating flows (common with the use of hit-and-run tactics) cause the meters as well as conventional pressure gauges to continually fluctuate, making accurate flow assessments difficult. To help alleviate this situation, nozzle crews must be trained to continually evaluate the hoseline behind them to prevent kinks from reducing flow.

Attempting To Evaluate Amount of Flow

In actual practice on the fireground, attempting to determine flows from any nozzle, automatic or conventional, is only a process of making estimates (figs. 5–8a and b). Using flowmeters on the pumper can accurately indicate flow, but it may be difficult to communicate

Figures 5–8a and b. As nozzle pressure changes and the automatic mechanism attempts to maintain constant pressure, crossover may occur. Crossover, sometimes a good indication that flow is being reduced, should be corrected by readjusting the pattern control to keep the water from fighting itself and reducing reach.

the flow to the crew working the business end of the line inside a building. One of the arguments heard again and again against the use of automatic is, "You don't know what you are flowing."

If a nozzle crew has no idea of the amount of water they are flowing with an automatic nozzle, chances are they won't know what they are flowing with a conventional nozzle either. Pre-fire estimates will give a department an excellent idea of what flow they might need in certain situations, but when the red stuff is blowing down a hallway, the only flow that counts is one that stops the fire.

An experienced and educated nozzle crew should be able to tell what effect their stream is having on the fire. If they feel they are not making progress, they can decide to increase the rate of water delivery or move the line to another spot and continue the attack, which might give them a better shot at the fire.

It may be that the best course of action is to take the line out of the building if the crew is endangered. If they decide to increase their water delivery, the automatic nozzle can make their life a little easier because all they have to do is call for the pump operator to increase the flow. Other decisions, such as where and how much water should be applied, are generally based on experience and quality of supervision.

While automatic nozzles cannot automatically simplify the solving of hydraulic calculation problems, they can help alleviate certain water supply problems by making it easier to generate streams with effective reach. They can also help improve overall firefighting efficiency by making it easier for the pump operator to provide increased per line water flows. However, they also present a set of operational handling problems that must be addressed before automatic nozzles can be effectively and safely deployed on the fireground.

It can be said that a person who complains about the ineffectiveness of automatic nozzles is probably someone who does not understand the principles of their operation.

Chapter 5 Review Questions

1. The first automatic nozzle wasn't developed as a nozzle. What was its original purpose?
2. What is the primary operating characteristic of an automatic nozzle?
3. What is the most efficient use of an automatic nozzle?
4. In the initial stages of master stream fire attack, what is the advantage of using an automatic nozzle?
5. What does stream crossover usually indicate?

6

Smoothbore Nozzles

The smoothbore nozzle was the first stream-shaping device to be used for fighting fires. Also called a solid-stream or straight-bore, it is the simplest of all nozzles, constructed of a threaded base on a length of material and bored with a tapered hole to form a tip. The size of the exit hole drilled into the tip and the geometry of the reducing area determine how much water will flow at a certain pressure.

Early stream-shaping devices had no shutoff. If the chief wanted the water flow to stop, he simply ordered the firefighters operating the handles or brakes on the hand pumper to cease pumping. In the early days of steamers, nozzles had no shutoffs either. Steam-pumpers had positive displacement pistons or rotary gear pumps, and until the relief valve was invented, shutting off a nozzle could have easily burst a hoseline or damaged the pumper.

In those early days, smoothbore nozzles were ideally suited for the firefighting tactics then in use. Since hoselines were short and protective clothing was nonexistent, a nozzle needed to provide a stream with good distance and good penetration for defensive operations. In addition, the smoothbore was simple and easy to manufacture with the machinery of the day.

Near the end of the 19th century, as steam engines took the place of hand pumpers and municipal water systems became more dependable, staffing was shifted from hand-pumping water to actual fire attack. The steady, powerful streams supplied by steamers along with improvements in hose construction allowed firefighters to work more easily both outside and inside fire buildings (fig. 6–1).

Since sending a firefighter out of a fire building to tell the steamer engineer to shut down a line was time-consuming and inconvenient, shutoff devices and engine-mounted relief valves came into general use around the turn of the 19th century. The most common type of shutoff used then was the plug type, simply a cylinder with a hole bored into it that rotated in the nozzle body. Sealed with strips of leather, these early devices were prone to leakage and, unless extreme care was taken by the manufacturer to machine and smooth the inside flow surfaces, turbulence caused by roughness and misaligned mating surfaces degraded the stream. Because of this degradation, it was common in large cities for pumpers to carry both shutoff nozzles for general use and smoothbore nozzles with no shutoffs (called street pipes) for maximum reach at defensive operations.

The *underwriter's playpipe*, a nozzle still being manufactured today for industrial firefighting, was developed in the late 1880s by an engineer named John R. Freeman for use not only

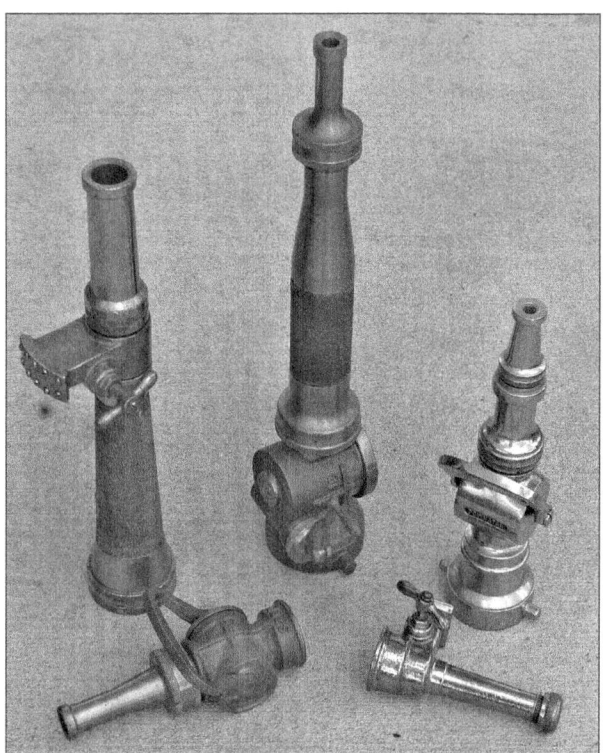

Figure 6–1. Here are nozzles made by Chicago and San Francisco as well as a nozzle made to New York City Fire Department (FDNY) specifications. Larger cities either made their own or had manufacturers fabricate smoothbore nozzles to their individual specification before the use of combination nozzles were common.

as a nozzle but as a measuring tool to determine the flow capacity of standpipe and sprinkler systems fed by fire pumps and gravity tanks. The device has 1⅛" and a 1¾" tips mounted on a copper-alloy tube wrapped in marlin cord to protect it and to improve the gripping surface when wet. The underwriter's playpipe served as a benchmark for many years when evaluating fire department nozzles because of its standardized output and reach when flowing water.

By measuring the output pressure from a smoothbore nozzle with a pitot gauge, then matching the indicated pressure with the tip size on a chart, the amount of water flowing through the nozzle can be easily and accurately determined. This over 100 year old method of measurement, sometimes using multiple nozzles, is still used by inspectors and engineers to calibrate flows of fire department pumpers and built-in fire protection systems, such as standpipe and sprinkler systems.

During firefighting's low-technology era, before the World War II, many big cities designed and manufactured nozzles in their own shops. For many years, apparatus in Chicago, San Francisco, Seattle, and other large cities carried a number of these devices, attesting to their robust construction and usefulness as firefighting tools.

From early firefighting up until about the time of World War II, smoothbore nozzles were common on interior hose racks. Their use was indicated because they were maintenance-free and were not prone to clogging with rust and scale. Shutoff devices were not desirable, principally because of their expense and the fact that shutting off a hose rack line was almost certain to burst the unlined linen hose then in use. If theft became a problem, the missing brass nozzles were replaced with pipes cast from iron.

In order for the stream to obtain maximum reach—an important consideration since most firefighting took place outside the building—early nozzles had machined tips attached to long playpipes. The purpose of the playpipe was to help shape the water to increase discharge distance; the long tube helped to reduce the exit turbulence by helping to straighten out the water flow as its velocity increased while traveling through the tapered bore to the tip. This smooth transition from pressure to velocity increased the stream's effectiveness for defensive exterior application, which was then the usual method of operation.

To help firefighters handle the nozzle reaction, large leather or rubber handles were later added to the playpipes to give firefighters a better grip on the nozzle. Elaborate stream-straightening devices such as vanes or grids were sometimes cast or inserted into the playpipe tube to help smooth out the water flow as it traveled through the tube. Their value was sometimes doubtful, but manufacturers, who had little else to help differentiate their product from that of the competition, loudly trumpeted the superiority of their designs in the literature of the day. A nozzle's effectiveness was measured in the distance its stream traveled and the compactness of its stream (fig. 6–2).

Today, shorter playpipes with their rounded rubber handles are still being sold and are probably specified more because of tradition than practical application. Playpipes were never designed for interior firefighting and can become difficult to handle while attempting to maneuver a hoseline inside a fire building. The theoretical reach enhancing advantage of a playpipe is outweighed today by the playpipe's weight, size, and handling disadvantages when used for interior attack.

Some playpipes are still being made with a device called a *ladder hook*. The hook was originally intended to secure and help control the nozzle when placed over the rungs of a ladder. In actual use, the nozzle is almost impossible to direct when hooked to the rungs, and the added strain of nozzle reaction could cause the rung to fail or the ladder to be pushed away from the building. The hook can also endanger personnel hand-holding a line if nozzle

Figure 6–2. For years, an assortment of different-sized tips were carried on pumpers to provide proper flow depending on a certain hydraulic situation. While this system worked well in the classroom, its needed hydraulic calculations and tip-swapping proved less than practical on the fireground. Most departments which consistently provided effective streams from smoothbore nozzles picked a standard tip size and pumped whatever pressure was necessary for proper stream operation.

reaction causes the nozzle to slip from the crew's hands. The hook, whipping backward along a firefighter's side or thigh, would cause the same damage as a jet-propelled pike pole let loose in a circus tent. Many departments preferring to use playpipes have ground off the hooks to prevent injuries (fig. 6–3).

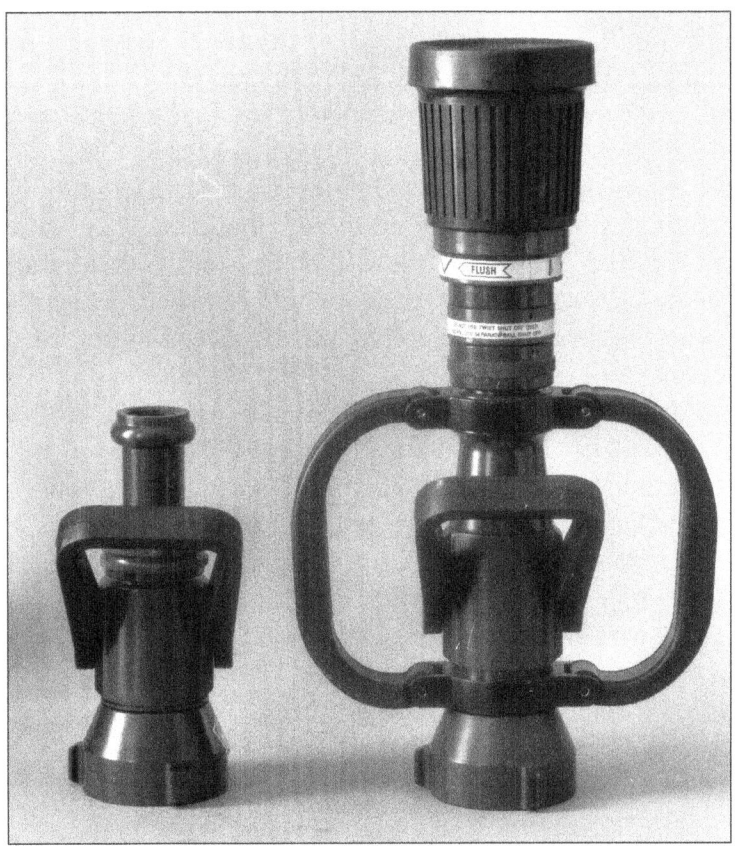

Figure 6–3. A side-by-side comparison of two 2½" smoothbore nozzles.

Nozzle Comparison

Flow	252 GPM at 45 PSI	250 GPM at 100 PSI
Reaction	89 lb. of force	126 lb. of force
Weight	2⅞ lb.	9½ lb.
Length	8½"	17½"
Price	$614	$1,155

Note. GPM = gallons per minute; PSI = pounds per square inch

Because contemporary playpipe design offers little advantage in the way of stream shaping, departments purchasing new smoothbore nozzles should consider eliminating playpipes altogether. The unit cost and bulk will be considerably less, the stream quality suffers little, and with just a shutoff and tip, the line can be maneuvered and the stream directed more easily inside a building.

For departments concerned with reducing firefighter stress through reduction of equipment weight, the shutoff and tip combination (less playpipe) is ideal. For example, the standard New York City 2½" nozzle, a shutoff with a 1⅛" tip, weighs less than 3 lb. compared with approximately 10 lb. for a standard playpipe and tip.

Smoothbore Nozzle Tips

Tips are the detachable ends of a solid-stream nozzle. Hydraulic science tells us there is only one size tip of similar geometry for each combination of flow and nozzle pressure. While it seems absurd nowadays to carry an assortment of tips in order to have the proper size on hand for a given situation, standard operating procedure years ago called for just that. In theory, changing tips should have solved reach and flow problems with weak and ineffective streams, but on the fire scene, firefighters rarely ran back to the pumper for the proper size tip. In retrospect, from the early- to mid-19th century, departments were more concerned with operating a stream with an effective-looking reach than supplying a flow rate with enough volume to stop a given amount of heat production. Fire service periodicals of the era were filled with stories describing complicated hydraulic formulas as the means of reaching what they called an *effective stream*. One magazine story in the early 1950s described a theoretical barn fire and how a department could lay a certain number of feet of hose to supply a ¾" nozzle flowing a total of 130 GPM at 60 PSI. By using present day rate of flow formulas and estimating the barn's size, this fire would require at least 1,000 GPM for knockdown and containment. Judging by what was written years ago, an effective stream was considered one that would hit the target, but not necessarily pack the flow punch for extinguishment.

Stacked tips were invented to help alleviate the problem of tip size management. Two or three different size tips were screwed together on the playpipe so the nozzle operator could select the proper size for the situation and remove the others. This design was great in theory but, again, rarely used in practice. A study of old fire publications showed hundreds of fire scene photographs with stiff lines, indicating adequate water supply, and tiny streams flowing out of the smallest tip in the stack. In the background, the building was usually burning merrily away (fig. 6–4).

Stacked tips can be an effective tool if they are used correctly. Unfortunately, in the heat of battle, firefighters forget to remove small tips when they have an ample water supply, resulting in inadequate, low flows. If you carry stacked tips or decide to buy them because some rating organization has them on a list, it is important to train the firefighters to use the largest tip possible, removing the smaller tips when they have the water available.

Departments which have standardized certain size tips and then pump the required pressure to supply the desired flow have the most success flowing good, working streams with smoothbore nozzles. In New York City, the standard issue 2½" nozzle has a 1⅛" tip, while Chicago has standardized an opening of 1¼", Seattle 1⅛", and Detroit 1¹³⁄₁₆".

Pump operators in those cities have developed charts showing pump discharge pressures for various hose lengths, making it easy to determine the proper engine pressure.

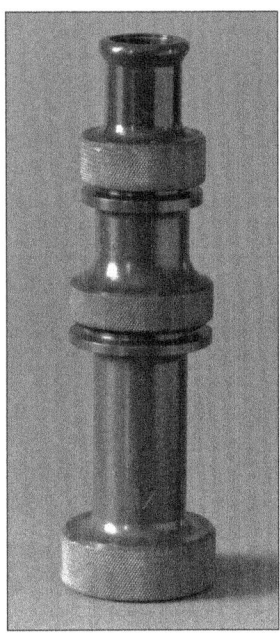

Figure 6–4. For years, stacked tips were required to be carried on every pumper. While in theory you could conveniently select the proper tip size to match incoming pressure and flow requirements, in actual practice it was rarely done. Remove the top two tips, the ones with the oddball threads, and put them in a drawer back in the shop. Run with the bottom tip which is 1¼" and has standard 1½" threads on the end if the line needs to be extended.

With any nozzle, seven variables enter the equation in order to produce an effective stream:

- Nozzle geometry and efficiency
- Desired flow
- Tip size
- Nozzle pressure
- Size of hose supplying nozzle
- Length of hose supplying nozzle
- Engine pressure

By eliminating the first five variables through standardization of hose and tip size, it becomes relatively easy to figure the engine pressure by counting the number of lengths of hose out of the bed and then consulting a chart for the proper engine pressure.

One of the biggest problems in properly utilizing smoothbore streams is supplying and maintaining proper tip pressure. Departments should evaluate the effect of tip pressure by flowing each brand of nozzle and tip in their inventory in order to observe the stream's reaction to pressure. With smoothbore nozzles, unlike combination nozzles (especially automatics), stream degradation due to gross under- or over-pressurization is readily apparent in both reaction force and visual characteristics.

For example, at a nozzle pressure of 50 PSI, a 1⅛" tip flows 266 GPM. If the nozzle pressure drops to 40 PSI, the flow will be reduced to 238 GPM. The reduction in reach and stream action will be almost unnoticeable to the eye, but a reduction in nozzle reaction should indicate to the crew the flow has decreased. A further reduction in nozzle pressure will cause extreme reduction in stream effectiveness, easily identified by a weak stream with decreasing range, visually indicating to the nozzle crew that the desired flow rate is certainly not being obtained.

Increasing nozzle pressure to 60 PSI on the same 1⅛" tip will cause the flow to move up to 291 GPM with a resultant increase in nozzle reaction. Other than a visible increase in the velocity of water exiting the tip, the effect on the stream range and firefighting action will be negligible. Over-pressurizing the nozzle beyond 65 PSI will result in increased flows, but the stream quality will degrade because of the increased internal nozzle turbulence. The increase in nozzle reaction will also make operation of the nozzle at the higher pressure unsafe to handle by a limited number of firefighters.

An over-pressurized nozzle will exhibit a stream that *rags* or *feathers* around the edges, caused by unequal velocity generated by water rubbing against the sides of the nozzle. As water passes through the nozzle's waterway, the liquid around the edges of the stream being shaped is subject to friction and turbulence as it contacts the sides of the nozzle interior. The water along the sides of the stream exits the tip moving at a slightly faster speed than the core of the stream and tends to fall away as it travels from the tip. The higher the nozzle pressure, the faster the velocity of the water moving through the opening and the greater the depth of turbulence along the sides of the stream. The greater the depth of side turbulence, the more water will fall from the stream on its way to the target.

If the nozzle is operating at a tip pressure up to about 60 PSI for handlines and 90 PSI for master streams, this ragging will be negligible. At higher tip pressures, the outside begins to peel away in greater quantity and will begin to reduce the reach of the stream (fig. 6–5a). In actual practice, over-pressuring's major problem will be in handling the stream's increased reaction. Over-pressurizing smoothbore tips on ground level master stream devices simply wastes water which falls away, never penetrating the fire (fig. 6–5b).

Over-pressurized streams on aerial devices can cause dangerous overloading of the device. Aerial ladders are designed to carry their load so that it pushes downward. The reaction of an aerial stream tends to load the ladder in the opposite direction of the direction of

Figure 6–5a. A nozzle with too little pressure

Figure 6–5b. A dispersed stream caused by too much pressure. Smoothbore nozzles supplied with too much or too little nozzle pressure allow visual assessment of proper operation.

its greatest strength. An increase in nozzle reaction that is greater than the ladder is designed to handle can cause catastrophic failure. Investigations of aerial device mishaps indicate that most failures have been caused by operator error. Applying an excessive load to an aerial device through the misuse of aerial streams is a dangerous practice.

Within a fire building, because of the relatively short distance the stream needs to travel to reach the fire, poor stream quality caused by excessive nozzle pressure will not be as much of a problem as the stress imposed on the crew attempting to handle the higher nozzle reaction. When over-pressurized smoothbore tips are used defensively outside the fire building, they simply waste water and the energy expended while attempting to control the line. Even though they are flowing more water, they could be delivering less onto the target than if operating at the proper pressure because of decreased stream quality.

Effect of Tip Size Selection

Tip sizes should be selected for practical application of the desired flow. For example, if a flow around 200 GPM is wanted, almost all tip sizes from ⅝" to 1¼" will supply that flow at various pressures. For the purposes of discussion, let's assume four different size tips are available, ¾", 1", 1⅛" and 1⅜". All will flow approximately 200 GPM but do so at widely varying pressures and nozzle reactions and will exhibit extremely different stream characteristics.

Tip Size	GPM	Nozzle Pressure	Nozzle Reaction	Reach
¾"	201	145 PSI	122 lb.	20'*
1"	210	50 PSI	79 lb.	61'
1⅛"	206	30 PSI	60 lb.	50'
1⅜"	217	15 PSI	45 lb.	38'

* Because the tip is being operated well above its operating pressure of 50 PSI, the stream exits the tip in the form of a fairly undefined spray which has little reach.

It can be seen from the chart that the 1" tip operating at 50 PSI nozzle pressure or the 1⅛" tip operating at 30 PSI delivers the desired flow with a reasonable working range while the reaction forces are low enough to allow the lines to be easily handled by the nozzle crews.

The ¾" tip delivers the flow, but the excessive nozzle pressure causes the stream to be extremely feathered and the high nozzle reaction would be extremely hard to handle. The 1⅜" tip delivers the flow with little nozzle reaction; however, the reach of the stream would be too short to be effective on most fires. It can be concluded that if a 200 GPM flow is desired, the flow can be most effectively delivered by a 1" or 1⅛" tip.

Theoretical discharge tables show wide ranges of gallonages and pressures for each tip size. However, for all practical purposes, when effective reach and nozzle reaction are considered, it is best to select a tip size that flows the desired gallonage rate at 40 PSI–55 PSI for handlines and 70 PSI–90 PSI for heavy-stream devices.

Because of friction loss in a given hose size, there is a limit to the amount of water that can be practically flowed from each size hose. Tests have shown that almost 250 GPM can be pushed through 100' of 1½" hose, but the hose is so soft at that flow that it can almost be pinched shut, and the discharge pressure is so high that the pumper could be described as *screaming*. As a general rule of thumb, the largest smoothbore nozzle that can be efficiently operated is equal to about one-half to three-quarters of the hose diameter. This translates to the following:

Hose Size	Tip Size	Flow at 50 PSI
1½"	⅞"	161 GPM
1¾"	15/16"	182 GPM
2"	1 1/16"	237 GPM
2½"	1¼"	328 GPM
3"	1½"	473 GPM

For all practical purposes, the 1¼" tip is normally the largest size used on handlines because the high nozzle reaction generated by larger tips at standard operating pressures makes safe control as a handline almost impossible.

Smoothbore Firefighting Tactics

As discussed in chapter 3, there are three basic ways to attack a fire. One is direct application of water to the burning material in order to cool it below its ignition temperature. The second is the indirect method where a spray stream is played into the heated overhead, generating steam that will theoretically absorb heat and eventually find its way to the burning material. The third is a transitional attack whereby the intensity of the fire is knocked down by the application of a straight stream from the outside of the building and then rapidly moving the line inside to accomplish extinguishment by direct attack.

In the municipal fire service, smoothbore nozzles were the universal stream-shaping devices used for direct attack until about the end of World War II, when indirect attack methods became popular. To quickly review, the pure indirect method of attack requires the building to remain unventilated and a fog stream played into the upper atmosphere of the fire room to generate steam and absorb heat. Indeed, ventilation is not a vital part of indirect attack; it must be delayed while the steam settles on the burning material in order to accomplish extinguishment.

In the early days of indirect attack, it was taught that smoothbore nozzles had no place in this method of firefighting, resulting in many thousands of nozzles being relegated to dark corners of storerooms as fog nozzles came into widespread use.

Despite the indirect attack fog frenzy of the 1950s and 1960s, many departments continued using smoothbore nozzles on attack lines, New York, Chicago, and San Francisco among them. All three cities are still standing, offering testimony to the fact that direct attack and smoothbore nozzles will extinguish fires quite efficiently.

Many officials in the fire service today are taking a hard look at the over 100-year-old methods of indirect fire attack and are reassessing the advantages of the smoothbore's stream qualities in modern interior fire attack tactics. These reassessments have been stimulated by a dramatic increase in overall fire intensity due to modern construction methods and increased use of synthetics in building contents. In addition, budget restrictions are forcing departments to provide increased protection from these more intense fires with fewer operating personnel.

The Smoothbore Nozzle's Operational Advantages

The smoothbore nozzle offers firefighters certain advantages when used in interior direct attack:

- The stream pattern greatly reduces generation of unwanted steam, maintaining visibility and greatly reducing the chances of thermal injury to the firefighters during interior attack operations.

- The stream is composed of more water mass and less air than a comparable flow combination nozzle, exhibiting greater penetration when the smoothbore stream is directed into high-heat areas. This water concentration allows more water to be applied to each square foot of burning material than if a spray stream was used.

- Because the smoothbore nozzle operates at half the exit pressure of a 100 PSI combination nozzle at the same flows, the smoothbore has less nozzle reaction.

- Since very little air is entrained in the stream and because the stream has a narrow cross-section, the possibility of pushing fire ahead of the nozzle is greatly reduced or eliminated altogether.

- Since the lines are carrying water at a lower pressure, they are easier to maneuver inside a building.

At the FDNY's Bureau of Research and Development, former head Steve Krupa says each engine company is issued an assortment of solid-stream and fog nozzles; their use is determined by the company officer. "We like to get inside the burning building and put the fire out," he said. "The smooth bore allows us to be more aggressive."

Captain Gil Moreno, former head of San Francisco's Bureau of Equipment, agrees with Krupa's observations. "Because of the nozzle's penetration, you do not have to get down on top of the fire to do some damage. And you can easily stay below the heat when operating with a forward tip," he said.

Captain Moreno said a previous fire department administration decided to remove the smoothbore tips from service and converted 100% to combination nozzles about 40 years ago. Within the next two years, smoothbore nozzles were reissued to the companies because fire experience during that time indicated their practicality and usefulness for interior attack.

The words *penetration, heat balance,* and *stress reduction* seem to be repeated over and over in discussions of smoothbore nozzles with experienced fire officers across the country.

In 1970, the American Insurance Association issued a paper describing spray stream operations. While describing the merits of spray stream application, they defined rather well the difference between spray and smoothbore streams.

> Spray streams, in contrast to smooth borer streams, generally absorb heat from hot atmospheres more rapidly, cover a greater area with water, and use less water; however, they require a higher discharge pressure, have a shorter reach, have less penetration, and are less effective in cooling and extinguishing fires involving sub-surface areas that have reached incandescence, such as deeply charred wood.

The paragraph, written 55 years ago, provides an excellent comparison of the two stream styles which still hold true today.

Given the knowledge gained from established studies, we know the absorption of heat by spray application results in the creation of steam. This can be a positive factor if extinguishment within confined areas only is considered. It can be a negative factor if the steam is generated when firefighters or victims are present in the environment. Since desired flow can be obtained from either combination or smoothbore nozzles, flow selection, not nozzle style, determines how much water is applied to a fire (figs. 6–6a and b).

Figures 6–6a and b. Comparing streams exiting from a combination nozzle (above) and a smoothbore nozzle (below), captured with the aid of a high-speed flash, shows the smoothbore stream to be more cohesive and have larger droplets than the stream exiting from the combination nozzle. Because the stream has to make two 90° turns inside combination nozzles, air is entrained in its stream, which separates the water molecules.

Penetration

The solid-stream nozzle's ability to penetrate into a fire area is a universal, positive observation. Experienced firefighters say the solid cylinder of water jetting from a smoothbore nozzle will easily reach the source of flame production within a structure without being vaporized by heat during delivery. Water exiting a standard combination nozzle is subject to extreme division of water particles because of higher exit pressure, hollow stream core, and the great amount of entrained air caused by the stream rapidly changing direction within a fog nozzle. This combination of entrained air and high nozzle exit pressure causes the water to vaporize more quickly while passing through heat on its way to the burning fuel because the water molecules are separated by air molecules. Fire officers experienced with the use of both direct and indirect methods of attack say the hotter the fire, the quicker a stream from a fog nozzle will evaporate, even when operated in the straight-stream position. This has the effect of causing less water to reach the fire even though the nozzle is discharging the required rate of flow.

With a smoothbore nozzle, less air is entrained in the exiting water than with a combination nozzle, so the water stays together in a more tightly moving mass. In stroboscopic photos, it actually looks more like a rope than a solid cylinder. This mass, along with the stream's lower exit velocity, makes the stream less subject to *shear*, or to unwantedly change direction due to air friction, wind, and gravity.

Two streams, one from a combination nozzle and one from a smoothbore nozzle, may appear to have the same quality and travel the same distance on the training ground; however, add a crosswind, heat, or both and the streams will react quite differently. The smoothbore stream generally continues forward on its course while the combination nozzle stream is bent in the direction of wind or thermal currents.

New, lightweight construction techniques, especially in multifamily dwellings, have created buildings described as lumber yards standing on end. When these buildings are involved in fire, a successful interior attack will depend greatly on the stream's ability to travel through the heat and hit the burning material. This lightweight, matchstick-style building construction, coupled with the high interior live loadings of synthetic materials common in today's dwellings, are causing fire officials to reassess the smoothbore nozzle's penetration qualities for fighting hot, fast-spreading fires up close.

Because of the smoothbore nozzle's performance and increased effective operating distance in high-heat conditions, the attack can begin as soon as the nozzle operator determines when the circular or rectangular movement of the stream can cool the compartment and hit base of the burning material.

The operator does not have to be close to the fire in order for the stream to accomplish efficient extinguishment. This effective operating distance helps keep the crew away from the thermal effects of heat and steam during extinguishment operations.

Managing Unwanted Steam

Another concern among fire officials is how a stream's reaction to heat will affect ventilation and the movement of products of combustion within a building. Since the stream from a smoothbore nozzle contains less air and operates at a lower exit pressure, it will not disturb the atmosphere within a fire building as much as a spray stream from a combination nozzle (figs. 6–7a and b).

There is little doubt the spray discharge from a combination nozzle into a fire area, especially into the overhead, will create a large amount of steam, and that steam will have to go somewhere. Every cubic foot of steam generated during firefighting operations must occupy a cubic foot of space in the area. The commonly accepted expansion ratio of water to steam is 1,700:1 at 212°F. But real fires have temperatures which far exceed 212°F—more in the range of 1,000°F–1,500°F. At 1,000°F, water has an expansion ratio of 4,200:1. The hotter the fire, the more steam volume created and the larger the area it must occupy.

In practical tests, it has been found that 1 gal. of water equals approximately 200 cu. ft. of steam at 212°F. With the increased use of high-flow handline nozzles, we can easily see

146 *Fire Stream Management Handbook* Second Edition

Figure 6–7a. When the stream from a smoothbore nozzle is worked off a ceiling, firefighting action is obtained from distribution of water over the burning material, not from turning into steam in the overhead. As can be seen, the water droplets are large, helping them penetrate through the fire's heat and reach the burning material below. Note how much water is available to soak the floor area. Some water does vaporize in the overhead, but it will tend to gradually reduce the heat rather than generate an immediate, violent cloud of steam. Also note that the water does not immediately angle downward but runs along the ceiling, dropping large water droplets along the way.

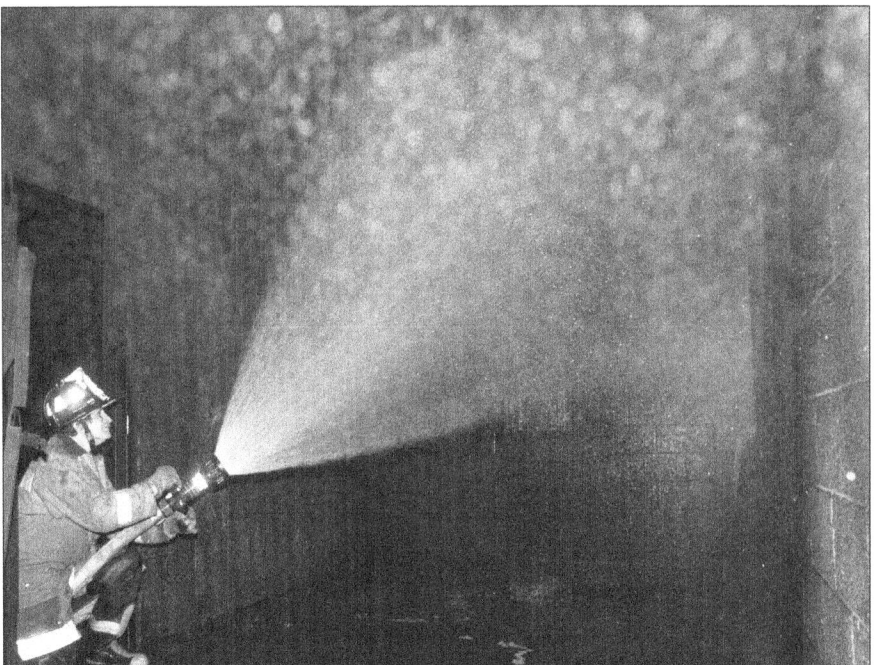

Figure 6–7b. A high-speed flash helps make a comparison of interior firefighting tactics. If a spray stream is worked into the overhead, as many have advocated in the past, most water remains in the room's atmosphere. If high heat was present, it would vaporize into clouds of steam, which would begin to cool the atmosphere, but in the process would drive steam and combustion products down on the nozzle operator. If the fire was located on floor level, little water would hit the fire.

why more firefighters are being burned by steam as they spray 150 GPM–200 GPM into an overhead atmosphere containing heat over 1,000°F.

During controlled tests measured by the National Institute of Standards and Technology, nozzle crews repeatedly inverted temperatures within a fire room when attacking test fires with flows from spray streams, forcing superheated steam, smoke, and products of combustion down to the floor while cooling the overhead to almost room temperature. The effect of 1,200°F of wet heat billowing downward can be extremely discomforting and, in many cases, dangerous to the nozzle crew, often proving harmful and in some cases fatal to any victims unfortunate enough to be trapped in the area.

To properly manage unwanted steam production within a fire building, the nozzle crew must accomplish the following:

- Keep as much finely divided water out of the superheated overhead as possible.
- Cool the compartment by using a rapidly moving straight stream, hitting the ceiling, sidewalls, and floor.
- Stop flame, heat, and vapor production as rapidly as possible at the flame/fuel interface by direct application of water.
- Quickly ventilate the expanding gases through selected openings, ideally before water is applied.

The difference between combination and smoothbore nozzles in radical steam production and air movement is determined by the individual stream's physical characteristics, how the streams react to heat, and how much air the stream moves. Any combination nozzle's spray stream will move a tremendous amount of air because the water is spread out by the nozzle design and atomized into small water particles separated by air particles. This air-enriched stream then picks up increased velocity upon discharge because of the combination nozzle's higher discharge pressures.

Water passing through a smoothbore nozzle encounters no baffles to severely deflect the flow at right angles inside the nozzle. Because the water flows forward, and because the stream operates at a lower discharge pressure, the stream itself does not carry a great amount of trapped air and does not tend to pick up as much air along the stream as a combination nozzle stream will.

A smoothbore nozzle's reduced air-mixing ability can be proven by dumping a gallon or two of dishwashing detergent in a booster tank. Discharge a stream parallel to the ground (to help eliminate aeration caused by water bouncing off the ground) from first a combination nozzle, then a smoothbore tip. The water-and-soap solution passing through a combination nozzle will pick up quite a bit of air and will discharge as a bubbly foam solution. The smoothbore's discharge will appear to be almost normal water because of the lack of entrained air. A combination nozzle's air-mixing ability will vary between individual nozzle designs and manufacturers.

The smoothbore stream, because of its minimum air movement qualities, is less likely than a combination nozzle to cause a steam-balloon temperature inversion when directed into burning material. All experienced firefighters have been enveloped in steam and smoke when making an interior attack, sometimes suffering burns on the ears and neck in the process.

After a demonstration of direct attack using smoothbore nozzles in a live fire situation, firefighters and officers are amazed at the relatively high comfort and low stress afforded personnel making a direct attack. A direct attack using smoothbore nozzles will not normally wrap the crew in heat, steam, and smoke.

In his book, *Fire Fighting Principles and Practices*, former FDNY Chief William Clark neatly sums up the theory of smoothbore nozzles when he says,

> In most cases, combination is no better, or not as good as, a straight stream. This can be explained when we ponder the fact that the finely divided spray cools the atmosphere of a confined space quite effectively, but we are not primarily interested in that. We realize that extinguishment is through cooling the fuel source to the point where it will stop generating fuel gases.

Smoothbore Nozzle Use with Standpipes

Depending on fire conditions, water flowing from a smoothbore nozzle can be distributed over a wide area simply by directing the stream into the overhead, floor, and side walls and deflecting the stream so that it is driven into the burning material. Some water will turn to steam as the solid stream breaks up, but most is formed into large heavy droplets that will do their job of cooling and extinguishment without being vaporized in the overhead.

Many chiefs assigned to high-rise districts in large cities strongly advocate the use of smoothbore nozzles when attacking fires in these buildings. It is rare that the fire in a high-rise occupancy is properly ventilated, and the use of the solid-stream nozzle will extinguish the fire without excessive pushing of flames or thermal balance upset. Former Chicago Battalion Chief Clarence Dixon, who commanded a district encompassing the downtown Loop area, thinks the smoothbore is ideal for most high-rise fires. He stated,

> You get immediate penetration without being enveloped in a cloud of steam. With proper ventilation and a combination stream, you can push the smoke and products of combustion away. In a high-rise building, the ventilation may not be ideal and the smoothbore disturbs the heat over your head less than a combination nozzle.

There are other advantages to the smoothbore nozzle as well when working off building standpipes. If low-pressure or fluctuating-pressure problems exist within the fire building's standpipe system, the lower operating pressure of the smoothbore nozzle will have a better chance of providing a workable stream where the combination nozzle, operating at double the pressure, may not. This reduction in the required nozzle pressure also means the hose will not be as stiff and is more easily maneuvered within a structure.

Another problem, especially in older occupancies, is clogging of the nozzle due to rust, scale, and trash dislodged as engine companies begin pumping into the standpipe system. Debris will easily flow through a smoothbore without having to stop the attack to flush or shut down to clean the tip. I have always stressed that the first criterion in evaluating a nozzle for standpipe use is, can a mouse pass through the nozzle without clogging it?

Reducing Stress

Nozzle reaction, or *kickback*, is generated by a combination of the weight of the water flowing and the pressure at which it exits the nozzle. The calculation of nozzle reaction is governed by laws of physics. A certain amount or weight of matter moving at a certain speed in a certain direction causes a reaction in the opposite direction. If a certain mass of water, for example 250 GPM, being discharged by a nozzle remains constant and the nozzle pressure is increased, the nozzle reaction will increase. Because smoothbore nozzles normally operate at 50 PSI, they dramatically reduce the effort necessary to counteract nozzle reaction over nozzles with the same flow rating operating at 100 PSI. Operating with limited suppression personnel, the lower nozzle reaction reduces operator fatigue caused by expending energy attempting to control the line's back pressure. This is especially apparent when using 1¾" or 2" attack lines with flows greater than 150 GPM. Since the operating pressure is reduced, the hose becomes less stiff and easier to handle and bend around corners when working inside a fire building (fig. 6–8).

Figure 6–8. A 1⅛" tip operating at 45 PSI nozzle pressure, discharging 252 GPM, and having a reaction force of 89 pounds of force being handled by a single firefighter. This is not the best practice, but let's face facts. At some point in every understaffed fire department, one person may have to operate a line for a short period of time to prevent a fire from spreading or endangering trapped victims.

For example, using standard nozzle reaction formulas, the following example shows back-pressures which can be expected from two nozzles, each flowing 185 GPM:

Combination nozzle flowing 185 GPM at 100 PSI = 93 lb. of force
$^{15}/_{16}$" smoothbore nozzle flowing 185 GPM at 50 PSI = 69 lb. of force

Some instructors use a rule of thumb which states that a firefighter can safely handle one-half their body weight in nozzle reaction force. If this value is used to estimate personnel requirements, a 186 lb. firefighter will be needed to safely direct the combination nozzle stream while a firefighter weighing 138 lb. could handle the stream from the smoothbore nozzle.

In addition to all the advantages of reducing stress and providing excellent stream characteristics, smoothbore tips require little maintenance and are less expensive.

If the interior or the tip edge of a solid-stream nozzle is damaged, burred, or dented, the stream quality is adversely affected. This quality is exhibited by a stream with poor reach, rapid break up, and uneven-looking flow along the stream's sides. Most smoothbore nozzles are manufactured with a counterbore on the inside edge of the nozzle tip to help protect the exit orifice from damage. If the shutoff bore is not matched with the tip bore, say by mixing products from different manufacturers, or if a gasket protrudes into the waterway, the stream will be of poor quality (fig. 6–9).

Figure 6–9. Modern smoothbore tips have a counterbore or step machined into the exit orifice which helps prevent damage to the opening which will degrade the quality of the stream.

Training

A firefighter wrapped in a protective envelope of boots, pants, coat, gloves, hood, breathing apparatus, and helmet cannot easily detect what is happening to the atmosphere in the fire building using the senses of touch, smell, sight, or hearing. Many firefighters have been injured from the effects of steam or direct flame contact caused by fire rolling around or over them as they advance on a fire. Because of visibility problems, it can be difficult for a nozzle crew to comprehend what is happening to the fire conditions and atmosphere ahead of the nozzle.

Under fire conditions, officials who are experienced in the use of both combination and smoothbore nozzles for interior attack, support the contention that a firefighter will get into less trouble with the atmosphere in the fire area using a smoothbore. All firefighters making interior attack must be fully trained, but because applying water to a fire is also an art which, unfortunately, can only be learned by actual experience, not every firefighter who handles a nozzle during attack will apply the water in the most efficient manner. If the nozzle operator has a low experience level, applying water from a smoothbore nozzle will cause fewer unwanted fire spread problems and will most certainly help reduce thermal injuries.

The crews must also be taught to shut down the stream immediately after flame production ceases. The smoothbore nozzle's advantage of creating less steam and increasing visibility is rendered useless if excess water continues to be vaporized by hot solids after the flames are knocked down.

Combination Stream Attack

Standard operating procedure, formerly used by a few departments, called for two lines to be used when attacking stubborn fires in stores, basements, or high-rises. One line employs a combination nozzle set at about 30° to be used strictly for pushing combustion products away from the crews. The second line has a smoothbore nozzle for penetration into the fire area far ahead of the spray stream. In this way, it was felt both nozzles were utilized to their best advantage.

Former Chief Ed Phipps of the San Francisco Fire Department, however, felt two streams from smoothbore nozzles was perfect for interior firefighting operations. "If you can put two lines together on a fire, you won't get hurt," he said. "The second line should be deflected off the ceiling and walls over the first line, with the nozzle swinging from side to side to keep the heat off the advancing first crew."

For the combination stream attack to work properly, there must be adequate ventilation openings over or beyond the fire area to vent combustion products. Fire and smoke cannot be successfully pushed into an unventilated area without reacting violently, blowing back into an area offering the least atmospheric pressure resistance. In most cases, this area is created around the nozzle crew by the effects of the flowing streams.

Twin Tips

One answer to the combination versus smoothbore nozzle question is to carry both on the same line. In 1986, the Miami Fire Department began to study its high-rise firefighting tactics. They wanted to carry a combination nozzle on their high-rise hose packs for general use but were concerned about performance in low-pressure situations. In addition, they wanted to address the problem of excess steam generation in enclosed areas during high-rise fires.

Working with a nozzle manufacturer, the *slug tip* was developed. These compact smoothbore tips were designed to be attached between the shutoff and the combination tip, so both nozzles are carried into a building as one unit. The combination tip is available for normal attack operations, but if fire conditions mandate or if the standpipe pressure drops, the combination tip can be easily removed and the attack continued using the smoothbore tip. Because the nozzle is carried into the building as a unit, the choice of tips can be made at the fire source, before or during attack. The decision to change tips when both are immediately available is not as major as having to stop the attack and run outside, rummage through compartments, then return to the nozzle with the desired smoothbore tip (fig. 6–10).

The removable smoothbore slug tip is no longer as popular as a shutoff with a permanently attached smoothbore tip. This makes the twin tip concept a bit more compact and lighter.

I see fewer departments today carrying this setup—what I called training wheels for smoothbore use.

A comparison of pressures and reaction forces can help in understanding the twin tip concept. Let's consider a 200', 1¾" preconnect line equipped with a low-pressure combination nozzle and a ⅞" smoothbore intermediate tip operating at an engine pressure of 144 PSI:

Combination Nozzle Flow/Pressure	Smoothbore Tip Flow/Pressure
150 GPM/75 PSI	170 GPM/63 PSI

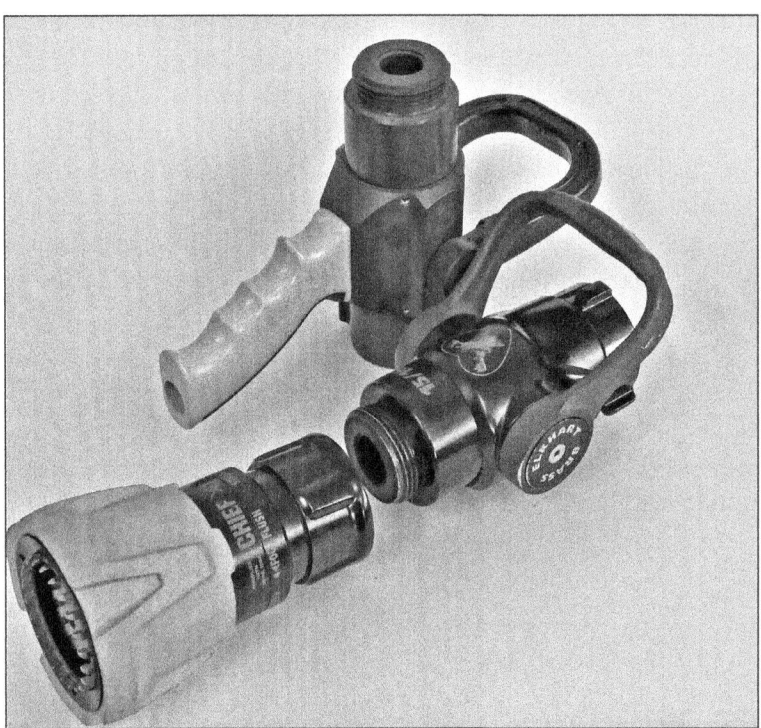

Figure 6–10. Growing in popularity is the design of having a smoothbore tip outlet provided on the exit side of a shutoff. A combination tip can be provided mounted on the outboard threads and be specified to flow less or match the flow rate of the smoothbore behind. Testing has proven that the shutoff tip provides exactly the same quality stream as a longer, detachable tip.

The smoothbore tip will flow 20 GPM more water than the combination nozzle at the same engine pressure while nozzle pressure is reduced somewhat. If the same layout is in use and the pressure is suddenly reduced to 100 PSI the following flows and reaction can be expected:

Combination Nozzle Flow/Pressure	Smoothbore Tip Flow/Pressure
125 GPM/52 PSI	140 GPM/39 PSI

As can be seen, the drop in engine pressure caused the combination nozzle stream to drop to a much lower flow while its reduced nozzle pressure rendered the stream reach less effective. With the smoothbore tip, while the gallonage was proportionally reduced, the reduction in nozzle pressure still allowed for an adequate firefighting stream with decent flow and reach to still be effective. If operating from a standpipe system, the effects of a drop in pressure become more pronounced. For example, let's assume a standpipe with 100 PSI available at the outlet is supplying 100' of 1¾" hose equipped with a 150 GPM at 75 PSI combination nozzle and a ⅞" smoothbore, intermediate tip. At the end of the line the following flows and nozzle pressures would be delivered:

Combination Nozzle Flow/Pressure	Smoothbore Tip Flow/Pressure
145 GPM/68 PSI	168 GPM/56 PSI

While both streams would be adequate for firefighting use, the stream from the smoothbore is well within the nozzle's normal operating range and has more effective reach and firefighting action than the stream from the combination nozzle.

If the pressure in the standpipe drops to 60 PSI, the following flows/nozzle pressures would be available:

Combination Nozzle Flow/Pressure	Smoothbore Tip Flow/Pressure
110 GPM/41 PSI	130 GPM/34 PSI

The reduced standpipe pressure would cause the stream from the combination nozzle to be relatively ineffective. While the smoothbore stream would not be as good as that flowed at normal pressures, it could still be used to hold or extinguish a fire until the standpipe pressure problems could be overcome. Remember that the older *National Fire Protection Association 14: Standard for the Installation of Standpipe and Hose Systems* which covered building standpipe systems called for only 65 PSI to be available at the system's uppermost outlet. While the newer version of the standard calls for 100 PSI, many buildings still have systems in place that only meet the older standard. This pressure is not enough to power an adequate flow through smaller-diameter handlines equipped with 100 PSI combination nozzles.

Because smoothbore streams operate effectively at pressures lower than needed to produce effective streams from combination nozzles, they can provide a higher level of protection for operating personnel if a pressure drop occurs during suppression operations (fig. 6–11).

Even if the department does not wish to use smoothbore nozzles for primary fire attack, the addition of a slug tip to the combination nozzle/shutoff assembly will provide a higher level of security for firefighters, if deployed when fireground conditions dictate their use.

Figure 6–11. A removable stream straightener can be used whenever using removable smoothbore tips. Keep in mind that approximately 3' of the hose behind a smoothbore shutoff is also part of the nozzle. Kink or severely bend this hose, and the resulting turbulence causes the stream to degrade. The stream straightener helps mitigate this. Because it also tends to trap debris, its use on high-rise nozzles is not recommended.

Twin tip/slug nozzle assemblies offer the following advantages:

- Flow can be increased at the nozzle without having to adjust engine pressures.
- The decision to utilize either a smoothbore or combination nozzle for attack does not have to be made in the street before entry is made but can be decided and easily implemented within the area of operation (fig 6–12).
- If pressures from standpipes or long hose lays prove inadequate for proper operation of combination nozzles, the slug tip can provide an immediate remedy, providing effective streams at nozzle pressures down to 20 PSI.
- If the combination becomes clogged with debris, it is an easy matter to spin it off and continue the attack with the smoothbore tip.

Threaded Tip Styles

Departments purchasing smoothbore tips, especially those on 2½" lines, should consider having the exit end threaded with the department's 1½" hose thread. Additional hose can then be attached without removing, and possibly losing, the tip. The Chicago Fire Department carry 150' of 1¾" line attached to the tip of a 1¼" smoothbore shutoff nozzle (fig. 6–13).

When needed, the line, which many call a *day line*, is stretched with enough 2½" line flaked out around the building for use if the fire proves too much for the smaller line. The smaller line is detached, and the larger flow is quickly available. In an attempt to increase fireground efficiency, some departments use a 1½" base, gated wye so a second line can be quickly attached to the large nozzle. This worked relatively well when the smaller lines flowed around 100 GPM, but with desired handline flow of around 150 GPM–160 GPM, trying to supply two lines being opened and shut at different times becomes almost impossible for the pump operator to control and allows dangerous reaction forces to develop, causing safety issues for nozzle operators.

Figure 6–12. The firefighter here, faced with a large amount of fire, opted to remove the combination tip, seen at the lower right, and continue his attack with the built-in smoothbore tip.

The use of a threaded tip provides more fireground flexibility than just a plain tip and depending on design, a number of auxiliary tips can be attached. For example, the FDNY uses a smoothbore tip that terminates with internal threads. A right-angle partition tip or a ½" 50 GPM overhaul tip can be quickly attached, depending on conditions (fig. 6–14).

I always recommend the use of a small stream shaper between the shutoff and nozzle tip. It helps reduce the incoming water turbulence and helps keep the stream straight when maneuvering the nozzle inside a fire compartment. The only place where the stream shaper is not recommended is in high-rise operation where it can become quickly clogged with debris during firefighting operations.

Figure 6–13. The threaded tip makes it easy to carry the initial attack line attached to the nozzle of a 2½" backup line to easily and quickly provide more flow if needed.

Figure 6–14. Depending on how the main tip is configured, a number of options can increase firefighting efficiency. From left, the partition tip threaded on the end of an FDNY style main tip, a 50 GPM overhaul tip attached to the end of an FDNY main tip, and a dual-size tip, useful for standpipe work.

Variations on Common Smoothbore Nozzles

In the quest of trying to provide a smoothbore nozzle with an integral combination nozzle, manufacturers are fielding nozzles that can provide both stream shapes. There were a number of attempts to create these in the past but problems with different gallonage and nozzle pressure settings and unwieldy controls limited their effectiveness.

Task Force Tips introduced the Vortex line that ranges from handline nozzles to master stream tips. The basic waterway is a smoothbore controlled by a lever shutoff. A turn of the barrel introduces internal fingers into the waterway that will convert the straight stream into an approximately 30° spray pattern. Tips are interchangeable, and use of the spray stream does not affect the flow rate (fig. 6–15).

Akron has a similar nozzle called the Ultrajet that operates in a similar manner. In place of interchangeable tips, flow rates are changed by the use of removable plates (fig. 6–16).

Both of these designs are relatively new, and time will tell how well they will be accepted into the marketplace.

The HEN nozzle has also been recently introduced into the fire service. It is a straight-through smoothbore nozzle; however, its outlet is square rather than round. Its primary feature is that has two stream positions: straight and fan. The straight stream setting behaves as a standard smoothbore, and the fan setting flattens out the straight stream into a thick, 30° wide fan pattern. This should not be confused with a circular spray pattern as the fan pattern exhibits heavy large droplets and is designed to help work the water around a compartment

Figure 6–15. A Vortex nozzle provides both a straight stream and can provide a 30° spray stream with a twist of the barrel ahead of the shutoff. It is designed so smoothbore tips can be switched out if needed.

during fire attack. Early users, especially in wildland work, have been enthusiastic about the effectiveness of the fan pattern and, again, time will tell how well the nozzle will be accepted in the marketplace (fig. 6–17).

The structural model of the HEN, designed to be used on a 1¾" handline, flows 160 GPM at 50 PSI, but smaller, lower-flow versions are offered for wildland firefighting where the fan pattern has been desirable for many years (figs. 6–18a and b).

Figure 6–16. Akron markets the Ultrajet nozzle that offers a smoothbore flow with a spray pattern with a twist of the barrel. Flow rates can be changed by replacing the front nozzle plate.

Figure 6–17. The HEN nozzle offers two stream selections, straight and fan. The fan stream, widely used in wildland firefighting, is now being introduced for structural use.

Figures 6–18a and b. The HEN offers a straight-stream pattern that is not round, but square. The fan stream is said to provide increased coverage when advancing into a compartment but is delivered in straight-stream-quality droplet size, providing less chance of generating unwanted steam.

Review

As with all fire service tools, there are advantages and disadvantages to using smoothbore tips. Like the leather helmet, smoothbore nozzles have developed something of a firefighting mystique along with a devoted, sometimes vocal, following.

A battering ram can be used to poke an inspection hole in a ceiling. A pike pole, however, might be a more practical tool for the job. Perhaps you could find a firefighter who carried a battering ram into battle with some success and now professes to use the tool for every task. While I can admire that firefighter's dedication, I would have to question their thought process.

Some fire service officials swear the smoothbore tip is the only nozzle they would ever use, and others say they would not allow one within a mile of the fireground. To adhere to either extreme is battering ram mentality.

Widely used by big cities, the smoothbore is proving more and more popular with increasing numbers of suburban and rural departments who appreciate its structural firefighting advantages of high flow with low nozzle reactions, and its superior penetration into high-heat areas.

A smoothbore nozzle is another very effective tool. We should learn its advantages and limitations, and then use it where it will do the job best.

Chapter 6 Review Questions

1. What determines the flow rate of a smoothbore nozzle?
2. What are the main advantages of a smoothbore nozzle over a combination nozzle operating at 100-PSI nozzle pressure?

3. What is the secondary use of an Underwriter's playpipe?

4. What is the main advantage in using a smoothbore nozzle for direct, offensive attack?

5. What is the prime indication that a smoothbore stream is overpressurized?

6. What is an advantage to using a smoothbore nozzle in older, standpipe-equipped buildings?

7

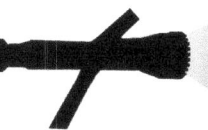

Low-Pressure Nozzles

Over the decades, the fire service has experienced radical changes in water delivery methods. More powerful apparatus with larger pumps and tanks, large-diameter supply hose, larger attack lines, and advances in high-flow nozzle design are hardware improvements that have helped fire departments deliver more water to a fire more efficiently than ever before.

Unfortunately, the advances in hose and water delivery appliances have occurred at the same time as a general trend of overall reduction in available fire scene suppression personnel. Pumpers carrying over 1,000' of large-diameter hose and capable of pushing out over 1,500 gallons per minute (GPM) at the fire scene are common today in the nation's fire service. Also common today, unfortunately, is the fact that these units, more often than not, arrive on the fire scene staffed with only two or three firefighters.

Recent surveys have shown that much of the offensive interior firefighting in this country is accomplished with two or three firefighters attacking a fire with a 1¾" medium-diameter hoseline flowing 130 GPM–160 GPM. While this flow range has proven generally adequate in the past, many departments are beginning to report an overall alarming trend of interior structure fires becoming more intense, generating more heat than lower gallonages can suppress. This increase in intensity is being traced by researchers back to three major causes:

- Increased loading of flammable contents within structures, especially dwellings. For example, a bedroom containing a dresser, nightstand, lamp, and bed may in today's world have a dramatically increased fire loading—a king-size foam rubber mattress, a plastic-covered home entertainment center containing a television, a hair dryer, and more.

- Extensive use of plastic and other synthetic materials for use as structural building materials, plumbing, and interior finishes. The use of plastics in furniture manufacturing, especially in low-cost items, is now commonplace. Synthetic furniture materials have been described by some researchers as "solid gasoline."

- Changes in building construction methods, especially the widespread use of lightweight roof and floor trusses, expose more surface area of the structural framing material to combustion than do older construction methods utilizing fewer structural members of larger dimensions.

Within the past few years, many departments have experimented with new tactics to combat this overall increase in fire volume. These new tactics generally require the application of greater amounts of water and planning from where that water will be applied. Traditional methods of simply increasing the number of operating hoselines to increase overall flow at a fire scene are not proving practical because of the common trend reducing the number of firefighters available to stretch and crew these additional lines. In career departments, personnel reduction is usually a result of budget cutting or renegotiated labor agreements, assigning fewer firefighters per vehicle than ever before.

Volunteer departments, too, are not immune to this general depletion of personnel. Many departments are finding their traditional sources of personnel shrinking as high living costs, especially in the eastern part of the country, are forcing active members as well as potential recruits to move out of the area in order to find affordable housing.

Over 30 years ago, as fire officials attempted to address the problem of fewer available fire scene personnel, experiments and testing by fire departments in Los Angeles, Chicago, and San Francisco attempted to address flow and personnel problems by finding solutions that utilize improvements in waterflow hardware.

Captain Gil Moreno, former head of San Francisco's bureau of equipment, said his department's testing of new hose and nozzles had a single goal: "We wanted a maximum of flow using the least amount of people."

Former Battalion Chief Claude Creasey headed the Los Angeles Fire Department's extensive testing program which addressed the same concerns. One of Los Angeles' goals was to evaluate the effect of smaller firefighters entering the department as a result of changes in physical requirements, due in part to affirmative action recruiting programs. Because discrimination based on height and weight was no longer allowed, more smaller-statured firefighters were being hired.

The safe control and operation of a traditional hoseline depends greatly on the collective weight of the firefighters holding the line, acting as an anchor, so to speak. In addition, as more and more firefighters already employed became concerned about physical fitness, many began weight control and reduction programs, which could cause a further reduction in an engine company's total available line-holding weight. Since an increasing number of firefighters had become smaller and lighter, traditional methods of controlling standard handlines began to prove less effective.

Beginning in the early 1980s, the Los Angeles Fire Department began a series of evaluations aimed at attempting to determine the maximum effective flow from a 2½" line that could be safely controlled by two firefighters. Tests were devised to simulate offensive firefighting as closely as possible. In one of the tests, using a 15" opening in a large panel as a target, crews were told to move in different directions while attempting to keep the stream within the target.

Firefighters of different sizes and with varying amounts of experience were rotated on the nozzle and the results were videotaped for later review. Other tests included calibrating the amount of nozzle reaction generated by various nozzles and pressures using a measuring device created by the department.

Chief Creasey and his evaluation group found that the most effective streams that two firefighters could handle were produced by lowering the nozzle pressure on a standard fog nozzle. He knew that flow was reduced by lowering the nozzle pressure; however, by working with nozzle manufacturers, nozzles were developed that would flow their rated 250 GPM capacity at 75 pounds per square inch (PSI) rather than the traditional 100 PSI. Success of the 2½" program in 1984 led to the purchase of new nozzle equipment and encouraged further testing utilizing the low-pressure nozzle concept on smaller attack lines.

About the same time these tests were taking place, the Chicago Fire Department evaluated a number of different nozzles in order to facilitate the adoption of a new attack program they called *quick water*. Historically, the Chicago Fire Department relied almost exclusively on reverse lays in fighting structure fires. In attempting to get water on the fire faster, the quick water program allowed engine companies to lead out preconnected lines in certain situations, backed up by a 3" or 4" supply line either hand-laid by the first engine or supplied by the second company. While performing evaluations, department officials found that they could flow more water by utilizing 1¾" hose and high-flow nozzles than they could with the 1½" they were then using without greatly increasing the physical effort expended by firefighters to stretch the line.

They experimented with automatic- and fixed-gallonage nozzles and rated each on firefighting effectiveness, maintenance considerations, unit purchase cost, and amount of training needed to effectively integrate the nozzle into their operation. The department determined that 150 GPM was an ideal baseline flow for a single attack line. This was based on an evaluation of the average interior volume of dwellings and other structure hazards found within the city.

In addition to the baseline flow requirements, it was determined the reaction caused by a combination nozzle flowing 150 GPM at 100 PSI was the maximum a single crew could effectively direct and safely handle in an interior fire situation.

At the suggestion of a nozzle manufacturer, the department tested a fixed-gallonage nozzle flowing the desired 150 GPM rate, but at 75 PSI nozzle pressure rather than the traditional 100 PSI. This pressure reduction had the desirable effect of reducing nozzle reaction as well as providing stream qualities ideal for their method of direct interior attack. The net effect was that the new hardware could deliver more water on the fire than the previous system while the considerable reduction in nozzle reaction had the desirable effect of reducing firefighter stress.

Additionally, both Los Angeles and Chicago were concerned with initial purchase price and maintenance considerations. Because fixed-gallonage nozzles have fewer internal moving parts than automatic- or adjustable-gallonage designs, they determined the simpler constructions would prove to have a lower repair frequency rate. The original low-pressure Los Angeles City nozzle, a modification of an industrial design, was constructed of only four parts, which they said was an important consideration during anticipated maintenance evaluation (fig. 7–1).

As a result of extensive evaluation and testing, Chicago, Boston, Memphis, Detroit and Los Angeles fire departments were among the first to utilize low-pressure nozzles on all attack handlines.

Figure 7–1. The original Los Angeles City low-pressure nozzle was a simple adaptation of an industrial nozzle that was National Fire Protection Agency (NFPA) rated at 230 GPM at 100 PSI, had a manufacturer rating of 200 GPM at 75 PSI, and in actual use flowed 160 GPM at 50 PSI.

High-Flow Handlines

At first, many departments wanting to flow more water using 1¾" or 2" handlines consider purchasing automatic nozzles along with the larger hose. While it is true that the automatic nozzle will easily allow for higher flows, it also has the drawbacks of high nozzle reaction, personnel misunderstanding their operational characteristics, the ability to create a good-looking stream at low flows, and problems in handling pressure surges. All these items can be easily addressed, and there were departments who began using automatic nozzles with quite a bit of success.

Then there were many others who were disappointed, primarily because they were flowing quite a bit less water than they had thought. High flows from 1¾" or 2" handlines can be easily obtained with nozzle styles other than automatics. In addition, it is important to realize that with standard-pressure combination nozzles, high flows also mean high nozzle reaction forces. In order to flow more water, either the number of personnel on the nozzle crew had to be increased or some way had to be found to reduce the reaction force while flowing the same amount of water.

As testing demonstrated, a department attempting to maximize attack potential with higher flows through larger hose and nozzles must take into consideration the fire-flow stress imposed on the firefighters operating the line in hostile environments. Not only must the crew contend with excessive heat and reduced visibility, but they must also physically exert themselves while advancing the water-loaded hoseline and then expend even more effort after opening the nozzle to counteract its reaction.

As discussed in chapter 6, one attempt to reduce this stress and to increase firefighting effectiveness is the use of smoothbore nozzles operating at 50 PSI tip pressure, rather than combination nozzles operating at 100 PSI tip pressure. The arguments advocating the use of certain design nozzles for interior fire attack are far from settled; however, each type has a number of distinct, positive advantages.

Smoothbore Nozzles

- Operate at less pressure at the same flows than a standard combination nozzle, reducing nozzle reaction

- Have greater penetration into high-heat areas because of stream characteristics

- Are less apt to cause undesirable problems with the overhead fire atmosphere because the stream contains less entrained air

- Have wide-open nozzle bores that are not prone to clogging by rust, scale, or stones

- Are less expensive to purchase and repair

- Do not entrain a great amount of air and do not tend to push a fire around as much as a combination nozzle

Fire officials using smoothbore nozzles on attack lines consider the nozzle's two most important characteristics to be its penetration power and low nozzle reaction. They feel a suppression crew can more effectively operate if they can apply water from a safe distance and do not have to continuously battle the effects of high nozzle reaction force.

Combination Nozzles

- Can absorb a wide range of nozzle pressures (over or under pressurization) and still provide a stream with effective reach

- Have spray pattern positions which can easily cover a large area more quickly for cut-off or holding operations

- Have spray positions which apply water more gently, causing less damage in close-in work

- Are more effective for fires involving flammable liquids because of spray patterns

- Can provide limited protection for the nozzle crew in certain fire situations with their wider spray pattern positions

- Are useful for indirect attack in attics or other confined areas using the spray pattern

- Can move a great volume of air with the spray stream, which is useful for forced ventilation operations

Most fire officers cite tradition as the primary reason for employing combination nozzles ("Well, we've always used them"). When questioned further, most state the combination nozzle's most important advantage is the myth they term "protection." As discussed in previous chapters, the protection many think they need may be to protect themselves from improper interior water application tactics.

At equal gallonage, streams from smoothbore nozzles operating at lower pressure than comparable combination nozzles cause less nozzle reaction and therefore reduce operator stress. Many departments, while appreciating the smoothbore nozzle's ability to reduce operating effort, do not want to give up the versatility of variable stream patterns offered by combination nozzles.

Retired District Chief Andrew O'Donnell, former director of training for the Chicago Fire Department, underscores this point:

> Basement fuel oil storage tanks are common in this city. If we make entry into a basement and find the fuel oil involved in fire, we need the spray pattern to be available immediately. We've found times we have to fight wind in high-rise ventilation, sometimes having to ventilate an area against the wind, and have found the air-moving ability of a spray stream extremely useful. In our operation, the little bit you lose in quality of stream with a combination nozzle over a smoothbore is not worth losing the combination nozzle's versatility.

One of the biggest drawbacks to using a combination nozzle is that the high 100 PSI nozzle pressure required for proper operation causes high nozzle reaction at greater flows. Many departments would like to reduce operator stress, but because of tradition or tactics, do not feel comfortable using smoothbore nozzles on all attack lines.

The History of Low-Pressure Combination Nozzles

To understand the development progression leading to the introduction of low-operating-pressure combination nozzles, we need to look back to the mid-1930s. As told to me in an interview with a retired Elkhart Brass engineer, the company brought a design from the Mystery Fog Nozzle Company in Germany to the United States. Elkhart constructed prototypes and then went to test them. In those days, flow was measured by how long it took to fill up 55 gal. drums. They charged the line at the standard 50 PSI and found that the flow rate was around 60 GPM, which was not the flow that they were looking for. They pondered about redesigning the nozzle and then one engineer suggested doubling the pressure to see what would happen. That is how the rated 1½" flow from the Mystery nozzle became 95 GPM at 100 PSI, not 100 GPM. It is also interesting that when other manufacturers started to manufacture competing nozzles, the 95 GPM rating was copied.

The first low-operating-pressure nozzle was placed in service in the Chicago Fire Department after World War II in the late 1940s. When Chicago purchased Mystery nozzles, they did not buy the 1½" version, but rather a 2½" version with 1½" threads designed to be mounted on a playpipe. This nozzle was rated 250 GPM at 100 PSI. Chicago's standard 1½" nozzle up until then was a ¾" smoothbore that flowed 118 GPM at 50 PSI (fig. 7–2).

Keep in mind that the early Mystery nozzles' gallonage flow varied depending on the pattern. The nozzle's rated gallonage was measured at a 30° fog. It flowed less in the straight-stream position and more in the 90° fog position.

In actual use, the nozzle was operated at 50 PSI in the straight-stream position, which gave them a flow of around 150 GPM. They had put into service the world's first low-pressure fog nozzle, although they didn't realize it at the time. No one knows why they purchased the larger nozzle for use on their 1½" lines. One possible answer could be that the nozzle was more robustly constructed than the 1½" version, but again, there is no evidence to back up this claim.

The nozzle was in use until the early 1960s. When they finally began to wear out, they were replaced with the then standard 1½" shutoff nozzle flowing 95 GPM at 100 PSI. The operating crews realized that they were not flowing the water they were used to when using the older Mystery nozzles, and in the late 1960s asked nozzle manufacturers to come up with a solution.

Experimenting with different stem designs, one company designed a nozzle with a flow rate of 150 GPM at 75 PSI operating pressure. From then on, the use of nozzles operating at lower pressures began to grow.

Figure 7–2. When the Chicago Fire Department equipped its engine companies with fog nozzles after World War II, for some reason unknown to this day, they used 2½" nozzles with 1½" thread. At 50 PSI working nozzle pressure, it flowed 150 GPM. Because it was a variable flow nozzle the actual flowing gallonage in the straight stream position was reduced to 180-GPM at 100-PSI even though the nozzle was rated at 250 GPM which it did flow in the wide fog position.

The Low-Pressure Age Begins

In attempting to find answers to the paradox of flowing more water while using less personnel, a number of widely separated departments who have performed evaluations reached remarkably similar conclusions almost simultaneously. Through testing, it has generally been determined that if 1½" hose is used, 150 GPM is the practical maximum flow for this size line. More water can be pushed through 1½" line by using high engine pressures. However, as the flows increase, friction loss becomes excessive, making the pumping operation very inefficient and the hoseline hard to handle.

Using 1¾" hose, a flow of 150 GPM–180 GPM can be easily obtained at normal engine pressures, and tests have shown that the reaction from those flows is about the maximum a one- or two-person combination nozzle crew can effectively control inside a fire building.

In the present day, a nationwide survey showed that the most common attack line in use flows 150 GPM at 50 PSI, commonly through a ⅞" smoothbore nozzle. If a department prefers a combination nozzle, a low-pressure nozzle can easily provide that flow rate and reaction force as well (fig. 7–3).

The lower nozzle operating pressure provides an additional advantage. In earlier chapters, it was described how airflow moves around a combination nozzle. As the water exits the nozzle, it picks up air from just ahead of the center baffle, creating a low-pressure area. In a 100 PSI pressure nozzle, the water exits at a high velocity and continues moving in a relatively straight line until it loses velocity and falls to the ground. In a low-pressure nozzle, the stream exits at a much lower velocity, and about 14" from the nozzle exit point, the vacuum actually pulls the

Figure 7–3. A matched set of nozzles, each flowing 150 GPM, demonstrates the difference that 25 PSI nozzle pressure can make. At left, the 100 PSI stream's small droplets mist off on the way to the target, and the stream forms no visible footprint at the end of its travel. The 75 PSI stream on the right delivers more water on target due to its larger droplet size.

hollow water stream back together, creating a more solid column of water. This action produces a stream that is similar to that which exits from a smoothbore nozzle, having equal range and similar stream qualities for nozzles with identical flow and pressure ratings (fig. 7–4).

Figure 7–4. The lower operating pressure generates less velocity, and the vacuum in the center of the stream, just ahead of the baffle, is able to pull the hollow pattern of water exiting the nozzle into a solid mass about 14" ahead of the device, rivaling that of a stream exiting a smoothbore.

Safety in Handling

For a high-volume handline to apply an efficient flow, the line must not only supply enough water to overcome heat production, but it must also be easily and safely handled by the available crew. This is the point missed by most departments when wanting to put high-flow handlines into service. If high operating pressure causes the line to be stiff or the nozzle reaction to be too high, the crew will take operational steps on the fireground to make the line more user-friendly. They will tell the pump operator to cut the pressure, adjust the nozzle to a lower gallonage setting, or close the shutoff part way. All of these steps reduce the flow rate. In order for high-flow handlines to function efficiently and safely on the fireground, operator stress must be reduced. There are two ways to accomplish this. One is to cut flow of a standard-pressure nozzle by reducing the operating pressure or reducing its gallonage. The second method is to utilize nozzles that deliver their rated flow at lower operating pressures.

Low-Pressure Nozzle Evaluations

Years ago, the fire company where I served as training officer evaluated a group of low-pressure nozzles to determine if they could add to their firefighting efficiency. Because of limited personnel availability, especially during the day, they sometimes experienced delays in quickly and effectively applying high-volume interior attack streams. The company responded to over 600 fire runs a year in a district including factory buildings of mill construction, three- and four-story balloon-frame apartments, Victorian-style single- and multiple-family dwellings, a railroad yard, and newer types of single-story mercantile and high-rise constructions. At that time, we were using automatic nozzles on our preconnected lines with acceptable results, so were able to approach the tests without having to prove a concept or to advance a cause.

Utilizing a portable flow meter and special line gauges, we began the evaluation. To provide a side-by-side comparison, one manufacturer supplied a matched set of 150 GPM nozzles with one calibrated to flow its rating at 100 PSI while other was set at 50 PSI. Testing at identical flows found the low-pressure nozzle exhibiting less nozzle reaction, making it much easier to handle. We rotated all members on each nozzle to verify this result (fig. 7–5).

We were surprised to find the droplet size exiting from the low-pressure nozzle was much larger than the size from the standard-pressure nozzle. The larger droplet size allowed more water to travel further than the standard-pressure nozzle, proving to be an advantage in actual firefighting use. The low-pressure nozzle's streams more efficiently penetrated into high-heat areas and were not degraded by thermal currents as much as the standard-pressure streams. When comparing streams from smoothbore and low-pressure nozzles side by side, we were impressed by their similarity in reach and compactness of the water column.

There was also a marked decrease in misting or peel-away in the low-pressure stream when operating in the straight-stream position. The footprint of the water delivered by the low-pressure nozzle was a distinct, easily visible oval. The standard-pressure nozzle dropped its water to the ground in a vague mist.

Nozzle operators detected a considerable difference in nozzle reaction between the matched nozzles flowing at different pressures. While both nozzles could be controlled by

Figure 7–5. The stream exiting a low-pressure combination nozzle (right) is remarkably similar to the stream from the smoothbore nozzle (left). Both of these streams are discharging 150 GPM, and both show compactness and ability to deliver most of their water within a small footprint area.

one person at their recommended discharge pressures, the operators stated the line operating at 50 PSI was much easier to control and felt much safer discharging water than the 100 PSI line.

The evaluations demonstrated the following points:

- The low-pressure nozzles exhibited considerably less reaction force and were much less stressful to operate at the same flows than automatic, adjustable-gallonage, or fixed-gallonage nozzles designed to operate at 100 PSI.

- Because the nozzles operated at less nozzle pressure, the engine delivery pressure was reduced, causing the hose to be less stiff and easier to maneuver in tight hallways and small rooms. It was also found that the reduction in operating pressure had a slight effect on increasing the tendency for the hose to kink. This proved true to a lesser extent when double-jacket, rubber-lined hose was used than with the double-jacket thermoplastic-lined hose.

- An important discovery was that the low-pressure nozzles' droplet size tended to be much larger than the droplets exiting a standard-pressure nozzle. When compared side by side with standard-pressure nozzles at equal flow, the larger droplets were not affected as greatly by wind and heat and allowed more water to land on target, especially in narrow fog positions. Later, in actual fire situations, it was found that more water was delivered to the flame/fuel interface, especially in high-heat situations.

- When compared side by side with smoothbore tips operating at the same flows, low-pressure fog nozzles in the straight-stream position exhibited similar characteristics of stream compactness and delivery of water on target. This observation was attributed in part to a noticeable decrease in the amount of air mixed with the exiting water which is common in all combination nozzles. This fact was verified by several nozzle manufacturers.

- When testing the low-pressure nozzle on live fires, they noted quite an increase in stream penetration over the 100 PSI nozzle, especially into high-heat areas during direct attack. In intense fires, the low-pressure combination nozzle delivered more water to the base of the burning material than the standard-pressure combination nozzle and achieved knockdown more quickly (fig. 7–6).

There are times when the nozzle must be worked across the compartment ceiling, sidewalls, and floor, not to create steam, but to distribute the water over a wider area. The company found this tactic more effective using the straight stream from a low-pressure nozzle than from a standard-pressure nozzle. The low-pressure nozzle's larger, heavier water droplets dropped through the heat to the burning material with very little being converted into unwanted steam. This tended to keep the thermal layering intact, which kept fire atmosphere disruption to a minimum.

This evaluation reinforced the results of tests by other departments and demonstrated the advantages of the low-pressure combination nozzle's ability to deliver a maximum amount of water with a minimum amount of firefighter stress, both physical and thermal, in an extremely effective form for safe and effective firefighting.

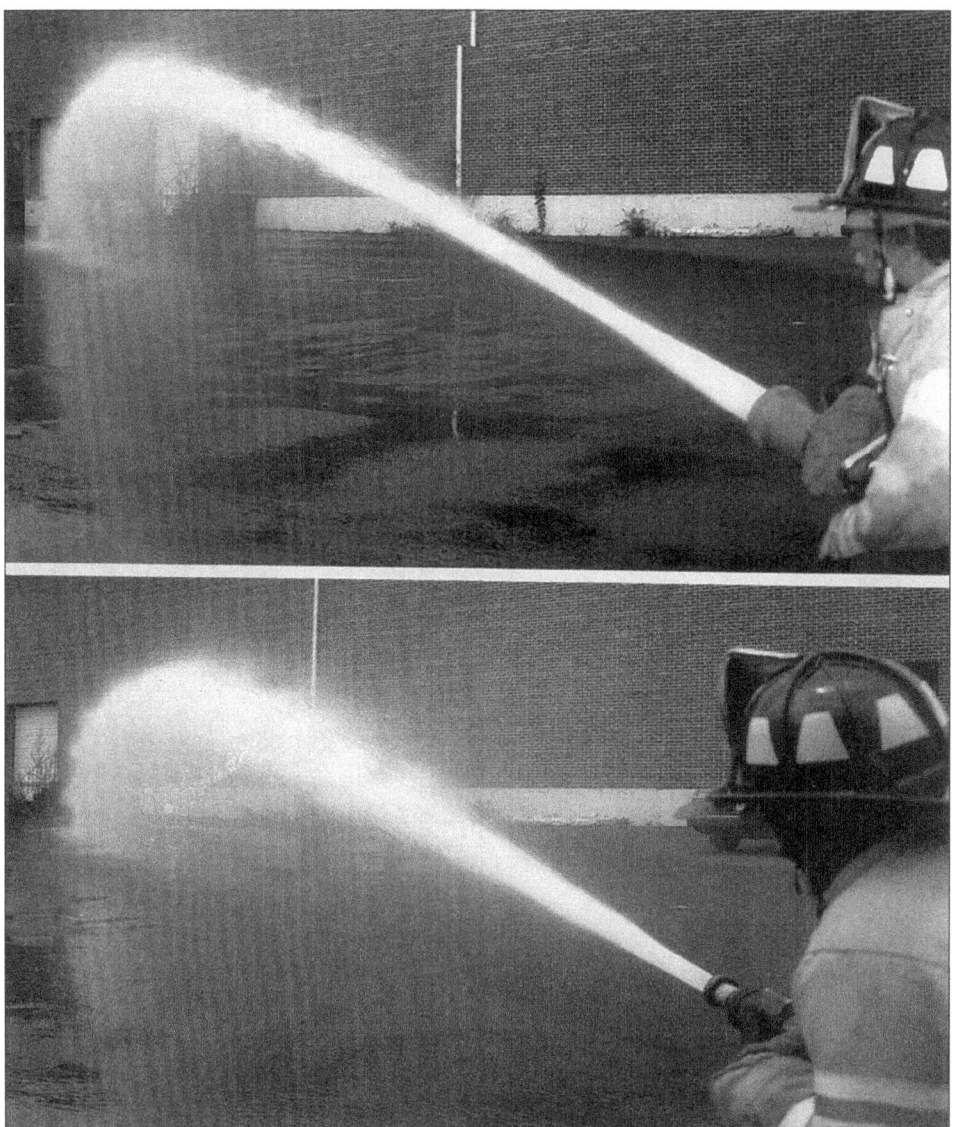

Figure 7–6. A comparison of remarkably similar streams from a low-pressure nozzle on top and a smoothbore nozzle below.

Converting Present Nozzles

Almost all fixed flow, constant-gallonage nozzles, adjustable-gallonage nozzles, or automatic nozzles now in service can be easily converted to high flow/low operating pressure use simply by changing the baffle stem, adding a baffle spacer, or returning the unit to the factory for adjustment and recalibration. In some cases, changing the nozzle configuration can be done by the department, but flow testing should be performed afterward to calibrate the modifications. A department can easily find out what their conversions will entail by contacting the nozzle's manufacturer.

All low-pressure nozzles are marked with their rated flow and operating pressure as per *NFPA 1964: Standard for Spray Nozzles and Appliances*. For many years, the standard required that combination nozzles be tested to supply their rated flow at 100 PSI; however, this was changed about eight years ago to allow more realistic flow rates and operating pressures (fig. 7–7).

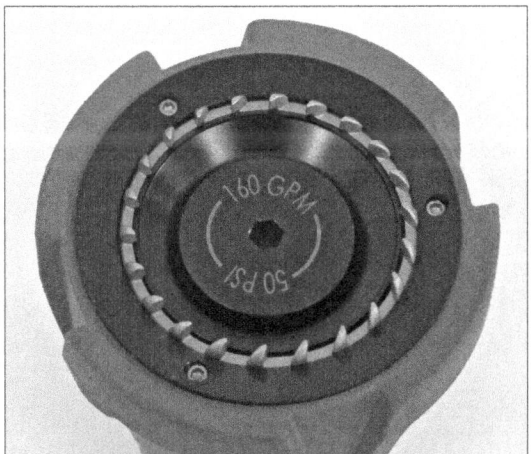

Figure 7–7. All nozzles produced today are marked with their rated flow and pressure at which that flow is obtained.

Explanation of Nozzle Reaction Forces

Arguments for and against the use of various nozzle designs often become nullified on the fireground as crews find they cannot safely operate lines which exhibit high nozzle reaction forces. Formerly, flow rate was the primary criterion when deciding attack line deployment. But if high flow rates are coupled with high nozzle operating pressures, high nozzle reaction force will result (fig. 7–8).

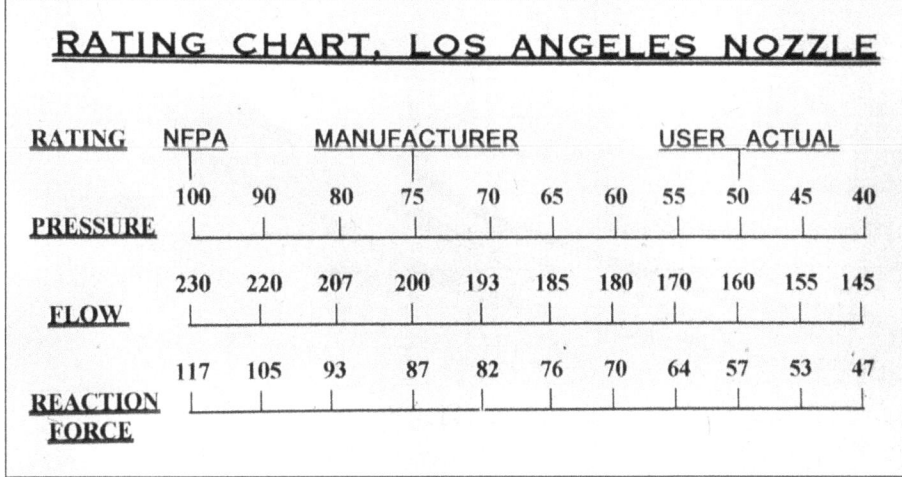

Figure 7–8. A simple chart explains the various ratings. When the first edition of this book was written, NFPA 1964 required all combination nozzles to be rated at 100 PSI nozzle pressure. The manufacturer could rate the nozzle at a different pressure, and then in actual use, the department would flow the nozzle to best fit its needs.

Newton's Third Law of Physics states that for every action there is an equal reaction in the opposite direction. As a stream flows from a shaping device, it creates a force in the opposite direction known as nozzle reaction. The amount of reaction depends on two things: the weight or amount of water flowing and the amount of pressure forcing it through the device.

Nozzle reaction forces can be mathematically calculated by using accepted formulas—one for smoothbore nozzles and one for combination nozzles.

The smoothbore reaction force formula is expressed as

$$RF = 1.57 \times (BD \times BD) \times NP$$

RF = Nozzle reaction in lb. of force
BD = Bore diameter in inches
NP = Nozzle pressure

The combination nozzle reaction force formula is expressed as

$$RF = GPM \times NP \times .0505$$

RF = Nozzle reaction in lb. of force
NP = Nozzle pressure (fig. 7–9).

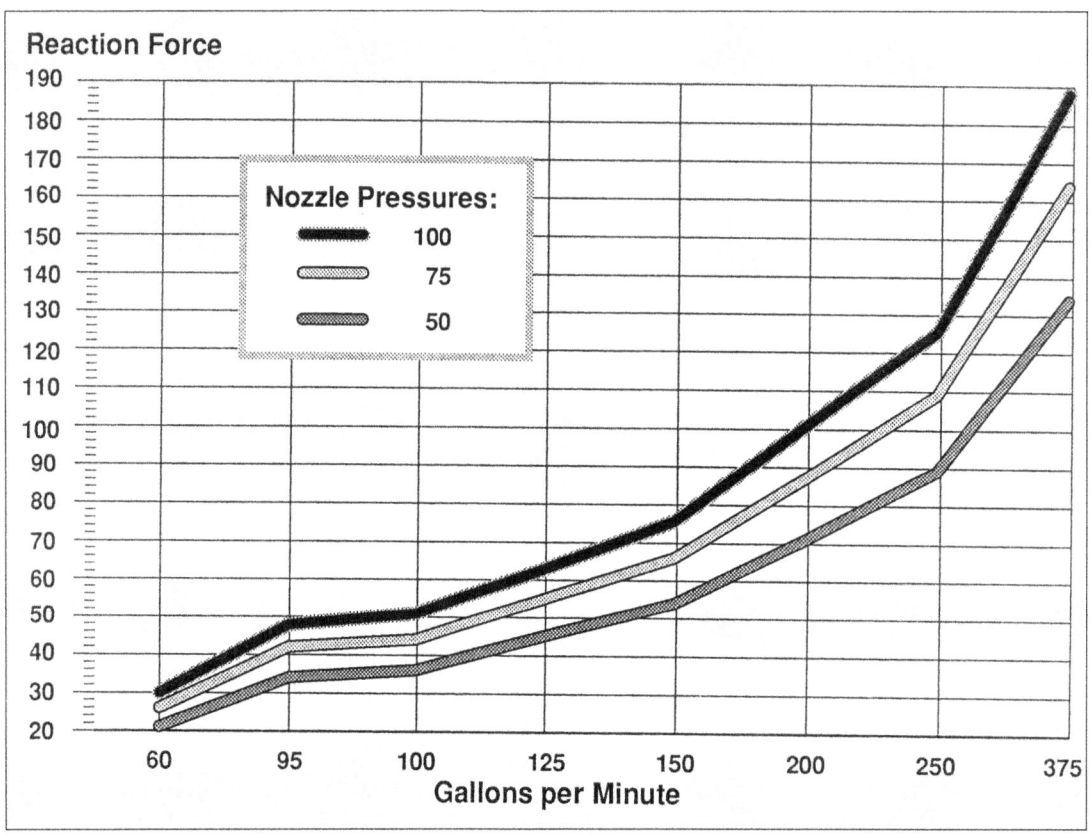

Figure 7–9. A chart showing the relative relationships of flow rate, rated nozzle pressure and generated reaction force. It is generally agreed that a reaction force of around 65 lb. of force can be generally handled by the average firefighter.

Chapter 7 Low-Pressure Nozzles   175

MASTER NOZZLE FLOW RATE AND REACTION CHART

STANDARD AND LOW-PRESSURE ATTACK NOZZLE STEM CHART

FLOW RATE (REACTION FORCE)

Nozzle Pressure	95-GPM @ 100-PSI	125-GPM @ 100-PSI	125-GPM @ 75-PSI	150-GPM @ 50-PSI	150-GPM @ 75-PSI	150-GPM @ 100-PSI	175-GPM @ 50-PSI	175-GPM @ 75-PSI	175-GPM @ 100-PSI	185-GPM @ 75-PSI	200-GPM @ 75-PSI	200-GPM @ 100-PSI	250-GPM @ 50-PSI	250-GPM @ 100-PSI	325-GPM @ 100-PSI
40	60 (19)	79 (25)	92 (29)	134 (43)	109 (35)	95 (30)	152 (49)	128 (41)	111 (35)	134 (43)	145 (47)	126 (40)	224 (72)	158 (50)	206 (66)
45	64 (22)	84 (28)	97 (33)	142 (48)	116 (39)	101 (34)	161 (55)	136 (46)	117 (40)	142 (48)	155 (53)	134 (45)	237 (80)	168 (57)	218 (74)
50	67 (24)	88 (31)	103 (37)	150 (54)	122 (44)	106 (38)	170 (62)	143 (51)	124 (44)	150 (54)	163 (58)	141 (50)	250 (85)	177 (73)	230 (82)
55	70 (26)	93 (35)	108 (40)	160 (59)	128 (48)	111 (42)	178 (67)	150 (56)	130 (49)	160 (59)	171 (64)	148 (55)	263 (98)	185 (69)	241 (90)
60	74 (29)	97 (38)	112 (44)	165 (64)	134 (52)	116 (45)	185 (72)	156 (61)	136 (53)	165 (64)	179 (70)	155 (61)	274 (107)	194 (76)	252 (99)
65	77 (31)	101 (41)	117 (48)	171 (70)	139 (57)	121 (49)	193 (79)	163 (66)	141 (57)	171 (70)	185 (76)	161 (66)	285 (116)	202 (82)	262 (107)
70	79 (33)	105 (44)	121 (51)	177 (75)	145 (61)	125 (53)	201 (85)	169 (71)	146 (62)	177 (75)	193 (82)	167 (71)	296 (125)	209 (88)	272 (115)
75	82 (36)	108 (47)	125 (55)	185 (80)	150 (66)	130 (57)	208 (91)	175 (77)	152 (66)	185 (80)	200 (87)	173 (76)	307 (134)	217 (95)	281 (123)
80	85 (38)	112 (51)	130 (59)	190 (86)	155 (70)	134 (61)	215 (97)	181 (82)	157 (71)	190 (86)	207 (93)	179 (81)	317 (143)	224 (101)	291 (131)
90	90 (43)	119 (57)	138 (66)	201 (96)	164 (79)	142 (68)	228 (109)	192 (92)	166 (80)	201 (96)	220 (105)	190 (91)	336 (161)	237 (114)	308 (148)
100	95 (48)	125 (63)	145 (73)	212 (107)	173 (87)	150 (76)	240 (121)	202 (102)	175 (88)	212 (107)	230 (117)	200 (101)	354 (179)	250 (126)	325 (164)
110	100 (53)	131 (69)	152 (81)	222 (118)	181 (76)	157 (83)	252 (133)	212 (112)	184 (97)	222 (118)	245 (130)	210 (111)	371 (196)	262 (139)	341 (181)

Streams having nozzle reaction forces under 45 pounds/force are considered as too weak for effective firefighting. Streams having nozzle reactions over 68 pounds/force are usually considered as generating too much reaction force to be safely handled by a single firefighter.

SMOOTHBORE NOZZLES
FLOW RATE (REACTION FORCE)

Nozzle Pressure	1/2" Overhaul Tip	3/4" Tip	7/8" Tip	15/16" Tip	1" Tip	1-1/16" Tip	1-1/8" Tip	1-3/16" Tip	1-1/4" Tip
35	—	99 (31)	135 (42)	154 (48)	176 (55)	198 (62)	222 (70)	248 (77)	275 (86)
40	47 (16)	106 (35)	144 (45)	165 (52)	188 (60)	212 (71)	238 (79)	265 (89)	294 (98)
45	50 (18)	112 (40)	153 (51)	175 (58)	199 (68)	225 (80)	252 (89)	281 (100)	311 (110)
50	53 (20)	118 (44)	161 (57)	182 (65)	210 (75)	237 (89)	266 (99)	296 (111)	328 (123)
55	55 (22)	124 (49)	167 (62)	191 (71)	220 (83)	249 (97)	279 (109)	311 (122)	344 (135)
60	58 (24)	129 (53)	174 (68)	199 (78)	230 (90)	260 (106)	291 (119)	324 (133)	360 (147)

Smooth Bore nozzles can generate effective streams at nozzle pressures as low as 30 PSI, but the streams will degrade rapidly at pressures over 60 PSI except for tips 1-1/4" and larger. Friction losses for attack monitor hose lines per 100': 400-GPM—2-1/2" 32 PSI, 3" 13 PSI; 500-GPM—2-1/2" 50 PSI, 3" 20 PSI

Review

If combination nozzles are preferred for interior attack, low-pressure nozzles offer the following advantages over combination nozzles operating at the standard 100 PSI pressure:

- They have less nozzle reaction force at the same flows than nozzles rated at 100 PSI operating pressure.

- There is less air mixed with the water stream, which tends to increase the stream's penetration into hot fire areas.

- The spray pattern water droplets are larger and tend not to vaporize as easily in undesired, overhead areas.

- Because of the lower operating pressure, the hose is not as stiff, allowing it to be more easily maneuvered inside buildings.

- Fixed-gallonage versions are relatively inexpensive and durable, and firefighters are more easily trained in their use than with automatic or adjustable-gallonage nozzles.

- In actual use, operators will flow the rated gallonage rather than attempting to reduce the reaction force by reducing pump pressure or by partially closing the shutoff. Both actions will considerably reduce the flow rate.

After extensive testing and evaluation, a growing number of departments are finding there is no compelling reason to purchase handline fog nozzles operating at a pressure above 75 PSI unless the operating crews enjoy making an interior attack on a hot fire while wrestling with high nozzle reaction force and generating billows of unwanted steam.

Chapter 7 Review Questions

1. What were the two major cities that were the first to develop low-pressure nozzles?
2. What is the main advantage of low-pressure combination nozzles over those operating at 100-PSI nozzle pressure?
3. Describe the straight stream characteristics of a low-pressure nozzle.
4. What are two advantages of utilizing a spray pattern?
5. What is the advantage of having large water droplets, such as those exiting from a low-pressure nozzle?

Heavy Streams

There are certain fires which, upon arrival of suppression forces, will have such a large area of involvement and are producing such a great volume of fire and heat that they cannot be handled by the flows available from handlines. There are also fires that spread rapidly after the initial attack despite the best efforts of the department, forcing firefighters to abandon interior positions and assume defensive positions outside the fire building to keep the inferno from spreading to other occupancies (fig. 8–1).

A similar situation is that of a superior army force overrunning one's troops. A military commander would take steps to protect their infantry; then, to help even the odds, would call in the big guns—heavy artillery and air strikes. Using this same strategy, a wise fire commander, recognizing a serious situation, should take steps to protect their forces, such as

Figure 8–1. At many fires, it will be evident upon arrival that a rapidly deployed, high-volume attack will be necessary to prevent further spread to nearby exposures. (Photo courtesy of Gordy Nord)

retreating from inside positions if necessary, regrouping labor-intensive handlines into heavy stream devices, and effectively using water applied from aerial apparatus and ground-level guns to prevent further spread and, finally, to extinguish the fire.

Heavy streams, sometimes referred to as *large-caliber* or *master streams*, are those which, because of their volume and operating pressure, cannot be hand-held when in use. The usual rule of thumb is to consider any stream flowing over 400 gallons per minute (GPM) as a heavy stream.

Heavy Stream Equipment

Prepiped guns were the first stream shaping devices used in firefighting. Mounted on the top of hand pumpers, they were the forerunner of the devices we use today for exactly the same purpose. During the steamer era, special wagon battery apparatus, sometimes called high-pressure units, carried large, heavy stream monitors for defensive operations. Water towers came into use early in the 20th century, allowing powerful streams to be played on the upper floors of burning buildings.

While certain wood aerial ladders could and did support ladder pipes, it wasn't until metal aerial ladders came into widespread use beginning around the late 1930s that heavy streams applied by aerial ladders became common. Water towers remained popular until the mid-1950s when most large cities phased them out in favor of the versatile aerial ladder. In 1958, the Chicago Fire Department put the first Snorkel unit into service, ushering in a new era in which aerial platforms, both straight and articulated, became the preferred means of safely applying heavy streams above ground level.

The first heavy stream devices designed for use off the apparatus looked like large playpipes attached to metal stands. While fairly cumbersome to place in position, they nonetheless allowed large-caliber streams to be applied from spots inaccessible to apparatus mounted monitors or high-pressure units. One of the first multiple-inlet guns on the market was the Grant multiversal, first manufactured by Akron Brass in the early 1920s.

Because of their weight, early portable guns were cumbersome and were rarely used in the offensive attack stages of large fires. Many larger cities mounted prepiped deck guns on pumpers. Since they were easily placed in service, they were used by some departments in the early stages of attack but rarely as the first stream in operation. Most departments employed both portable and mounted guns strictly as a defensive weapon used on large, out-of-control fires. The prepiped deck gun's popularity declined through the 1960s and early 1970s in favor of top-mounted portable guns. The reasoning behind their use was that they could be operated from the pumper, being fed by short lengths of hose, or they could be removed from the rig for action elsewhere.

Demounting the devices, which weighed around 75 lb.–125 lb., could not be accomplished rapidly or with any degree of firefighter safety. Responding to user demands, manufacturers began developing portable units made of aluminum, considerably reducing operating weight, and introduced more compact, two-inlet lightweight devices that have proven ideal for offensive operations.

Older portable monitor designs split the stream at the joint, which allowed vertical travel, then reunited it just before it entered the stream shaper. The split design balanced the reaction force of the water as it flowed through the device. Newer designs developed in the 1970s have a single, formed waterway resembling a pretzel which more effectively keeps the stream together. The bends balance flowing water reaction force, improving stream quality and reducing the gun's weight. In addition, the geometry of the design directs the reaction forces downward to make the portable device more stable when flowing water. The majority of heavy stream nozzles now purchased by the fire service are based on this bent loop design. Besides keeping the stream together, they are easier to manufacture and generally less expensive than split waterway designs.

Removable, mounted units that became popular about 20 years ago can be operated either as a deck gun or demounted, installed on a base, and operated as a portable. While the price of a single removable gun is less than that of two pieces (a permanently piped gun and a portable gun), the removable gun can only be used in one place at one time. If a large fire demands flows from both a deck gun and a portable gun in different locations, an engine equipped with a single removable unit cannot supply both. Unless price is a critical consideration, it is more effective to equip an engine with a permanently prepiped deck gun that can flow at least 1,250 GPM, as well as providing a lightweight portable monitor for ground operation remote from the pumper.

Older portable guns were supplied with two, three, or four 2½" clappered inlets. Now, however, departments have a choice of inlet number as well as size. With the increasing use of large-diameter hose, many departments are purchasing devices with 4" or 5" Storz inlets to enable the gun to be supplied directly with the large hose. If smaller hose is all that is available at the fire scene, a lightweight, clappered Siamese can be snapped into place and the gun fed with multiple smaller lines.

Combination Aerial and Pumper Apparatus

In recent years, there has been a tremendous increase in the popularity of aerial devices mounted on engine companies. These range in size from 50' telescoping and articulating booms to 75'–100' aerial ladders. These devices multiply the effectiveness of attack engines by allowing heavy aerial streams to be put into service quickly with a minimum of personnel.

While the fire service is an important market, most of the master stream equipment sold today is to industry. Hundreds of guns mounted permanently on supply piping or hydrants are common in many petrochemical, offshore, and chemical operations, ready for immediate use to help contain a fast-spreading fire involving product storage or process equipment.

Heavy Stream Nozzles

Before choosing a nozzle for a heavy stream device, consider its intended use and water supply. In most cases, both smoothbore and combination nozzles should be available in order to generate the most effective stream for a particular situation.

Automatic or selectable-gallonage combination nozzles are particularly effective for master stream operation during the initial attack phase because of their positive operating characteristics under conditions of varying supply. Water supply is a major problem with flowing effective heavy streams, especially in the early stages of a large fire. On the fireground, it may be impossible at the beginning of operations to supply a master stream with more than one supply line. If the tip flow is not matched to the water supply, a weak stream with insufficient reach will result. If a properly sized smoothbore tip is matched to the limited water supply and additional lines added later, the gun will have to be shut down and the tip replaced in order to take advantage of this increased water supply. In actual practice, most companies keep the smallest tip of a stack-tip assembly mounted on the gun but rarely remove small tips in the heat of battle even when additional lines become available.

Automatic and selectable-gallonage nozzles can help solve these problems without the inconvenience of swapping tips. In the early stages of attack when water supplies may be weak, the automatic nozzle's baffle closes to match the lower nozzle pressures created by a small amount of water flow. As the flow is increased, the tip automatically adjusts the nozzle opening to accommodate the extra water. This means you will be able to hit your target with at least some type of stream at the start and you will efficiently utilize the additional flow as more water becomes available. For most heavy stream work during initial attack, automatic gallonage nozzles have proven to be useful and convenient tools to use. Although the nozzles provide a variety of patterns, a straight stream is the most effective pattern position for heavy streams because of its maximum reach and penetration qualities.

Many officers prefer the characteristics of smoothbore nozzles when flowing heavy streams. A set of stacked tips will usually do a good job of matching streams to the water supply at much less cost than a combination automatic; however, training will be needed to educate officers, pump operators, and firefighters in the proper use of the tips. Each time a line is added, they should shut down and change tips to flow the added water supply effectively. If they forget to do this, it will be more efficient to leave the automatic combination nozzle in place.

When using smoothbore tips or fixed-gallonage combination nozzles on master stream appliances, remember that if the nozzle pressure is increased, both the flow and the nozzle reaction will increase. If nozzle pressures of over 100 pounds per square inch are being supplied, the reaction force could become excessive and could cause a portable unit to break loose. Because automatic nozzles keep the nozzle pressure constant as the flow increases, the reaction force can be less than that of a fixed tip overflowing at high nozzle pressures.

Using Heavy Streams Effectively

Until fairly recently, the use of heavy streams was considered a tactic for use only at out-of-control fires, a last defensive resort to help contain the spread and protect exposures. Unfortunately, utilizing large streams only for defensive operations is a waste of a valuable offensive attack resource (fig. 8–2).

Many departments are finding the heavy hit from a master stream device a potent and effective weapon and are stopping many fires that would have been considered hopeless losses in past years with these streams. A heavy stream should be considered a viable option for immediate use during initial attack on a large fire (fig. 8–3). When used offensively, a

Chapter 8 **Heavy Streams** 181

Figure 8–2. The first master streams to be deployed should be placed to stop the spread of the fire to exposures. In doing so, plan for the fire increasing in intensity before the streams become operational.

Figure 8–3. For a blitz attack to be successful, the water must be applied directly to the burning fuel, should be applied at the highest flow rate possible, and should be applied from an area to either create a gateway through which handlines can be advanced or to protect an exposure. This attempt at a blitz attack using tank water is doomed to fail. The stream is not being directed at the burning fuel, the stream from the combination nozzle is not of sufficient volume or reach to penetrate the fire area, and it is being directed from an area that neither protects exposures nor creates any headway for firefighters advancing with handlines. (Photo courtesy of Alan Chaniewski)

heavy stream device can effectively stop an extremely large volume of fire if the following criteria are met:

- It is used for attack as soon as possible in the early stages of a large fire.
- The flow rate is of sufficient volume to stop or slow production of heat.
- The stream is able to reach the fuel being burned.

Every plan for strategic firefighting tactics should include a quick, high-volume initial attack made with heavy streams. Immediately upon arrival at the fireground, an engine company should be capable of putting streams in service that can flow 500 GPM–1,000 GPM using only the personnel riding on the engine. This can be most easily accomplished with a top-mounted prepiped deluge gun monitor. A prepiped gun can be put into operation rapidly because little work is required other than sending one firefighter topside to operate the stream. In many cases, the tank water might be all that is needed to break the back of a fast-spreading fire. To operate effectively, the gun should be aimed in the general direction of the intended target, and the pump should be throttled up to operating pressure before the gun's discharge valve is slowly opened. Handwheel discharge controls are excellent for this use. Since the gun is not being immediately supplied with the proper operating pressure, this will help eliminate wasting water if the stream misses its target or falls short. Some departments are now fitting a ball or gate shutoff on the discharge tube or below the deck gun, so that the gun operator can more effectively control the flow of water.

If an engine company normally responds with a limited number of firefighters, the use of a prepiped gun can help protect personnel when used to quickly knock down vehicle, rubbish, or certain types of structure fires—garages, barns, or storage buildings, for example—before advancing a handline. With practice, the intensity of fires in these situations can be immediately reduced with a 10 second–20 second burst from the gun while a handline is being stretched. Training is a key element of this tactic. Water should be used wisely, and care must be taken to have the gun positioned and aimed directly at the fire before flowing water. As soon as the fire begins to darken, the gun should be shut down to avoid wasting water.

Blitz Attack Tactics

Many departments, especially those protecting rural areas, are now carrying portable ground monitors or guns preconnected to the pump with 150'–300' of 2½" or larger supply line. This enables a high-volume attack to be started immediately from the doorway or other exterior openings in barns or other large structures (fig. 8–4).

Former Chief Gregory Kozey of the Eastford, CT, fire department called the preconnected gun his "invisible fireman." He said that if the fire volume demands a large flow, the gun is easy to position and easy to operate with limited personnel. When the number of on-scene personnel is limited, the preconnected gun line can be used in place of a high-flow handline, even inside a building. Tests have shown that a well-involved barn fire can be controlled in less than a minute, and usually using less than 500 gal. of water, by flowing a 500 GPM–1,000 GPM blitz attack through a portable gun. In practice, the gun is positioned

Figure 8–4. Attacking a large amount of fire with the tank water may make the difference between saving or losing the exposure.

as close to the fire as possible and the straight stream aimed high into the structure, sweeping from side to side. Here, an automatic master stream nozzle is an ideal tip, able to effectively discharge as much water as the pump can in addition to having good reach and a range of patterns to fit the water discharge to the situation.

Mounting

Unless mounted topside, over the pump, and connected for immediate use as a monitor, portable guns are best stored where they can be removed from the apparatus without climbing or using excessive physical effort. A portable but unpiped gun mounted over the pump may look impressive in a parade, but precious minutes may be lost as personnel swarm over the rig at a fire to remove it for use at ground level. Departments without prepiped guns may choose to mount their portable over the pump and make connections with short sections of 2½" or 3" hose. The decision of where to store the device should be made based on where the department thinks the gun would do the most good in the first stages of fireground operation (fig. 8–5 and fig. 8–6).

In a high-volume attack, the flow rate should be as much as the pumper can supply through the heavy stream device. If the fire is of such magnitude on arrival that a high-volume attack is desired, it is good practice to hit the fire with everything you've got to knock it down as quickly as possible. If it is decided to mount a high-volume attack, the gun should be positioned as close to the fire as possible and the stream properly directed to cover as much of the fire area as possible.

Figure 8–5. This properly mounted monitor, permanently piped on the pump on this New York Fire Department (FDNY) engine, allows for 360° coverage. The tiller-type handle makes it easy for a firefighter to quickly and accurately aim the stream.

Figure 8–6. The poor mounting location of this monitor does not allow for 360° coverage and the stream cannot be depressed over the side to reach the inside of first floor occupancies. In cases like this, manufacturers offer extension devices that allow for low-profile storage but can be quickly extended upward so the stream can be directly where needed.

While concentrating efforts on an offensive blitz attack, it must be remembered that a second line of defense must be formed in case the initial attack is unsuccessful. As offensive operations are taking place, a continuous water supply should be established, and streams should be positioned to further protect exposures (fig. 8–7). If the fire building is beyond saving, then the initial attack should be concentrated on containment tactics and exposure protection, not necessarily on the main body of fire.

One limitation in supplying the amount of water needed for a successful blitz attack, especially in older apparatus, is the size of the tank to pump line. A 2½" tank suction line, such as commonly found on pre-1990 pumpers, will supply only about 250 GPM or less, depending on the number of elbows in the piping. A single 3" line, common in apparatus built since 1990, can supply about 600 GPM. Departments using the blitz attack as standard procedure

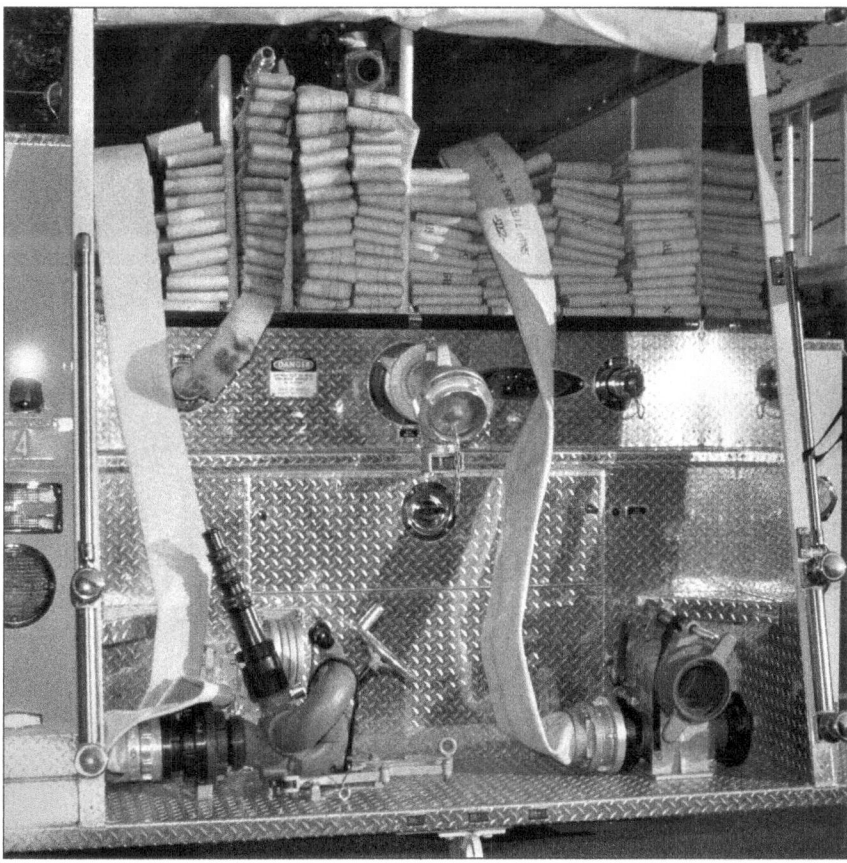

Figure 8–7. Too few city departments take advantage of preconnected portable monitors. An exception is Rocky Hill, CT, which mounts preconnected guns attached to 500' of 4" hose. The pumper is supplied by a bed of 5" hose (at right).

should consider installing 3½" or 4" tank suction lines in order to provide higher flows when using on-board water for attack.

When used for offensive attack, the heavy stream must be able to penetrate the fire effectively in order to slow or stop its progress. When positioning apparatus for use of a prepiped monitor, or when placing ground devices, make sure they will be able to not only hit the fire, but also cover an exposure if the offensive attack proves unsuccessful (fig. 8–8). It is extremely difficult and time-consuming to move heavy stream devices once water is flowing. In the confusion during the early stages of a large fire, many commanders, taking comfort in the fact that water is flowing toward the flames, are reluctant to shut down streams even if they are ineffective and could be better used elsewhere.

While stream placement is important, consideration must be given when positioning vehicles to the safety of the unit and its personnel if threatened by fire suddenly blowing out of the building or if the structure should show signs of collapse. A pumper or aerial device can easily become an $800,000 or $2 million exposure if conditions change and if the rig becomes threatened by the fire or collapse. It is not very cost-effective to risk destroying a piece of equipment worth more money than the burning building. Steps should be taken to stretch lines to protect the apparatus from heat or flames before they are actually needed. A length or two of 1¾" line with a combination nozzle is easily handled and provides enough flow to disperse quite a bit of heat (fig. 8–9).

Figure 8–8. A heavy stream attack here knocked down all visible fire in these upper floor apartments, buying time for companies stretching attack lines in the interior.

Figure 8–9. The possibility of collapse must be taken into consideration when positioning apparatus at the scene of a large fire. Tied together by a fire escape, this wall carried its full height outward, striking the Snorkel unit, killing one firefighter and seriously injuring another. The old rule that a wall will fall only one third of its height can sometimes prove false.

Years ago, especially in New England mill towns containing large, timber-framed factories, it was common practice to carry a large knife on the apparatus to quickly sever hoselines so the vehicles could be rapidly moved out of danger. In the fire's early stages, proper positioning of apparatus should take the fire's probable path into consideration in case it burns out of control. Officers and apparatus operators should constantly monitor the fire's progress and should begin the process of relocating threatened apparatus long before the movement becomes an emergency. Apparatus arriving on later alarms should be careful not to take positions that could block the evacuation and relocation routes of apparatus that had arrived earlier (fig. 8–10).

Sometimes, despite all efforts poured forth in an aggressive, hard-hitting, and rapid attack, there will be fires that defy all extinguishing attempts, turning the building into a total loss. The fire may have had too much headway before arrival, or the building area may have been so large that streams could not reach the burning material. These types of fires are what the late Chief Emanuel Fried called "two-handed fires." Two-handed fires are those requiring the chief to use both hands to steady themself on the fender of their car to keep from collapsing after they get out and sees what is confronting them.

Large volume, out-of-control fires require rapid implementation of defensive operations in order to effectively and safely contain the loss. When planning defensive strategy, one basic principle is to consider what has already burned or is burning as history. What has already happened is not as critical as the fire's potential—what it is going to do and where is it going to go. Large fire tactics call for thinking and planning further ahead than what is considered normal at common fires. A good rule of thumb is to ask yourself, "What will this fire look like in 20 minutes if I do nothing?"

As a fire's size increases, its spread becomes more rapid. A common mistake made by commanders inexperienced with handling large fires is underestimating the speed at which the fire can move both upward and laterally.

Figure 8–10. The possibility of collapse must be uppermost in the minds of commanders when positioning personnel and apparatus. Even a partial collapse such as this roof overhang can prove fatal if steps are not taken to keep personnel out of the danger zone.

A word of advice. Many otherwise capable fireground commanders earn reputations of always crying wolf when they transmit multiple alarms too quickly, bringing extra equipment to the scene before all the facts are known. In many cases, a building enveloped in heavy smoke looks like extra alarm material before it is properly ventilated and the first line goes into operation on the seat of the fire. I've worked with many big-city chiefs whose standard procedure has been to wait until the first line was hitting the fire before deciding to call in help. Most of the time, the fire darkened down immediately, and a relatively minor fire was revealed when the smoke lifted. On arrival, even though the fire gave the appearance of being very serious and spreading rapidly, these chiefs had confidence in their forces and allowed them a little time to do their work before calling in reinforcements which, in most cases, were not needed. All of these chiefs, however, would never delay transmitting extra alarms if there was a question of life safety or if the fire volume was obviously beyond the capabilities of the first alarm companies. They just factored in a little common sense tempered with experience before making decisions about how much help was needed on the scene (fig. 8–11).

Throughout my entire career, I have never criticized an incident commander for asking for more help. Extra units can always be returned, but not having help arrive in a timely manner can materially increase fire losses and endanger occupant and firefighter safety. Pull the hook if you need to. We can always talk about it later.

Figure 8–11. Lack of training in supplying and operating high-flow streams becomes evident on the fireground. The simple remedy of shutting off one of the monitors to help supply an effective stream on the other certainly was not covered in this department's training.

Another common problem with attempting a high-volume attack is a department's lack of practice with high-caliber streams, which slows deployment time. Because high-caliber streams are not used every day, fire officers will often forget they are available. Over and over, across the country, I have seen fire companies stop in front of the fire building and stretch the small-caliber handlines they are familiar with, even if the fire's intensity dictates otherwise. This relates to what I call the *theory of recency*. If a tool or tactic has not been used recently, firefighters forget the equipment is carried or forget how to use it.

Stretching the wrong sized line leads to a step-up situation where each size line is abandoned as it proves inadequate, and the next size line stretched. Master stream devices are finally put into service through an indecisive trial-and-error process. This action is extremely labor intensive and is very time-consuming—time that the department may not have to waste if it is serious about saving the building or the people inside. Frequent training with heavy streams will help make their availability more familiar to the operating forces and increases their chances of actually being used when needed (fig. 8–12).

Heavy streams and the tactics used to apply them should show positive fire suppression results within a relatively short period of time. This is one reason high-volume attack works well in rural areas with limited water supply—quick results in heavy fire situations.

If flames are not being darkened within a reasonable time, officers should consider changing tactics or stream position quickly to better suit the situation. The fire service has a tendency to put a large stream into operation, then not order it shut down until long after it has ceased being effective. Conversely, a master stream should be given time to apply enough water to cool the burning material. We have all seen videos of a master stream being rapidly wiggled from side to side while not enough gallons-per-square foot are being allowed to flow into the burning material and do its job. With offensive handlines, it is desirable to rapidly move a stream operating on the interior to help cool the compartment and to make sure water is being distributed to the burning material. With a master stream, go a bit slower and give the mass of water time to work (fig. 8–13).

Figure 8–12. Instead of stopping before the fire building was reached, a second-arriving engine laid its supply lines past the involved building. As the fire developed, the lines melted and burned, rendering a 1,000 GPM pumper useless.

Figure 8–13. Many departments, through lack of planning and training, perform the same operations no matter how large a fire is encountered. This fire was attacked in a similar manner in which the department attacks a one-room dwelling fire. A single 2½" supply line was lain from a hydrant and two 1½" preconnects were deployed even though fire volume dictated otherwise.

There is no reason that a master stream cannot be effectively used to hit and move in an operation similar to advancing a smaller attack line. When heavy streams are used on a large fire, all officers must be capable of quickly recognizing operational deficiencies and then take effective action to overcome them. If the streams are not hitting any fire, shut them down. This will save wear and tear on the apparatus and personnel and can help the commander reevaluate the situation and position streams more effectively.

Large Fire Combat Tactics

Most textbooks on firefighting tactics and for that matter, instructors, explain methods of initial attack in some detail, yet give very little guidance on how to combat an extremely large fire—one that could threaten structures over a wide area (fig. 8–14).

Luckily, most departments will never experience a fire that sweeps through many large, closely-spaced frame buildings, or consumes a timber-framed mill building spread out over a two-block area, burning hoselines and apparatus in its path, igniting additional exposures faster than lines can be laid to protect them, and depositing 10 lb. chunks of flaming roofing material a mile from the fire's center, or have a brush fire consume acres of wildland before high winds push it through a residential area, destroying every house in its path.

Rare? Maybe, but a large fire could confront any department, getting out of control amazingly quickly and easily. Transport trucks could spill their load of flammable liquid on a downtown street, a train derailment can occur in a built-up area, a gas pipeline could develop a major leak, a plane could crash into an apartment building, or an arsonist could strike a match on a hot, windy day. Conflagration-type fires can start small yet spread fast, overwhelming initial attack forces who underestimated their potential, or they can start large, set off by an explosion in a hazardous manufacturing process, for example (fig. 8–15).

Figure 8–14. It makes little sense to pump water 75' into the air if it is needed at ground level. After the threat to exposures was mitigated, this stream could have been shut down and ground-level handlines used for extinguishment. Many times, aerial streams are put into action with little thought given to their effectiveness on the fire at hand.

Figure 8–15. Here, a heavy stream knocks down a fire in a one-story commercial building. The design of modern, heavy-duty aerial devices allows their streams to flow without restriction anywhere they are capable of being aimed. (Photo courtesy of Steve Redick)

One reliable indicator of how rapidly fire conditions are deteriorating is how fast the initial attack companies are being forced back. Are hoselines being relocated to get a better shot at the fire, or are they being pulled back because the heat and smoke are becoming untenable? Is firefighter safety being jeopardized by unsafe structural conditions? A wise incident commander will continually assess conditions and take immediate action to change tactics when indications point to a losing battle (fig. 8–16).

Many command officers help evaluate large fire tactics by using the 20 minute rule of thumb. If firefighting efforts make no appreciable headway on a fire in 20 minutes, it is time

Figure 8–16. At any large, open-burning fire, always be mindful of flying brands that can be spread, depending on the wind, many blocks around. If flying brands are present, it is wise to deploy additional companies downwind to control the possibility of additional blazes igniting.

to pull personnel out of the building and change to defensive tactics. Sometimes, it is hard to estimate how long the fire had been burning before the start of initial fire suppression efforts. However, a commander should keep track of the time spent on the scene after arrival and time withdrawal actions accordingly. In some cases, the 20 minute rule may be too long. A commander must be able to read a fire building and its construction, knowing the approximate location of the fire, how long the building will remain standing, where the fire is headed, and how fast it will get there. Any uncertainty, such as discovery of signs of structural weakness, should cause the commander to seriously consider discontinuing offensive attack and changing to defensive operations (fig. 8–17).

In many large cities, if no progress is reported on a fire within an hour, or if the all-out or strike-out signal is not given in that time, the commander who would normally respond on the next higher alarm is dispatched to evaluate the situation. Once it becomes apparent that offensive attack is not making any headway or that a fire cannot be held, the following steps are necessary:

- If the fire demands, call for help. A common mistake is to underestimate needs. It is better to have companies on the scene and available for immediate deployment than to lose time having them respond from quarters when their presence is indicated.

Figure 8–17. At large fires, the type of occupancy can dictate the use of defensive tactics at an earlier stage than would otherwise be used. This fire in an automobile paint shop made conditions hazardous for personnel due to exploding flammable liquid containers.

- Determine a point at which it may be practical to cut off the fire's spread and prepare to make a stand. This cutoff point should be located so equipment and personnel will be in place and operating long before the fire arrives.

- Deploy additional companies as a second line of defense—a fallback line—in case the front lines fail to hold the fire.

- Rapidly develop heavy streams and don't be afraid to use them.

- Have companies arriving on greater alarms find sources of water distant and independent from companies already operating.

- Rapidly but thoroughly remove citizens from the path of the fire. The middle of a firestorm is a lousy place to attempt a rescue.

- Prepare for possible flying brands additional fires downwind of the original fire.

- Immediately establish an incident command system in accordance with up-to-date procedures, and make sure that all operating and incoming fire companies know that this will be a defensive operation (fig. 8–18).

Figure 8–18. Companies arriving on additional alarms should take steps to locate engines at the water source for better control, and also take steps to maximize the water available. A Chicago engine here connects to the hydrant with its hard suction hose, mandated by the Chicago Fire Department standard operating procedures, as well as a second 5" line, assuring that every drop of water that is available can be utilized. (Photo courtesy of Jim Regan)

When the fire reaches such a point that the commander, after careful fireground evaluation, decides to call for help, it is much better to overestimate what help might be needed than to run out of companies when a gap develops. Time spent in getting the additional companies into operation could be a critical factor in limiting the fire's spread. Better to have the help on hand, held in a nearby staging area, than to have companies respond all the way from their quarters. In rural areas, travel time can become critical if departments are located some distance apart (fig. 8–19).

The use of the incident command system to deploy arriving companies is critical. Many fires which could have been stopped with an early, organized effort have caused great losses because of command confusion and indecision. Most large-city chiefs agree that resources, efficiency, and fire scene organization can rapidly deteriorate when the amount of fire exceeds the capabilities of the normal initial response for a given area, causing unfamiliar companies and officers to be summoned. How much it deteriorates depends on how quickly the commanders can positively respond and adapt to the unfamiliar working environment.

Staging Area

A staging area should be established near the fire scene early in the action and as soon as additional alarms are sounded. A shopping center parking lot can be an excellent location, providing room for incoming apparatus to maneuver and park. When dispatched, the extra

Figure 8–19. Rainshower application of heavy streams over a free burning, vented fire rarely accomplishes extinguishment. This is because the lightweight water droplets are carried upward by the fire's thermal currents before they can cool the burning fuel. Also, the application rate per square foot is too low to cool great masses of burning fuel. It is more effective to concentrate the stream into a compact water column which will penetrate through high-heat areas and cool the source of flame production.

alarm companies should be given the location of the staging area before they leave quarters. A staff officer should be stationed in the staging area to relay operational orders to companies waiting for assignments. As companies are deployed, it is an easy task for the staging officer to determine how many are available in the staging area. When communicated properly, the incident commander can use this number as a guideline to determine when to call the next greater alarm.

It is important that the staging area officer keeps close track of not only the number of companies staged, but also evaluate the type of companies that might be needed. Three ladder companies in staging will be of little use if tankers are needed to provide proper water supply operations.

If the first-line crews are driven from their positions by intensifying fire, a strategic retreat should not turn into a rout. Crews should back out slowly, fighting if possible. Officers should monitor conditions and should be able to order crews to change positions long before an emergency exists. Of course, unforeseen circumstances, threat of collapse, or rapid fire spread should dictate immediate evacuation as rapidly as possible (fig. 8–20).

Every department should have a prearranged signal for immediate building evacuation, such as repeated blasts on apparatus air horns. In the heat of battle, it has happened that radio evacuation messages went unheard or were misunderstood. Firefighters' lives were lost as a result. There is no mistaking the urgency of evacuation when a number of air horns sound at the fire scene.

All firefighters, including personnel from other departments called in for mutual aid, should be made aware of and thoroughly understand the fireground signal for immediate evacuation. As personnel pull back because of fire intensity, you must insist that the crews regroup quickly as soon as they are out of danger and account for all members. It may be

Figure 8–20. It is good practice not to set the outriggers and elevate the aerial device until water is provided in sufficient quantity to mount an effective attack. At this fire, rapidly spreading flames caused the aerial device to be moved five times before it was considered in a safe position from which to operate.

possible for these crews to put lines rapidly back into operation in other positions, but firefighter safety trumps all other actions. Provide enough fresh personnel to maneuver lines and to relieve those subjected to high heat and smoke conditions frequently.

It is good practice to distribute the number of lines among engine companies so a single engine will not be supplying both master streams and handlines. The pressure and volume requirements of large-caliber streams can make handline operation from the same engine dangerous if the large streams are suddenly opened or shut down, even if the pressure controller is set.

The second line of defense should be established using heavy stream equipment and staffed with fresh personnel. Lines should be stretched and pumpers connected to a water source, ready for instant charging when necessary. If the first-line crews get pushed back, the second-line firefighters should take immediate action to allow first lines to safely escape, disconnect, and move apparatus. The second line should then work to contain the advancing fire. If tactics require the activation of the second line, it then becomes the first line, and will need to be backed up by its own second line of defense (fig. 8–21).

When repositioning apparatus, enough time must be allowed to bed aerial devices properly. At one large fire, I watched as a truck company driver attempted to move the apparatus with the ladder extended over 60' in the air. Luckily, when the driver began to raise the hydraulic jacks, he saw the unit tipping, realized his mistake, and was able to stabilize the unit with hydraulic down pressure on the jacks. If that particular truck had mechanical jacks, or if the operator had not noticed the unit tipping, it would have collapsed on its side and been destroyed in the oncoming fire.

Figure 8–21. When the first line of defense is overrun by rapidly advancing fire, a second line should be in place to allow the first line to be evacuated safely and to provide a continued extinguishment or containment effort. In this case, a second line of defense could have been provided by portable deluge guns operated from the opposite curb line.

When setting up aerial devices to apply water at a large fire, it is a good practice to position the aerial unit and activate the hydraulic pump power take-off, but not to set the jacks or raise the device until after water has arrived. At a rapidly spreading mill fire, I watched as a large-city department moved a platform unit five times until a safe operating position was found before flowing any water, as they tried to keep ahead of fast-spreading flames. The first two times, the crew set the jacks and raised the booms, only to be ordered to move before the supply lines could be charged. The last three times, the crew readied the unit for operation but did not set the jacks or raise the device, making the apparatus much easier to reposition in a hurry if needed.

At all times, and especially in the early attack stages where building condition cannot be easily evaluated, the possibility of structural collapse must be considered when locating firefighters, apparatus, and streams. A wise fire scene commander will appoint a safety officer as soon as possible at a large fire to help keep personnel and equipment out of danger.

Considering the Effectiveness of Smoothbore Streams

As soon as it becomes apparent that an aggressive initial attack is not producing the desired results, many departments that use combination master stream nozzles for initial attack should consider a change to smoothbore tips. Combination master stream nozzles are designed to put as much water in contact with heat in the shortest possible time. With a large volume of fire, however, peripheral heat will tend to evaporate a fog stream before it can reach the burning material than a stream from a smoothbore master stream nozzle.

As we have discussed in preceding chapters, the water exiting from a combination nozzle, because of its rapid change of direction within the nozzle, contains a great amount of entrained air. The stream is therefore lighter in weight than the stream from a smoothbore nozzle. This light weight could cause the stream to lose reach because of bending effects of wind or high heat. Because of the stream's relative lightness, it can also be easily pulled into the rising thermal column by the fire's air currents without hitting any burning material (figs. 8–22a and b).

A combination nozzle can be effective in some high-volume situations. When applying water to certain exposures, the gentler impact of a spray pattern will tend not to tear away sheathing materials or break windows as easily as a smoothbore stream. If water supplies are erratic, automatic nozzles can help maintain stream reach under varying pressure conditions by adjusting the nozzle pressure to suit the supply situation. It's usually much more effective to have a stream from a combination nozzle hitting an exposure than to have a stream from a smoothbore dribbling in the street because of problems in supplying enough volume.

A common misuse of a combination nozzle when combating large fires is to apply a wide spray over the top of the fire. This rainshower tactic cannot deliver enough water per square foot to an unvented, free burning fire to accomplish extinguishment. In addition, thermal currents carry most of the lightweight water particles upward before they have a chance to cool the source of heat production (fig. 8–23).

A straight stream from a combination nozzle or the stream from a smoothbore nozzle will deliver most of its water in an area of about 4 sq. ft. when delivered within effective range. If the nozzle is flowing 750 GPM for example, it will be delivering about 188 GPM per square

Figures 8–22a and b. The effect of wind is demonstrated by these streams from a smoothbore nozzle (a) and a combination nozzle (b). Both streams are flowing 800 GPM but the hollow-core stream from the combination nozzle is being carried away by the effects of wind currents. This action will most likely occur when a stream enters the vertical thermal currents of a large fire.

Figure 8–23. These streams from combination nozzles are having no effect on this fire in a motel. Their limited range allows only a fine mist to drift into the flame/fuel interface area. The aerial platform stream at right is accomplishing nothing but keeping the third floor wet. Surround-and-drown can be an effective tactic for out-of-control fires. Surround-and-mist should be considered ineffective. (Photo courtesy of Becky Gerard)

foot. If a 30° spray pattern is used, it will cover an area of approximately 400 sq. ft., about 60' from the nozzle. This gives a volume of about 1.9 GPM per square foot. The great amount of heat generated by a free burning, upwardly vented fire cannot be extinguished by a gentle rainshower. The water column being applied from a master stream device must be concentrated as much as possible to utilize the water's weight for penetration through high-heat conditions, allowing it to strike the burning fuel. Once it hits the fuel, the high GPM rate can be effectively utilized to cool the heat source and extinguish the fire.

Water that does not reach the burning material at the flame/fuel interface extinguishes no fire. In order for a spray stream to absorb heat without striking the burning material, the fire and the resulting steam must be contained within the building. This cannot take place if the fire is venting outside the structure (fig. 8–24).

The only positive point about the rainshower style of application is that it does look good to the uninformed when viewed on TV during news programs. The television news business is based on visual impact—the more flames that are shown on the tape, the longer the airtime will be. A wide fog pattern being swept back and forth over a burning building, which is having no effect on suppressing flame production, will look visually impressive during a news broadcast. An extinguished fire may not receive much airtime but it sure makes more sense to the building's owner and occupants and to the city's tax base (fig. 8–25).

Unless the fire is confined within the building and not venting to the outside, spray streams do little more than waste water. To accomplish extinguishment, the stream must hit the base of the fire, not ineffectively tickle tongues of flames around the fire's edges. If proving unproductive, weak handline streams should be consolidated into more powerful master streams. It is a wiser decision to save as much as possible with fewer high-caliber streams than to lose everything because of the limited large fire-extinguishing capability of handlines.

Figure 8–24. An example of a *sky shot*, a stream directed at the sky in hopes it will fall back to earth and extinguish the fire. This stream is accomplishing no extinguishment because the spray stream is being vaporized before reaching the flame/fuel interface. (Photo courtesy of Steve Redick)

Figure 8–25. Master streams should be applied as close as possible to the fire for the best effectiveness. The stream from the Squirt unit in this picture is being applied from inside a window opening and is being directed by the officer on a ladder below. The ladder pipe stream is being applied directly into the fire through a burned-out roof hole from overhead.

Ground-Attack Monitors

It is my belief that one of the most game-changing and effective devices introduced into fire stream hardware in the past 75 years was the introduction of the ground-attack monitor.

Smaller than a conventional monitor, these devices are designed to safely flow up to 500 GPM, have an integral shutoff for hit-and-move operations, and allow one person to safely control that flow rate while easily maneuvering the stream to penetrate a large amount of fire (fig. 8–26).

They are lightweight and many departments are finding them ideal as one-person, high-volume preconnected line extremely effective for initial attack (fig. 8–27).

The use of a ground-attack monitor allows a deployed stream to continue in use, even if the crew has to leave because of heat or flame exposure. Officers should not hesitate to take ground-attack monitors guns inside buildings to more easily provide the volume needed to stop a rapidly spreading fire. I remember our department saving an office area of a large, totally involved factory by placing two portable guns in the connecting access doorways. Only two people were needed to apply 1,400 GPM in the successful stop. Ground-attack monitors are also excellent for use inside warehouses or churches to reach extremely high ceilings (fig. 8–28).

Evacuate all civilians, both residents and spectators, from the area in which you plan to operate. Do this early and you will have one less worry if the fire starts leaping over your forces.

Due to the tremendous upward draft created by the fire's thermal column, it is common at large fires for all kinds of loose material to be sucked high into the sky, including large embers and chunks of burning roofing material. This debris will be deposited downwind at distances from a few feet to a few miles. Provisions should be made to provide extra companies to combat the possibility of additional fires downwind resulting from flying brands.

Figure 8–26. The ground-attack monitor can easily allow the safe application of a 500 GPM stream to assist in knocking down a large body of fire. When needed, it can be shut off at the monitor and quickly moved to another vantage point.

Figure 8–27. A ground monitor preconnected to a 250' 2½" line on this rural pumper. Having a master stream immediately available, being safely flowed from a lightweight monitor that is easy to position and reposition, can materially contribute to a save.

Figure 8–28. All nozzle manufacturers market ground monitors. It is wise to test out those being considered for purchase under your own department's conditions before writing the check.

Fighting a Losing Battle

At some point, the incident commander may have to resign themself to the fact that the original building(s) has been lost and that operational tactics will have to be drastically altered (fig. 8–29).

Except for containing the spread of large fires, heavy streams poured on large, self-vented conflagrations are used largely for public relations purposes. It's a fact. Unless you can apply water in enough quantity to overcome the heat being generated, you cannot extinguish the fire. If the critical flow rate is 20,000 GPM, consider that there are very few departments that could supply the water and extinguish the fire at that stage. Sometimes, no matter how much effort and water are expended in battle, the department will have to wait until the fire itself consumes its own fuel. All fires will eventually burn down to a size that can be handled by the available streams.

Master streams, if not used wisely, simply waste staffing and time by slowing down the burning process, keeping companies on the scene longer than if the fire was contained and allowed to rapidly burn up its fuel. That's not to say the fire should not be fought, but common sense should indicate how much effort should be expended on a lost cause (fig. 8–30).

How many times have we seen a large occupancy totally involved with the roof collapsing and flames shooting everywhere? Thousands of gallons of water per minute are being applied by the fire department with seemingly no effect; suddenly, the main body of fire

Figure 8–29. A building this involved on arrival calls for covering the exposures and making sure the firefighting personnel are operating from a safe distance as within a short period of time, this building will start to collapse. Risk nothing to save nothing certainly applies here. (Photo courtesy of Bill Harp)

Figure 8–30. No fire department on earth could muster enough flow to extinguish this fire. When confronted with a fire of this magnitude, it is senseless to continue to send endless extra alarms in an attempt to extinguish what is already burning. Additional alarms may be needed to help contain the fire's spread. In this case, proper resource management calls for removing apparatus and operating personnel from danger and adopting defensive tactics to keep the blaze from spreading. There is nothing for the incident commander to save in this situation except firefighters' lives.

begins to darken and is then extinguished in a relatively short time frame. The fire has simply burned itself down into a size that was then rapidly controlled by the flow being applied. How long it takes for a department to extinguish a large fire is usually determined by how long it takes for the decrease in heat production to equal the extinguishing capabilities of the water being applied.

Once all exposures are protected, it's time to take stock and determine what further action is really needed. A decision to commit extra companies should not be based on emotions but on the stark realization that the fire won this battle.

On the scene of a large conflagration, a smart commander will know when to say when. It makes no sense to send extra alarms and commit additional companies when their work will have absolutely no effect on the final outcome of the fire. Besides exposing more personnel to injury, it strips adjoining response areas of companies which may be needed if a second fire breaks out. It takes only a few sticks to effectively beat a dead snake.

Protecting Exposures

When used defensively, a heavy stream's primary purpose is to prevent the extension of fire by protecting exposures. Remember, exposures can be located both internally and externally.

A good tactic to use when protecting an exposure is to reduce the amount of heat being absorbed by the exposed material. Obviously, this is most easily accomplished by keeping the exposed face wet with a stream. A good nozzle operator will make sure the exposed surface is continually being soaked and will play the stream back and forth between the exposure and the fire, attempting to reduce the fire's heat while keeping the exposure cool (fig. 8–31).

A stream of water washing the face of an exposed building reduces the heat building up in the exposure's surface material and deposits a film of water on that surface. If the fire continues to generate great amounts of radiant heat, the water will evaporate into steam, further cooling the exposed material. Be careful not to rip off sheathing or to break any windows when directing a stream on an exposure. If an opening is made in the building's outside skin, then the exposure becomes internal in that spot and the outside stream may not be able to cover the internal exposure because of the stream's location. The use of heavy-caliber combination nozzles may be effective in distributing water over an exposure's surface if care is taken to assure that the stream is reaching all parts of the exposed surfaces, including the roof. A spray stream merely directed between a burning building and an exposure is not as effective as placing the water directly on the exposure. To prevent unwanted fire spread, it is important that the water be deposited on the exposure's surface. Tests have proven that at

Figure 8–31. In this fire in the Bronx, New York, heavy streams protected exposures until a ruptured gas main could be shut down, allowing companies to safely enter exposed structures. There was no attempt made to enter threatened structures until the gas was shut off, ensuring a safer operating environment in the threatened exposures. (Photo courtesy of Jim Estrin)

least 80% of the radiant heat generated by a large fire will pass through a water curtain only to be rapidly absorbed by a dry exposure. While self-operating water curtain nozzles may appear to be doing an effective job, they are actually doing very little to reduce surface temperatures if they don't soak the building.

Sometimes the best defense may be a strong offense. Knocking down the source of heat production does wonders for protecting exposures. A good exposure protection stream should be able to drench the entire exposure as well as also being able to penetrate the threatening fire.

Do not depend strictly on outside lines to protect a threatened occupancy. Lines should be stretched inside exposed buildings to quickly cover any internal outbreak of fire. Radiant heat will easily travel through glass, igniting materials inside without the glass having been broken. If an exposure is equipped with a sprinkler system, lines should be stretched to the exposure's fire department connection and immediately charged to prevent the system from being overwhelmed if a large number of heads open at once.

Directing Heavy Streams

If the fire is venting itself upward, be careful in letting large overhead streams push the fire back down into the building. One of the most common mistakes of large fire offensive attack is to order an aerial stream directed into a ventilation hole cut or burned in the roof. These streams rarely accomplish much extinguishment, but will reverse the natural course of the fire, taking what needs to be vented to the atmosphere (the flames and products of combustion) and stuffing them back into the building. I never cease to be amazed at firefighters who do not understand that the flames leaping from a hole in a roof will continue the burn with the same intensity inside a building if a stream is directed to keep them there (fig. 8–32).

If the roof has collapsed or the fire has burned a large hole, an aerial stream will aid extinguishment as it will more easily reach inside the structure. But as long as the roof is intact, the roof will do what it was designed to do—keep water out of the building. A much more effective practice is to apply streams from below the roof, horizontally and upward through windows or other openings, to help penetrate cockloft and attic areas and push flames out of the building through the vertical ventilation hole. Application should start at the lowest point where fire is showing and work upward as extinguishment takes place. It makes no sense to work on extinguishing fire in the top floor as flames eat away at lower structural supports. The top floor is almost certain to light up again when it collapses into the burning basement (fig. 8–33).

To be effective, streams applied from ground level should enter the building before passing over the edge of the roofline. A stream directed toward the roof and passing over the roof's edge is practically useless. Thermal currents will whip the water upward before it has any effect on the base of the fire. These sky shots are nothing but a waste of water and effort. In the heat of battle, many nozzle operators tend to aim streams at the flame tongues rather than the flame's base, thinking they are doing some good. If the desired result is to extinguish the fire, officers spotting these streams should have them immediately redirected to where they can actually hit some burning material, not uselessly tickle the tips of flames (fig. 8–34).

Chapter 8 **Heavy Streams** 207

Figure 8–32. The means of stopping this fire is being mistakenly directed 20' too far to the right in a misguided attempt to protect an exposure which is not yet threatened. There are no lines stretched to the interior of the structure. The building was destroyed after four alarms were sounded. Extinguishing the fire removes the threat to the exposure. (Photo courtesy of Jo L. Keener)

Figure 8–33. This stream is accomplishing nothing because the water is not penetrating where the fire is burning—under the roof. Notice the lack of ventilation—not one window on the second floor was opened and cleaned so streams could penetrate the building's interior. (Photo courtesy of Jo L. Keener)

Figure 8–34. Streams directed over the edge of the roofline accomplish little extinguishment. What water clears the roof's edge is quickly sucked upward into the fire's thermal column and carried away. (Photo courtesy of Gordy Nord)

When applied from ground level, heavy streams will be effective up to about the third floor. The farther away from the building the heavy stream device is positioned, the more penetration will be obtained on the upper floors. Aerial devices must be used if it is necessary to effectively apply large amounts of water at higher levels and have it effectively penetrate into the structure.

Heavy Stream Safety

It is critical that all personnel be removed from a building before heavy streams are applied from the exterior. These streams flow tons of water per minute and could help aggravate structural weaknesses or be absorbed by stored materials, increasing floor loading to dangerous limits and possibly leading to collapse. Tons of water flowing out of a building can sweep firefighters along their path, causing serious injury. Heavy streams, especially spray streams, applied from the exterior of a building can rapidly push heat and combustion products to other parts of the structure. Anyone who has ever been caught inside a building when a heavy stream was applied will tell you it is not a pleasant or safe experience. Be sure the building is cleared of personnel before directing heavy streams into the building from the outside (fig. 8–35).

Figure 8–35. When a fire has progressed to a point where inside operations are unsafe, external streams must be operated at a safe distance and, if smoke obscures the target, must be aimed using senses of sound and sight.

The same dangers could exist when operating heavy streams from portable ground-attack monitors inside a building, although they will not be as severe if the operating crew directs the stream properly. Before inside monitors are charged, it is good practice to clear operating personnel from the area between the monitor and the point of exit ahead of it for ventilating smoke and heat.

One of the primary causes of collapse on newer aerial apparatus devices is failure to properly set the outriggers. It only takes a few more minutes to accomplish the set-up job properly. Any set-up time saved is quickly forgotten when the ladder or boom carrying personnel crashes to the ground, or worse, into the fire. Luckily, aerial apparatus being produced today have interlocks to prevent misuse, such as rotating an aerial device toward a side where the jacks are not properly extended. But if the jacks are short set, it may be that the aerial device cannot be placed in the position desired. A good practice is to fully extend all jacks to allow for 360° aerial device operation (fig. 8–36).

A large fire is not something that happens every day. As a matter of fact, most firefighters will face only a few in their entire careers, but it is something you cannot plan enough for as any commander who has watched buildings crumble and the department pushed to the limit will tell you.

Figure 8–36. An aerial stream being properly applied to knock down a heavy volume of fire in the top floor and cockloft area. Too many departments would have extended the basket over the roof and attacked the fire venting out the roof, pushing it back inside the building. When using master streams, if the fire is venting through a selected channel, attack operations that drive the fire out the channel will cause less property loss than stream placement that drives fire back into the interior of the building. (Photo courtesy of Alan Chaniewski)

Chapter 8 Review Questions

1. What is the definition of a heavy stream?
2. What are one of the ways to maintain effective stream reach under fluctuating water supply conditions?
3. What are the advantages of a preconnected hose line with an attached heavy stream device?
4. What is an important point when implementing a department's evacuation signal?
5. What is the main consideration when making the decision to use a heavy stream?

9

Fire Hose Basics

Fire hose is another of those critical but generally taken for granted tools used in the water delivery system. Far from being a simple water conducting device, fire hose is a carefully constructed balance of jackets, inner liner, and outside protection. Each of the three components has a specific function, the efficiency and durability of which is dependent on quality of materials and construction methods used in their manufacture.

Until the mid-1800s, most fires were fought by water transported to the scene in buckets. Original hand pumpers discharged their water through a small pipe or monitor attached to the top of the pump tub. It was not until the late 1860s that hose became available to convey water more easily from the hand pumps, and later steam pumpers, to the fire. Early fire hose was fabricated of leather, fastened together with copper rivets and washers and, as can be imagined, it was heavy, stiff, and commonly leaked.

Around 1895, unlined fire hose made of circular woven linen yarns began to replace leather hose. It was certainly much lighter. As the hose fibers, commonly made of flax, became wet, they swelled up and tightened the weave, causing the hose to become watertight. Unlined linen hose, because of its lack of durability, was rapidly replaced with rubber hose in municipal fire service use. Unlined hose continued to be specified, however, for use on interior hoselines and hose racks until the 1960s and is still used in some areas for forestry applications.

Following the invention of the vulcanizing process as a means of curing raw, soft rubber into a harder, more useful product, the fire service slowly made the transition from bulky and unreliable leather hose to unlined linen hose, then to a multi-layer, rubber-lined and -coated hose with an interior fabric reinforcement. This rubber hose was as bulky, heavy, and stiff as leather hose, but was not as prone to leaking. It also proved more durable than unlined linen hose. Its wrapped construction resembled some hose used today by industry; for example, fuel delivery hose used to service airliners.

Along with the development of more efficient steam pumpers, which were rated to deliver their capacity at 100 pounds per square inch (PSI), a demand was created for hose that could safely contain the new, higher operating pressures. Manufacturers responded by designing a seamless, circular, woven cotton jacket hose with a rubber liner. It was constructed in much the same manner as modern fire hose, although the all-cotton jackets woven from the heavy yarns needed for strength made it bulky and heavy. Because of limitations in then available rubber compounds, the tube had thick walls, adding to the hose's weight and stiffness.

Nevertheless, cotton-jacketed hose was quite an improvement over previous leather, unlined linen, and rubber hoses and was rapidly and universally adopted by the fire service.

Until well after World War II, most fire hose was of circular woven cotton jacketed construction. While practically all of this hose had double jackets, some large cities specified hose with triple jackets, the intent being to increase the operating pressure and extend the service life of the product. Having three outer jackets in place of one would help to counter the effects of wear and abrasion. One can imagine the great weight and stiffness of triple-jacketed hose and why it was unpopular with firefighters in the cities that used it.

Remarkable advances were made in the synthetic fiber industry following World War II, and important discoveries were made in the development of thinner yet stronger man-made fibers. DuPont introduced such synthetic fibers as Dacron along with other filament and spun polyesters. Because of their strength and weight advantages, these fibers quickly found their way into fire hose construction. These fibers, along with similar synthetic materials from other manufacturers, allowed hose jackets to be woven that were lighter, more flexible, and less bulky than previous all-cotton jackets. Originally, a combination of synthetic and cotton yarns was used to provide a hose having higher test pressures than all-cotton hose yet taking about a third less space in storage.

One major problem with the use of cotton was it was prone to mildew forming and rotting when stored wet. A number of treatments were developed to slow down the cotton's tendency to deteriorate that included the repeated application of wax. One upside of using cotton was that it resisted heat relatively well; however, the quality of a cotton jacket varied greatly depending on the quality of the cotton used. Generally, in those days, the higher the per-foot price, the longer lasting the hose.

Hose Construction

Except for some specialized interior and forestry applications, practically all fire hose used today consist of three major components:

- Inner liner or tube
- Inner reinforcement or strength jacket
- Outer jacket or protective rubber covering

Inner Liner

All municipal fire hose has an inner liner. The liner can be constructed of synthetic rubber or a special thermoplastic tube material. The inner liner is the hose component that contains water within the hose assembly. Years ago, the liner was calendared and formed by gluing together numerous strips of rubber cut from sheets. Today, all inner liners are extruded from machines, assuring an even thickness of the tube's walls and making the tube stronger and lighter than calendared construction (fig. 9–1).

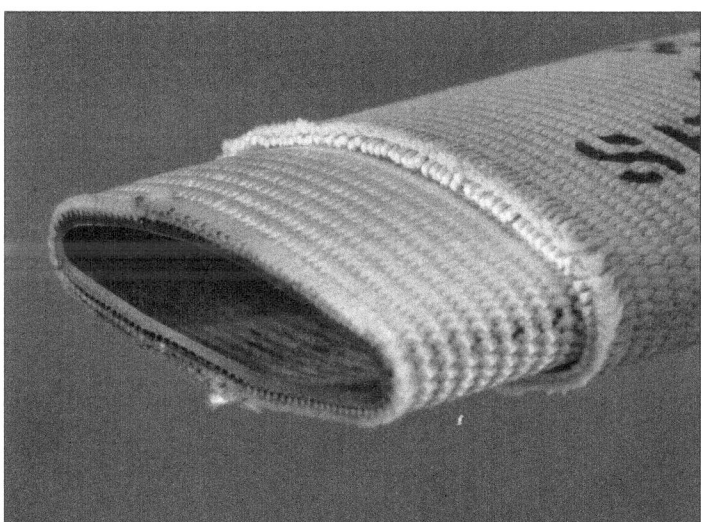

Figure 9–1. Hose used in the fire service is constructed of an inner liner or tube, an inner reinforcement or strength jacket, and an outer jacket or covering to provide additional strength and protection from wear.

Woven Inner Jacket

Since the tube is simply a thin cylinder of rubber or thermoplastic, it requires reinforcement in order to contain pressure. To reinforce the inner liner, a woven jacket is wrapped around the tube to provide the additional strength needed to handle water at fire service pressures.

During woven hose construction, the rubber tube is coated with an adhesive compound, inserted into the woven jacket, and then cured or vulcanized using steam pressure. This forces the tube to become permanently bonded to the woven jacket. A synthetic tube uses adhesive to attach itself to the reinforcing jacket. The result at this stage is single-jacket hose. Rarely used in municipal fire service, single-jacket hose is widely utilized in industrial and military applications where the hose is stored more than it is used, and its lack of abrasion resistance does not present a problem. While single-jacket hose certainly possesses the strength to contain its rated pressure, the exposed woven reinforcement jacket is extremely susceptible to damage from use in fire environments. If worn or cut, water pressure can cause it to rupture.

Woven Outer Jacket

A second jacket of woven material or rubber is normally used to protect the inner assembly, shielding the inner reinforcement jacket from the effects of abrasion, cutting, and heat frequently encountered in fire department use. If the outer jacket fits properly, it will also help increase the strength of the total hose assembly, contributing to higher pressure ratings.

A woven jacket is constructed of two types of yarn strands: warp and filler. The filler strands go around the hose in a circular pattern to provide most of the jacket's strength and normally cannot be seen. The warp strands run the length of the hose from coupling to coupling. Their main purpose is to provide protection for the filler strands. The warp strands are the ones you can see on the outside of the hose.

Fire Hose Yarns

As more post-war advances were made in fiber technology, and as the long staple cotton yarns became more difficult and expensive to obtain beginning in the early 1970s, hose constructed of 100% synthetic yarns began to dominate the municipal fire hose marketplace. As personnel reductions became common into the 1980s and departments took on other duties such as medical care and hazardous material response, less time was made available for hose maintenance. Since all-synthetic hose products require no drying and little care, hose of 100% man-made fiber construction has become the standard in the fire service. There are some disadvantages to all-synthetic construction, mainly centering on its poor heat resistance characteristics. Despite this problem, most departments will agree that all-synthetic hose provide a high level of serviceability while requiring much less attention and maintenance than that which must be given to cotton-blend construction (fig. 9–2 and fig. 9–3).

While most hose appears similar at first glance, not all synthetic hose construction is alike in wearability or heat resistance. It is wise to be familiar both with the materials utilized in hose construction and the methods used to bring these materials together into a completed assembly. The combination of materials and construction methods determines the hose's durability, length of service, return on investment, and, of course, safety available within the fire environment.

Fire hose jackets are constructed of woven yarns. This yarn is constructed of a number of individual threads, each made up of fibers. Cotton yarn is made up of millions of short fibers twisted or spun together, binding the fibers together by friction. To gain strength, quite a bit of material is needed, and binding friction is increased somewhat by spinning the fibers more tightly together.

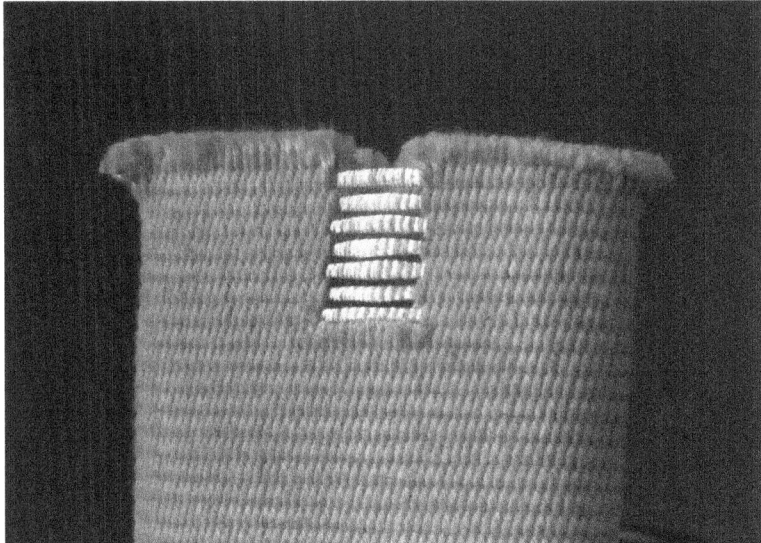

Figure 9–2. Woven hose is constructed of two types of yarn strands: warp and filler. The filler yarns provide strength and are wrapped around the assembly in a circular pattern. They are the lighter colored yarns shown in this cut-away sample. The warp yarns run lengthwise and provide protection for the filler yarns.

Figure 9–3. This close-up shows how the warp yarns are woven around the filler yarns to provide protection. Filler yarns in all-synthetic construction are usually filament type, which can be recognized by their flat appearance. The warp yarns are usually of spun type, which provide superior insulation and wearing qualities. They are round in shape.

Since cotton fibers are short, all-cotton yarns used in fire hose construction must be spun to provide the necessary length and strength. Synthetic fibers, on the other hand, are extruded from machines in continuous strands. Because length can be specified at will, synthetic yarns used in fire hose construction can be of two basic types: filament and spun.

Filament yarns are composed of a number of continuous extruded filaments, each resembling a strand of silk and lain side by side. Because the filaments are unbroken for the entire length of the yarn, the yarn possesses great strength. Unfortunately, this strength can be easily compromised if individual filaments are cut, broken, or abraded. This susceptibility to wear can be compared to that in nylon pantyhose, which are also woven from single, continuous filaments. If the garment is snagged during wear, filaments will break, causing an unsightly run down the full length of the filament. If unprotected filament yarn is used where it can be damaged, cutting, or abrading the individual filaments can cause rapid loss of strength. Because of its strength and its need to be protected, filament yarn is used mostly in the filler or internal cords where it is covered by the bulkier and longer lasting spun yarns.

Spun synthetic yarns use short, chopped filaments twisted tightly together to gain strength and bulk. This construction is similar to yarns made of cotton, comprised of millions of fibers that vary from ½"–1¼" in length. Spun yarn appears round while filament yarn appears flat. Because spun yarn contains many fibers, the damage from a snag or abraded spot tends be localized as other nearby fibers take on the load. Unless the entire yarn is cut completely through, not all the yarn's strength or thickness will be lost. Filament yarn has no way of transferring the load to other fibers within the yarn if subjected to the same amount of abrasion or cutting, so once individual filament fibers are cut, the yarn will rapidly lose its rated strength.

Some hose manufactured today use a blended synthetic yarn that has entangled filaments. This yarn usually contains the same amount of material as spun yarn, but more bulk is provided without increasing the total weight of the yarn because the material is entangled, twisted, and fluffed-up similar to the material on the tip of a cotton swab. Depending on the hose manufacturer's specifications, entangled yarn is usually constructed with a high percentage of nylon material. Nylon can stretch more than polyester or cotton yarns while still providing its designed strength, contributing some positive pressure absorption characteristics. Nylon has poor heat resistance and must be protected by other types of fibers or by a polymer protective coating. Because of its increased surface area, entangled yarn can absorb and effectively retain the polymer treatment material, widely used as a heat resistance and abrasion treatment (fig. 9–4). Entangled yarn construction tends to be looser and more open than spun yarn and, for that reason, is rarely found in municipal hose construction without a protective polymer outer jacket treatment. This treatment helps bind the fibers together and keeps them from separating during use.

Most woven jacket fire hose uses a combination of yarns, each to its best advantage. Because of its strength, filament yarn is used in the filler or strength cords. This yarn is then covered by spun or entangled yarns, providing the necessary bulk for abrasion and heat protection. How the yarns are blended and woven can determine how well the hose resists kinking and how well it resists the effects of abrasion and snagging. If the outer jacket is tightly woven, more material will be contained within a certain area, increasing abrasion resistance; however, a tighter weave may also make the hose stiffer. It is important to have the manufacturer's salesperson demonstrate how each type of construction is designed. This is most easily done using short samples in which all components are visible.

Figure 9–4. All synthetic woven jacketed hose is sensitive to damage by excessive heat. This hose was damaged by falling embers. A charged hose line is less susceptible to heat damage than an uncharged line. There is hose available in the market that is designed to have higher heat resistance and meets the radiant heat test outlined in the UL 19 standard; however, it is extremely expensive and its acceptance has been slow. (Photo courtesy of Alan Zale)

There are some hoses on the market that have a high percentage of filament yarns in the outer jacket. This makes the hose glide somewhat easier when being stretched or advanced, but also makes it prone to unwantedly slide out of the hose storage bed. A number of accidents caused the *National Fire Protection Agency (NFPA) 1901: Standard for Automotive Fire Apparatus* o now include hosebed restraint devices.

A design of hose that is increasing in popularity has an inner tube consisting of a circular woven inner reinforcement with an extruded rubber covering forming an integral tube and inner jacket. This hose is relatively resistant to kinking and eliminates the problem of the tube delaminating from the inner jacket. This type of construction also provides more heat resistance than conventional double jacket hose styles.

Fire Hose Sizes

Thirty years ago, attack line sizes were generally 1½" and 2½". After the New York Fire Department (FDNY) adopted 1¾" hose, 2" and now 2¼" sizes are finding their way to the front lines (fig. 9–5). Each size has its advantages and drawbacks, and selecting attack line size takes much research and hands-on trial with the actual products. Don't be afraid to have your dealer or manufacturer's representative bring you hose and nozzle equipment so you can properly evaluate the hardware in your own operating environment.

With the advent of thermoplastic hose liners, manufacturers facing an increase in rough-liner friction loss decided to fractionally increase hoseline diameter to compensate. Then, when pitted against each other in the marketplace, manufacturers kept increasing their hose size so they could boast about their offering having less friction loss than the competitors. Consider this: You decide to use a standardized friction loss chart to compute the necessary pump pressure to supply a preconnected attack line. But, with all the different manufacturers' hose diameters, the standardized chart recommendations may not accurately reflect the

Figure 9–5. Five fully involved barns and a 40-mph wind pushed this fire, threatening the farmhouse. A 250' 2" preconnect flowing 250 gallons per minute (GPM) stopped the fire's spread in its tracks, and the house was saved. (Photo courtesy of Mike Ullery)

proper and safe pressure needed. In these cases, the only way to accurately determine the exact pump pressure needed is to screw the preconnect together and flow the line using a calibrated flow meter to determine the required pressure. But then when you reload using a hose made by different manufacturers, your calculated pressure will not be accurate. A few years ago, at firefighters' insistence, manufacturers began to market hose made to true diameters. So inside, a length of 1¾" hose actually measured an accurate 1¾", which certainly simplifies pump pressure calculations. If a manufacturer wanted to deviate from the nominal size measure, for example FDNY's 1¾" hose diameter which actually measures 1.88", they told us. Not all hose is true to its advertised diameter, so when purchasing, you will have to specify the true size you want.

Fire Hose Pressure Ratings

Considerable confusion is generated when attempting to compare fire hose pressure ratings. Most fire hose manufacturers follow the guidelines contained in *NFPA 1961: Standard on Fire Hose*, which addresses fire hose manufacturing, and *NFPA 1962: Standard for the Care, Use, Inspection, Service Testing, and Replacement of Fire Hose, Couplings, Nozzles, and Fire Hose Appliances*, which addresses fire hose care and usage. Among other items, these standards describe three basic pressure ratings for municipal fire hose:

- Acceptance test pressure
- Burst pressure
- Service test pressure

Each pressure and test is designed for a specific purpose, and anyone using or specifying fire hose should be familiar with each.

Acceptance or *proof test* pressure is a one-time test, performed at the hose factory on a special test table after the hose is coupled and is ready for delivery. Keep in mind that if the hose is delivered to a distributor or department uncoupled, the hose is usually subjected to an acceptance test. Hose is pressurized to its acceptance test pressure for 15 seconds, after which it is inspected for any leaks or broken yarns. It is then drained, coiled, and boxed for shipment. This test ensures that the couplings are properly attached and that there is no major structural defect in the hose. End users should clearly specify that all hose purchased be tested to its rated proof pressure by the manufacturer. Years ago, cotton fire hose was normally rated by this test in various pressures from 200 PSI–600 PSI. Because the amount and quality of the cotton material had to be increased to withstand higher proof pressures, the higher pressure rating usually indicated that the hose was of better construction and contained heavier materials.

With the introduction of synthetic materials, rating hose by acceptance pressure alone is not as good an indicator as it was in the days when all-cotton hose was common. Almost all cheap, poorly constructed, and all-synthetic hose can withstand an acceptance test of 15 seconds at high pressures when new. The main difference between poorly-constructed hose and high-quality hose is in length of service life—how long it will continue to contain that pressure when in use.

The short length minimum *burst pressure test* is performed by a special testing device at the manufacturer's factory. It is the benchmark test from which all other pressure ratings—proof, service test, and standard working pressure—are mathematically calculated. In this test, a 3' section of hose is pressurized until it bursts open, and the pressure recorded. Burst pressure can serve as another evaluation guideline; however, it simply indicates that the hose must withstand a pressure one and a half times the acceptance test pressure or three times the service test pressure. This test should be performed twice: once on a straight hose sample and then on a kinked hose sample. The kink test is important because it indicates any weakness in the warp yarns. If the jackets are of extremely open or loose weave, the hose will have a greater tendency to fail when the yarns are flexed on a bend. The burst test is performed on random samples from a batch of hose, not on pieces from each length.

The annual *service pressure test* is performed by the fire department or third-party testing organization to assure that hose has enough strength and integrity to remain in service. The service test pressure is usually stenciled on the hose and is half the acceptance pressure. NFPA 1961 calls for a minimum service test pressure of 300 PSI for attack hose and 200 PSI for supply hose.

The working pressure recommendation is a safety guideline for the department and is computed as 85% of the annual service test pressure.

The following chart will help clarify how the different pressures relate to each other:

Proof	Burst	Annual	Working
450	700	225	191
600	900	300	255
800	1,200	400	340

Proof and burst pressures are most useful when making evaluation and purchasing decisions. Annual tests determine working pressures and will determine if the hose is safe to use.

There are other performance evaluations used by manufacturers, such as rise, elongation, twist, and kink tests. Unless they are witnessed by qualified inspectors, their usefulness to an individual department is not as important as the three major tests.

Countering the Effects of Heat and Wear

The outer jacket or cover has a function similar to that of a firefighter's turnout coat: to provide protection (to the assembly, either human or synthetic) from the fire environment. As with protective clothing, one of the outer jacket's main protection functions is to combat the effects of excessive heat. Unfortunately, the polyester fibers used in a great percentage of hose construction have low resistance to high temperatures and can rapidly lose strength or melt if enough heat is present. One manufacturer's marketing official explained, "Remember back when everyone smoked? Someone would be at a party and an ash would drop on the leg of his all-polyester leisure suit, immediately burning a hole right through to his thigh." All-polyester fire hoses have the same problem as an all-polyester leisure suit; it melts rather quickly at relatively low firefighting environment temperatures, not necessarily from air temperatures, but more from coming into contact with hot surfaces or embers. Flammability standards for children's sleepwear, in fact, are higher than the melting point of polyester hose jackets. There are a number of things that can make fire hose more heat resistant, but no matter what is done at the factory to treat the hose, the most effective method to keep hose from failing in use is proper care in its deployment and maintenance.

Consider for a moment the example of a nozzle crew working its way down a hallway. Wrapped up in a protective envelope consisting of NFPA-standard turnout coat and pants, boots, gloves, hood, helmet, and breathing apparatus, the firefighters are seemingly well protected. As entry is made into the fire area, the firefighters must be aware that the hose they are pulling has the least heat resistance of any equipment in the interior firefighting chain and that it can soften and fail from the fire's heat long before the firefighter feels any discomfort.

Tests have shown that water flowing through a hoseline has little to no effect in reducing the probability of heat damage by helping to keep the yarns cool, despite previous beliefs. A water-soaked jacket can provide a bit of protection but should not be relied upon. The best insurance against hose failure is having the interior crews help minimize damage by being careful not to drag the line through embers or across window frames filled with glass shards. Unavoidably, embers can fall on the hose and cause heat damage. If water is flowing and the hose is wet, their effect can be minimized. It must be remembered, however, that any heat source coming into direct contact with the outer jacket of any fire hose, especially untreated, all-polyester jackets, can cause failure.

Two firefighter deaths in a fatal fire in a large city were partially blamed on hose deterioration due to high heat; however, it was surmised that there was little or no water in the hoseline at the time. Tests by the Bureau of Alcohol, Tobacco, and Firearms afterward proved that this had little effect on the outcome.

There have been calls to increase attack hose outer jacket heat resistance and there have been products introduced into the marketplace that have high temperature–resistant fibers blended into the outer jacket construction. Initial testing has shown that these fibers exhibit practically no elasticity, so, in use, the outer jacket expands and becomes quite loose. The major hurtle in more departments adopting this type of hose is its extremely high price tag. Mildew is a form of bacteria attack in which microbes consume the cellulose fibers within the yarn material. Visually, mildew appears as black or brown spots on the hose fibers. In its advanced state, it appears as a white or light green fuzz. Although mildew does not materially affect the strength of synthetic fibers, it is rather unsightly and can make the hose difficult to handle. Good practice is to wash hose after use and, in freezing weather, dry it as well.

Rubber-Covered Hose

Rubber-covered hose provides high-heat damage resistance while offering the extremely positive advantage of needing practically no maintenance. For years, rubber-covered construction has been the most widely specified for large-diameter supply hose. Its advantages, which include being able to be repacked immediately after use and its ability to be thin yet strong, have helped increase the widespread acceptance and use of large-diameter hose.

Rubber-covered hose available to the fire service is constructed in one of two methods: plied and extruded (or through-the-weave).

Internally, *plied construction* resembles the assembly of the double-jacket hose. An inner tube is vulcanized to a woven jacket and is similar in appearance to standard-construction hose. In place of a woven outer jacket, an extruded rubber outer jacket is provided, giving resistance to abrasion and heat as well as extra strength. All plies are then vulcanized together, providing a unitized hose that tends to be a bit more flexible than extruded designs (fig. 9–6 and fig. 9–7).

Extruded construction encapsulates a woven inner reinforcement, similar in appearance to a woven jacket, within a liner and cover of extruded rubber. This is accomplished in a single process as it is extruded through a machine, the rubber compound forming both the inner liner and outer jacket. The molten rubber compound flows through the woven reinforcement jacket, creating a single, unitized assembly (fig. 9–8).

There are advantages and disadvantages to both methods of construction, the main differences being that the extruded construction has slightly more resistance to kinking while

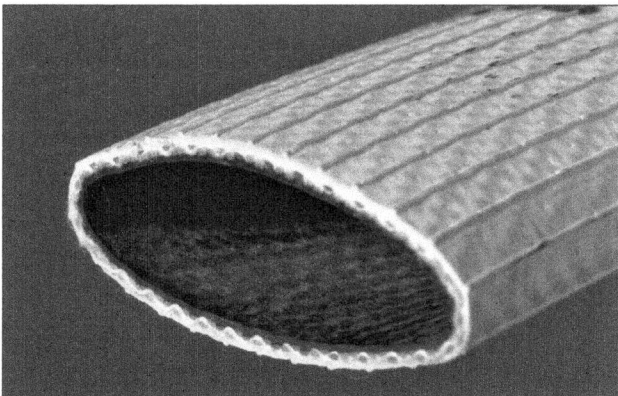

Figure 9–6. An example of rubber-covered hose using plied construction. This style of hose can be recognized by the black inner liner.

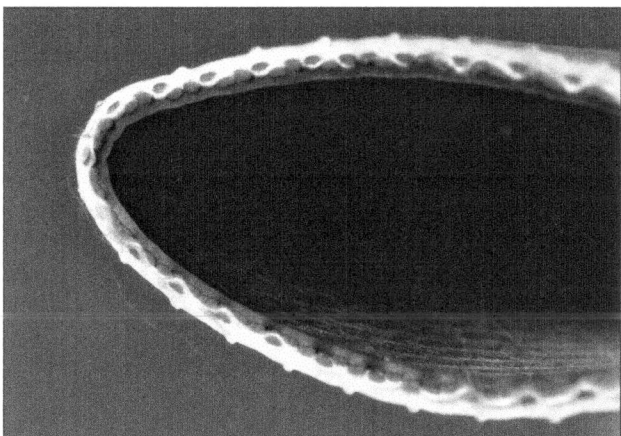

Figure 9–7. A close-up of plied style rubber-covered hose. This design encapsulates the reinforcement liner.

Figure 9–8. An example of rubber-covered hose constructed by the extrusion method. Note how the reinforcement jacket is encapsulated within the construction.

the plied construction packs more compactly. The importance of each style of construction seems to vary depending on which salesperson is making a presentation. The best way to evaluate each construction is to contact departments using each style and ask for their opinions on flexibility, wearability, and deployment characteristics. It is also wise to test individual lengths with your own personnel and apparatus to determine which construction is most suitable for a specific operating environment.

Close inspection of the surface of rubber-covered hose will reveal that it is punctured with thousands of microscopic holes called weep vents. If the inner liner is damaged and water is leaking through the woven strength jacket, it will seep from the hose through these holes. If water is discovered dripping from the outer jacket of rubber-covered hose, it is a good bet the inner liner is seriously damaged, requiring immediate inspection and repair.

While rubber-covered hose has been historically the most popular style for large-diameter supply hose and industrial applications, there appear to be a great number of arguments both for and against the use of rubber-covered hose for attack line use. Rubber-covered hose is either loved or hated probably more than any other type of fire equipment. There seems to be no middle ground when rubber versus conventional woven hose construction is discussed. While tradition seems to be the largest obstacle to more widespread use of rubber-covered hose for attack lines, there are many more practical comparisons which should be considered before a purchase decision is made.

Some of the advantages in the use of traditional double-jacket fire hose construction are

- It is easily folded for storage.

- It is easier to deploy because it slides much easier and creates less friction when being dragged.

- It is usually less expensive than rubber-covered hose.

- It is not as susceptible to cuts and punctures as rubber-covered hose.

- It is generally available with higher proof pressures than rubber-covered hose.

Some of the advantages in the use of rubber-covered fire hose construction are

- Some constructions will usually unkink themselves when charged, getting water to the nozzle more quickly.

- It will leak if punctured, but the rubber holds the reinforcement in place and it does not tend to split open as double-jacket hose can tend to do, allowing the hose to continue to supply water to the fire.

- The rubber covering is easily cleaned and can be more easily decontaminated if soiled during a hazmat incident or after being dragged through a rural barnyard.

- Rubber-covered hose has high resistance to heat.

- It will not absorb water in use so it will not freeze as easily as water-soaked woven jackets at winter fires. It will also not smell or suffer the effects of bacterial action if packed wet. Its weight will not increase in use as no water is retained inside the jackets.

- Because it can be wiped down with a rag and then repacked without drying, the need to keep an inventory of spare hose is reduced and, in most cases, the hose can be repacked and the pumper can be put back into service at the fire scene.

By some manufacturers' estimates, it accounts for about 20%–25% of the total footage of attack line sold. It is about twice as expensive as double-jacket hose, but it does offer advantages: kink and heat resistance, the ability to be repacked immediately, and the resulting reduction in spare hose inventory. These advantages may help justify its higher purchase price depending on a particular department's method of operation.

According to a survey of users, double-jacket hose dominates the municipal market because of cost, tradition, and ability to resist tearing and punctures. In addition, it is usually lighter in weight than rubber hose, folds more compactly in the bed, and is more flexible.

Other Methods of Protecting Hose

One of the most widely used methods of increasing the heat and abrasion of synthetic double-jacket hose is coating the outer jacket with a liquid polymer or elastomer compound. These treatments, called Hypalon, Encap, and other names, are applied by dipping the completed outer jacket in the coating solution after the hose is assembled. How effective this treatment is in preventing heat and abrasion damage depends on the quality of the treatment material, how it is applied, and how well the yarns are able to absorb the coating material. Unless the desired treatment is specified by name, some hose companies have been known to color the hose with dye or use a dilute solution of marine boat paint as their treatment. Just because an all-synthetic hose is yellow in color does not indicate it has been properly treated.

Because it has more internal surface area, spun or entangled filament yarns can retain the treatment material in a more efficient manner than continuous filament yarns. The real difference in efficiency is how long the treatment sticks to the yarns. The longer it is retained by the jacket, the longer the hose's service life should be.

Forestry Hose

Hose utilized for forestry service differs somewhat from that used in municipal fire service. Backpacking hose for miles to reach a remote burn dictates that the hose be much more lightweight in construction than municipal hose. Dragging lines through burned-out areas during mop-up operations exposes the jacket to a great amount of heat damage potential from burning stumps, embers, and smoldering logs, so forestry hose must be somewhat heat resistant. Almost all the forestry hose constructions available today are of 100% synthetic construction with some type of protective coating. Forestry hose construction is balanced to be lightweight and heat resistant while not sacrificing wearability and low maintenance requirements.

Some forestry hose is manufactured so that the inner liner *weeps*, or seeps water when in use. This form of self-wetting soaks the outer jacket for extra protection in high-heat environments. While certainly useful in wildland firefighting, this construction is rarely seen outside of brush and forested areas.

A Discussion About Lightweight Lined Hose

Beginning about 30 years ago, a new style hose was marketed which, in spite of double-jacket construction, appeared extremely lightweight and compact. The major difference between this new construction and older hose constructions is in the liner. Rather than having an extruded synthetic rubber tube as in conventional constructions, the lightweight hose sports a tube fabricated of an elastomeric-polymer thermoplastic. While this tube is extremely thin, it has a higher tensile strength of the heavier and thicker rubber tube, is extremely resistant to aging, and will remain flexible at temperatures as low as −60°F. In contrast, synthetic rubber used in conventional hose linings becomes embrittled around 20°F, at which time will tend will snap and break if bent (fig. 9–9).

One of the problems with this liner is its lack of bulk or wall thickness which, with conventional tubes, tends to smooth out the liner interior. When attached to the inner jacket, the thin tube exhibits a rough interior appearance caused by thin tube clinging closely to the weave of the jacket. This appearance, called *washboarding*, causes increased friction loss due to increased turbulence as the water rubs against the interior walls of the tube. This friction loss is reduced by manufacturers by making the hose fractionally larger in diameter, or using a finer, flat twill weave on the inner jacket, or a combination of both (figs. 9–10a and b).

While the hose looks great when packed, the lack of wall thickness also helps contribute to increased kinking. This is especially evident when the jackets are also reduced in bulk

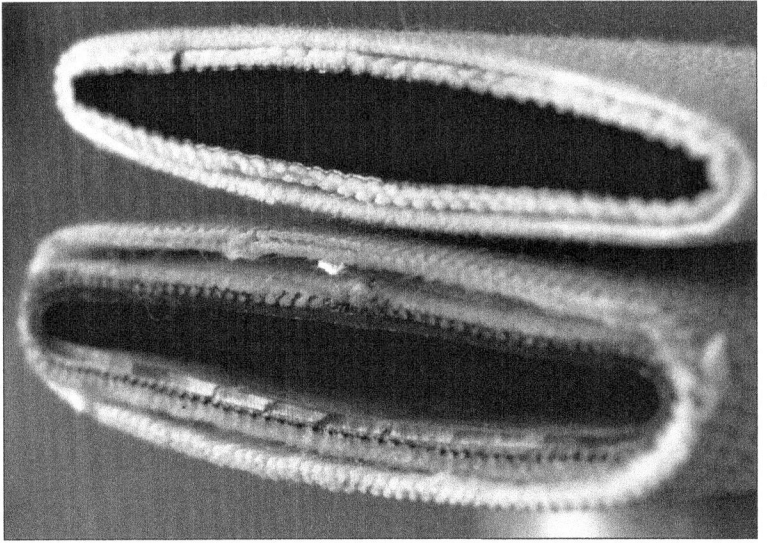

Figure 9–9. A comparison of the end cuts shows the relative thickness of the rubber tube compared to the thermoplastic liner (top). The lightweight liner's thickness is shown by the reflected, light white line.

to produce a lightweight, high-rise hose. Turbulence caused by the rougher surface area of the liner can also have an effect on stream quality when smoothbore nozzles are used. For practical purposes, the more ragged stream that could be produced by thermoplastic-lined hose has only a limited effect on the stream's firefighting ability; however, when placed side by side with a similar smoothbore nozzle being supplied by rubber-lined hose, the difference will be noticed.

Originally marketed for use in high-rise hose packs, the thermoplastic-lined hose's kinking tendencies have reduced its desirability when lines are operated in a cramped stairwell or at low pressures. If using smoothbore or low-pressure nozzles, testing any hose before purchase is a must to ensure that the hose will perform to expectations using an individual department's engine and nozzle pressure operating recommendations.

Most manufacturers are providing inner and outer woven jackets on thermoplastic-lined hose that are almost identical to rubber-lined hose, equalizing durability and heat resistances. The major difference between the two hose designs is in the inner liner (fig. 9–11); the thermoplastic is thinner, and the hose is more likely kink when operating at lower pressures than with the rubber (fig. 9–12).

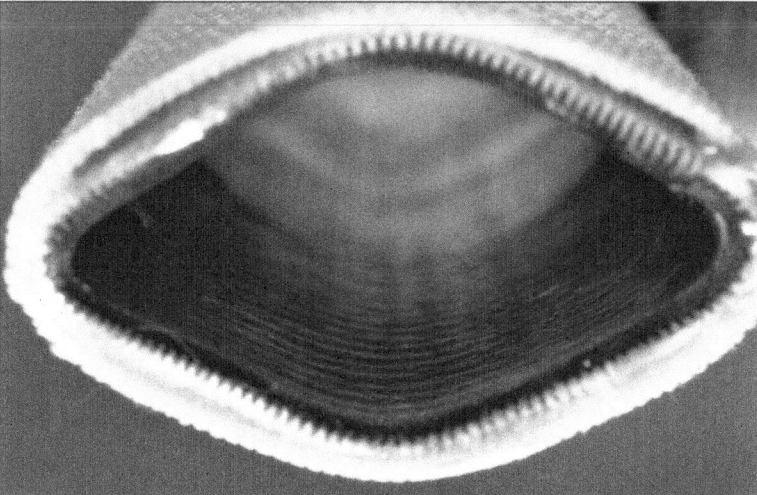

Figures 9–10a and b. Rubber liners (bottom) are relatively smooth when compared to thermoplastic liners. Friction loss caused by the rough interior (top) is easily counteracted by manufacturing the hose fractionally larger in diameter.

Figure 9–11. Newer-style hose construction consists of an inner liner of woven reinforcement and extruded rubber, with a conventional woven jacket over. It is relatively kink resistant and is ideal for low-pressure attack line operations.

Figure 9–12. A comparison of the relative thicknesses of various hose styles. At the bottom is the rubber-covered construction, then cotton-synthetic, 100% synthetic, and all-synthetic with thermoplastic liner on top.

Hose for High-Rise Use

In the past some departments attempted to use hose designed for different applications other than municipal firefighting when equipping for high-rise suppression operations. A desire to reduce weight seemed to be their primary motivation. Reducing weight is important, especially if hose packs must be carried up many stories to reach the point of operation. However, there are other considerations besides weight reduction that must be carefully evaluated when purchasing hose for high-rise use (fig 9–13).

Operating pressures will generally be lower when operating hose streams from standpipe connections than when operating directly from pumpers. Operating at lower pressures generally increases the tendency for hose to kink. It is important that each hose style's kinking characteristics be carefully evaluated before purchase.

Durability and heat resistance are also important qualities that hoses need for high-rise applications. It is a dangerous practice to use woven, single-jacket hose for aggressive attack operations. One little ember can put a nozzle crew out of business in seconds. And it is extremely time-consuming to run down 20 floors to pull a replacement length from the engine.

Figure 9–13. Hose for high-rise firefighting is normally carried in hose packs at 100' per firefighter. Hose packs should be designed so a blindfolded firefighter can release the hose, as anyone who has operated in an almost black stairwell can attest.

Finally, the hose selected should be capable of flowing at least 200 GPM from standpipe connections. Delays of 30 minutes to 1 hour are not uncommon to get a stream into operation on the fire floor in high-rise operations. Delays give the fire more time to build up heat that cannot be easily relieved by common ventilation techniques. When the stream finally gets into operation, it must have the power to penetrate into blast-furnace-like conditions. Departments experienced in high-rise operations say that nothing less than 2½" hose should be used, but there are viable alternatives.

The 2" hose has been gaining in popularity for high-rise attack line for the following reasons (fig 9–14):

- It is compact and lighter than 2½" hose but can still flow 250 GPM 200' from the standpipe valve.
- If equipped with 1½" hose fittings, two lines can be stretched from the first standpipe valve using a gated wye.
- It is lighter to carry and when filled with water is lighter to maneuver than 2½" hose.
- It is equipped with 2½" couplings, negating the use of adapters when connecting to a standpipe valve.
- It will easily flow 250 GPM 200' from the standpipe valve.

Recently, manufacturers have introduced 2¼" hose designed for standpipe use. Using standard 2½" thread couplings, it can achieve flows of 225–250 GPM and is intended for use with a 1⅛" smoothbore tip. At the time of this writing, it is too early to gauge its widespread acceptance.

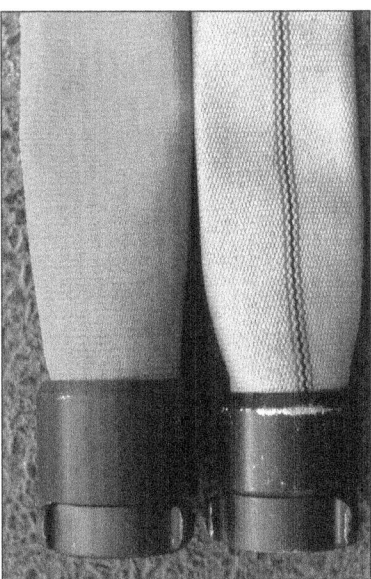

Figure 9–14. Only fractionally larger than 1¾" hose, 2" hose can easily flow 250 GPM and is normally coupled with 1½" thread. Can you tell which size is which? The slightly larger coupling bowl size of the hose on the left is the tip-off it is 2".

No department would consider changing into lightweight, less protective turnout gear or using 10-minute breathing apparatus in an attempt to reduce weight during high-rise operations. Hoselines will be a department's only means of fire extinguishment on upper floors of high-rise occupancies. Excellent double-jacket and rubber-covered hose is available are kink-resistant and durable. It is dangerous thinking indeed to consider small-diameter, thin-walled hose effective or safe when working from standpipe connections in high-rise operations. If you would not use the hose and nozzle on a ground level line, you should not use it in a high-rise pack.

Large-Diameter Hose

As the general trend of fewer on-scene firefighting personnel becomes more widespread, the use of large-diameter hose is increasing as departments attempt to find a less labor-intensive method of efficiently moving large quantities of water. Departments are able to move more water to the fire by laying a single large-diameter line than was formerly the case with multiple lines of 2½" or 3" hose. This makes firefighting operations more efficient, since large amounts of water are rapidly made available on the fire scene. Large-diameter hose construction is similar to that used in smaller lines intended for attack use. The major difference is that proof and service test pressures are reduced as the need to keep the hose light and flexible dictates the use of lighter materials.

Attack hose must have the ability to counter the operational effects of wear and heat. This requirement can make their construction rather bulky. Supply hose, on the other hand, needs to be light for ease of deployment and should pack compactly so the required amount can be easily carried in standard pumper beds.

As hose diameter increases, the need for high pressure ratings decreases. To move 1,000 GPM 500', only 40 PSI engine pressure is needed with 5" hose while 400 PSI would be needed with 3". As matter of fact, a 1,500 GPM pumper, supplying its rated capacity at

150 PSI, can move that capacity over 800' through 5" hose. If the pressure is increased, the volume decreases. However, since the pumper's highest rated pressure is 250 PSI, it is not practical for the pumper to exceed that pressure in actual operation. With this limitation in mind, we can see why high proof pressure ratings are not as necessary for supply lines as they are for attack lines or smaller diameter supply lines.

Probably the most common coupling for large-diameter hose is the one-third turn Storz device. Invented in Germany years ago, this coupling was quickly adopted by the fire service on this side of the Atlantic because it offers a number of operating advantages not found in traditional screw thread couplings. Its first advantage is that it is sexless. Each coupling will fit into another, eliminating the need for double male or female adapters. It offers the advantage of being coupled and uncoupled quickly without the need to lift the heavy water-filled hose up off the ground, as is needed with screw thread couplings.

Large-diameter hose is manufactured in three basic constructions: woven double jacket with rubber or thermoplastic liner, plied rubber covered, and extruded rubber covered. The choice of which hose to purchase will depend mainly on packing characteristics, dealer and manufacturer support, and price.

The easiest hose to pack and the flattest folding is the double-jacket construction. This advantage tends to be offset somewhat by its wearability and heat-resistance abilities, which may be somewhat less than rubber-covered hose depending on construction and type of protective coating. These may or may not be considerations depending on the hose's intended use (fig. 9–15 and fig. 9–16).

Figure 9–15. The fireground is not the place to learn how to handle large-diameter hose. Keep in mind that kinks drastically reduce available flow. (Photo courtesy of Jim Regan)

Figure 9–16. And the pump operator wondered why he was not getting enough flow. All operating personnel must be trained in and then continually practice fireground use of large-diameter hose and appliances. (Photo courtesy of Jim Regan)

Plied construction is more flexible and folds flatter than extruded construction. Both rubber constructions are approximately equal in wearability, and both contain the previously described advantages and disadvantages inherent in all rubber-covered hose.

In actual practice, the selection of large-diameter hose construction usually hinges on price and dealer support. It is wise to evaluate all three styles before making a purchase. Surveys have indicated all three types can provide satisfactory service.

Large-Diameter Flow Systems

A large-diameter hose itself must be considered as only a single component of a high-volume supply system. This system also consists of adapters, valves, and pressure relief devices, all of which must be matched to insure proper performance and safety. These components are separated into two categories: supply and discharge. In the supply category are those fittings needed to complete a hydrant to fire lay. A large-diameter hose is used to designate either Storz or threaded fittings. They include

- Hydrant fittings, both large steamer to large-diameter hose and 2½" to large-diameter hose. A relay valve can be used here if needed.
- Special spanner and hydrant wrenches.
- Incoming relief valve on the pumper. This valve is usually gated and the most important component of this fitting is the relief valve.

On the discharge side, the following fittings will be needed:

- 3" or larger piped discharge from pump
- Installed or removable adapters from the Storz connection to threaded connections on the device being fed

- Distributing device, portable hydrant, or multiple discharge gated wye, preferably with relief valve

The inclusion of relief valves is the most important consideration when designing a large-diameter hose system. Great weights of water travel at rather high speeds when large-diameter hose is in operation, much more than if 2½" or 3" hose is used. If this water is suddenly stopped, say when a common quarter turn butterfly inlet valve is slammed shut, tons of water must come to an instant halt. For example, if 1,500 GPM is flowing 1,000', 8,508 lb. of water (over two tons) is moving at a speed of 17 mph through the hose. It certainly cannot be expected that about ³⁄₁₆" of hose wall will safely contain that water if it suddenly and violently wants out because a valve was slammed shut, creating an enormous water hammer.

Built-in pumper relief valves can relieve only up to about 750 GPM. When they operate, that flow is passed around from the discharge side of the pump to the inlet side. While pumper relief valves will provide some help relieving pressure on the discharge side of the pump, they are of relatively little value if tons of pressure and water weight are suddenly applied to the intake side. Governors and pressure controllers simply throttle down the engine to cope with pressure surges. Since they do not physically relieve any excess intake pressure, the entire pump can rapidly become over pressurized (fig. 9–17).

NFPA 1901 requires that pumpers have intake relief valves permanently installed in the supply piping. These offer excellent protection if they are adjusted properly. In actual use, many departments, annoyed by their constant dribbling, tend adjust the pressure setting upward so that the valve will only open after dangerous pressures have already developed.

Figure 9–17. During operations with large-diameter hose, water will flow from the vent port of the relief valve to relieve excess pressures which could damage hose or pumps. It is a good idea to mount this valve away from the pump operator's panel so the operator will not have to straddle the large hose or be injured by the vented water.

It is important that these valves are set only slightly higher than normal hydrant or incoming supply pressures.

Engines using large-diameter hose systems should always be provided with incoming gated relief valves on the pump's large suction inlets. When large-diameter supply lines are in use, and if a large enough water volume is being discharged, they will stay shut. If only a small amount of water is being moved, they will, of course, open to relieve excess incoming pressure. This venting should not be viewed as a nuisance but as a reminder that dangerous intake pressures could develop if the valve is not operating properly.

Since they will rapidly act and dump excess water overboard, they do a fine job of keeping a water hammer from ever reaching the pump. Because of the damage to personnel, hose, and apparatus that water hammer can cause, the use of incoming gated relief valves on pumpers and discharge relief valves on manifold devices should considered as mandatory.

When planning for a large-diameter hose system, it is important to plan not only for the hose, but also for all the fittings that will be needed to make the system operational.

Inspection and Maintenance

No matter what materials are used in the construction of double-jacket hose, it should be cleaned and dried after use to help prevent mildew formation and to help prevent reduced service life from dirt, grit, and glass bits abrading the hose's outer surfaces. Mildew will not attack synthetic yarns, but it will cause foul odors and may cause the fibers to hold on to dirt, which, in turn, will increase internal wear. If rubber-covered hose is not wiped down, water lying between folds can generate a slime which, while not being damaging to the hose, can make it slippery.

Because it forms a substantial part of the firefighter's life-safety support system, hose should be afforded the same care as protective clothing and breathing apparatus. It should be cleaned and inspected after each use, and all hose should be tested annually in accordance with the procedures described in NFPA 1962 (fig 9–18).

All lengths should be inspected during and after use for any tipped or loose couplings and for indications of water weeping through the outer jacket. This weeping, appearing as tiny drops that reappear when wiped away, indicate that the tube may be damaged. Rubber-covered constructions contain tiny weep holes in the outer jacket designed to vent any water leaking from the inner jacket to the outside of the hose. Hose with tipped or cocked couplings or showing signs of weeping should be taken out of service. Many fire officials are quick to condemn a length of hose because of a worn or frayed outer jacket, which has little effect on the hose's performance, while ignoring signs of serious internal failure.

There appears to be a trend away from departments repairing damaged hose because of liability problems in case of subsequent rupture or coupling loss at a fire. Many departments are now returning damaged hose to the manufacturer or a qualified third party for recoupling and testing rather than attempting the job themselves (fig. 9–19).

Departments are wisely having aerial ladders tested and recertified yearly. If problems are found, the ladder is returned to the manufacturer or qualified dealer to have the problem corrected or is permanently taken out of service. Should hose, the failure of which can endanger suppression crews, cause serious injury, or loss of life, be treated any differently?

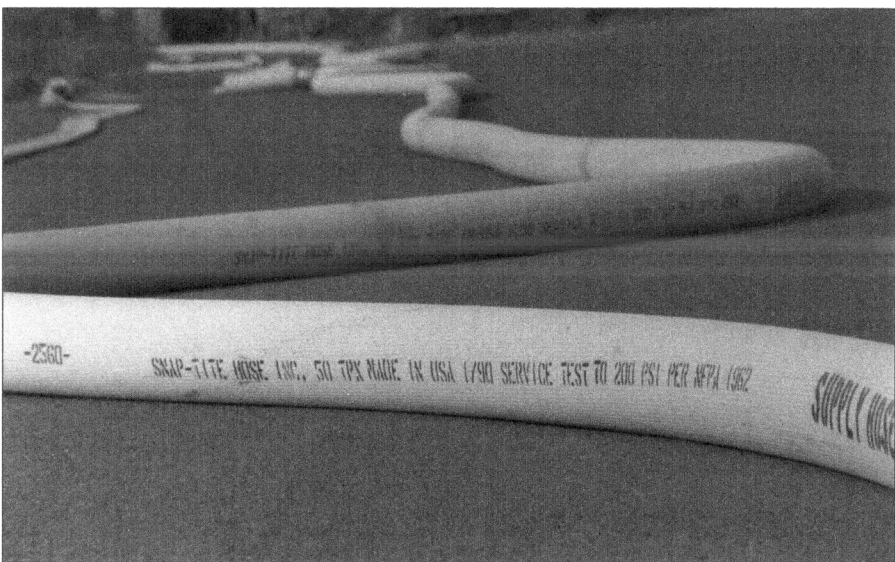

Figure 9–18. All new hose has the service test pressure rating marked on the hose. In this case, the hose should be tested annually to no less than 200 PSI. The service test pressure rating is half the acceptance test pressure rating.

Figure 9–19. While mildew will not attack synthetic yarns, it will degrade hose strength over time and, besides being unsightly, becomes slippery when wet. Here is a classic case of a department not properly maintaining its life-saving equipment.

I think not. We have the responsibility of creating the safest environment possible for suppression crews engaged in firefighting operations. With thousands of dollars spent on protective clothing and breathing apparatus for each firefighter, does it make sense to waste all this protection when the crew has to depend on a length of hose pieced together with a homemade repair (fig. 9–20)?

Figure 9–20. Hose must be tested each year to help determine if it is capable of supplying water for firefighting. Careful inspection should be made of outer surfaces and couplings for any signs of damage, slippage, wear, or leaking. Suspect hose should be taken out of service and either repaired by the manufacturer or replaced. Just because a length holds service test pressure does not necessarily mean it is fit for continued frontline service.

Chapter 9 Review Questions

1. What are the three basic components comprising municipal fire hose?
2. What is the purpose of the inner jacket?
3. What is the purpose of the warp yarns?
4. How is fire hose burst pressure determined?
5. What is the most common method of increasing the protection of the outer jacket?
6. What are the main disadvantages of a lightweight, thermoplastic tube?
7. What are the three types of construction used in large diameter hose?
8. What is the NFPA standard outlining hose testing procedures?

10

Class A Foam

Back in 1977, a new chemical intended strictly for Class A firefighting was introduced into service in rural Texas. It has since evolved into what is proving to be one of the most dynamic advances in wildland fire extinguishment since the development of woven jacket hose. This new generation of firefighting chemicals is collectively called Class A foam, a somewhat misleading term because when the agents are used for interior firefighting with standard hose and nozzles, they actually produce little foam. When Class A agents are added to water, adherents say that the resulting solution increases knockdown and holding potential anywhere from 3 to 15 times over plain water alone. Sound too good to be true? Read on.

As Class A agents have come into widespread use in the past 30 years, radical changes have been made in wildland firefighting tactics. Structures, in the past doomed to destruction in the path of uncontrolled brush and forest fires, are surviving cloaked with a thick, insulating cover of white, fluffy agent. Clearing firebreaks through brush and duff, once an extremely labor-intensive activity, is being replaced with a wide strip of Class A foam, which is proving almost as effective as bare ground in preventing fire spread (fig. 10–1).

In a presentation at a wildland firefighting symposium in 1987, Ron Rochna and Paul Schiobohm, fire service specialists at the Department of the Interior's Interagency Fire Center and two of the then-most experienced experts in the development and use of Class A foam, reported on the results of their extensive research.

Performance evaluations were made for direct attack, indirect attack, and mop-up. Class A was determined to suppress and repel fire in situations where water would not. Cost comparisons of mop-up work showed that straight water to be significantly more expensive than foam as more of it had to be used to obtain the same results as a Class A/water mixture. They predicted that Class A's future would be to replace all the then-current plain water applications and present new suppression opportunities to the fire management communities.

Basically, Class A agents extend the life and effectiveness of water by reducing the surface tension of water molecules enabling greater penetration into solid materials.

The notion that water is free is a fallacy. The Bureau of Land Management fights most of its fires where water sources are miles away. They made a statement that $20 can make 500 gal. of water into 5,000 gal. of effective water when the additive was used.

In a report in 1988, after more testing and actual experience using Class A agents on actual fires, they said, "No evidence exists that water by itself is the best way to extinguish fire."

Figure 10–1. A test fire being attacked with Class A agent using a compressed air foam system (CAFS) handline. Knockdown is considered average to good, depending on who you are talking to.

Water has been perceived for years to be cheap or free. This is not true. Hydrants are not free. Time and equipment spent hauling water is not free. The public has been misled for years to believe that a fire that cannot be put out with water is more fire than people can handle. This is not true, and we should not continue to support this misunderstanding.

While first perceived as useful only in wildland firefighting, Class A agents are now proving their worth for structural fire attack as well. In August of 1990, a series of controlled tests were conducted by Missouri fire instructors in a vacant motel at Sikeston. Fires were set in four identical rooms in the one-story building to compare the results of attack with both plain water and Class A agent solution. The building provided an excellent comparative proving ground as each fire was identical in instrumentation and included identical furniture loading and placement within each room.

The fires were allowed to develop in intensity until the ceiling temperatures caused flashover at approximately 1,000°F. The fires were then attacked until temperature readings were reduced to approximately 150°F. The fires were not completely extinguished to allow evaluation of rekindle times. To more fully evaluate Class A agents' effectiveness, low flow rates of between 30 gallons per minute (GPM) and 48 GPM were used to prevent high water application rates from influencing knockdown times (fig. 10–2).

Researchers found that by using Class A agents, the fires were knocked down in 29% to 52% less time than when using plain water alone. More importantly, especially for rural firefighting where water is scarce, the total amount of water needed for knockdown and overhaul was drastically reduced when Class A agents were used—242 gal. of plain water compared to 77 gal. of treated water, for example—in one test fire.

Figure 10–2. In August of 1990, a series of tests using Class A agent on a number of identical fires was conducted in Sikeston, MO. The agent was applied using standard nozzles and tactics. In the photo, the coating ability of the agent is evident on the shingled overhead. By painting the exposed area, the fire's spread was prevented before direct attack was made on the room. (Photo courtesy of Gary D. Rawlings)

History

To better understand how the agents achieve their superior firefighting qualities, it is helpful to trace Class A extinguishing agent development from its beginnings.

Around the mid-1970s, the Texas Forest Service began experimenting with a byproduct of the paper manufacturing process called soap-skim. This foamable residue is skimmed from kraft paper liquor vats and comes from the mills as a dark brown, sticky material resembling axle grease. It must be diluted with water to make a pourable concentrate. The mills were looking for a way of getting rid of the unwanted goo about the same time Texas officials were looking for a method of improving wildfire and rural fire protection across their vast state.

It was found that if this solution was added to water in about an 8%–9% concentration, similar to the amount needed to produce Class B foams, the water's wetting ability was increased, proving useful for penetrating and soaking brush, wood, and charred surfaces. It was also discovered that if compressed air was injected into the hoseline, the water–concentrate mixture would foam up and exit the nozzle appearing like snow. The first units to be put into service were called the Texas Snow Job.

The first operational units delivered their agent by utilizing the force of compressed air pressurizing a premix agent tank. This unit appeared and operated almost exactly like a large dry chemical fire extinguisher. Later devices used an engine-driven air compressor and positive-displacement pump to deliver the water/concentrate/air solution to the nozzle.

Live fire testing in all types of structures indicated Class A agents could provide quicker knockdown times than plain water and can keep the fire from reflashing for long periods after knockdown, making entry into the fire area much safer for interior crews.

Reports from firefighters indicated that the delivered product exhibited increased firefighting ability characteristics when compared to those of plain water. Most notable was the finding that the solution penetrated burning materials more thoroughly and more rapidly than plain water. The foam, which resembled shaving cream, would cling to vertical surfaces, allowing time for the water it contained to penetrate the surface underneath. Because the foam entrains quite a bit of air in its bubble structure, it formed a thick insulation blanket that proved quite effective in preventing a fire's spread to surfaces it covers. Another advantage was that knockdown times were achieved more rapidly than when plain water was used, multiplying the effectiveness of limited water supplies.

While initial results were encouraging, there were some drawbacks. The original systems developed in Texas used premix water–concentrate solutions in either pressurized or booster-type tanks. One limitation to their use in the field was that they had to be batch recharged after use, similar to recharging a fire extinguisher. This made it difficult to put the unit back into service without returning to quarters.

Because a large amount of agent was needed to treat the water, transportation and storage could become a problem during an extended operation in the field. While soap-skim was readily available in areas surrounding paper mills, it is not common in other parts of the country. In addition, while the soap-skim from kraft paper operations is biodegradable, the skim from white paper operations contains chemicals that make it toxic to the environment, precluding its widespread use for wildland firefighting.

As word of the soap-skim/compressed air combination's effectiveness was made known throughout the wildland firefighting community, the Bureau of Land Management, U.S. Forest Service, and California Department of Forestry began extensive testing and experimentation with soap-skim as well as a number of similar agents and delivery methods. These early tests with soap solutions and, later, detergent additives, proved instrumental in the development of the present generation Class A agents.

Development of a Suitable Agent

The Texas project demonstrated the dramatic increase in fire-killing effectiveness when agents were used on Class A fires. However, the soap-skim agent had additional drawbacks that needed to be overcome before more efficient application equipment could be developed.

Two of the soap-skim's major drawbacks were that it would not form an effective firefighting foam unless compressed air was injected into the stream, and a high concentration of agent (6% or more) had to be mixed with water to form an effective working solution.

Foam manufacturers introduced a new generation of Class A foam concentrates in the mid-1980s that contained detergents in place of soap to increase the water's wetting ability. Foaming agents were also added that allowed the solution to entrain air without the need for compressed air injection. Concentration levels of the desired chemicals were increased so the new agents could provide an effective firefighting solution when mixed at a rate of 0.2%–1% of concentrate to water. This means that for normal firefighting, 1½ gal. of Class A agent will treat a 500 gal. booster tank, as compared to 30 gal. of 6% concentrate. This means less concentrate needs to be carried on board and that the cost per treated gallon is extremely low compared with other agents.

These new concentrates easily formed a lightweight foam with conventional structural water application equipment without the need to inject compressed air. The newer liquids were also fully biodegradable and caused no damage to the environment.

What Are Class A Agents?

When Class A agent concentrates are mixed with water, the resulting solutions can be described as water in an improved form. Class A foams are quite different from Class B foams in the way they chemically suppress fires. Class B foams use water as a carrier to get the foam solution to the fire. In the case of aqueous film-forming foam (AFFF), when the foam contacts the surface of hydrocarbon products (i.e., gasoline, fuel oil), a film is formed over the surface of the liquid to prevent the further release of flammable vapors. In the case of protein foams, the foam blanket itself acts as the sealing barrier. With both types of Class B foams, water is needed to provide the proper foam solution ratio and to provide the carrying medium to get the foam solution to the fire. Water provides a cooling benefit, but it is the expanded foams themselves that suppress the release of flammable vapors.

With Class A agents, the reverse is true. Water, not the foam concentrate, is the major extinguishing medium. The water does the work, but the agent helps it to get where it needs to go and keeps it there long enough to do its job. Class A bubbles increase the heat-absorbing surface area of the water droplets, and the water is fortified by the addition of powerful detergent wetting agents called *surfactants* which reduce the solution's surface tension, allowing it to penetrate deeply into solid materials.

As commonly used for structural firefighting, applied with conventional nozzle equipment, Class A agents actually generate little foam. However, standard nozzles provide enough air mixing to make the solution effective. The present generation of wetting agents used in Class A concentrates are unlike the older wetting agents. They are designed to form effective solutions using less concentrate and have much greater penetration power. These penetrating agents, blended with additives designed to produce stabilized foam when mixed with air, can be effectively used at low temperatures, are biodegradable, and are environmentally safe.

One key to the agents' effectiveness is the reduction of surface tension within water molecules. When added to water, the water molecules are spread apart by the Class A concentrate's surfactant molecules. When air is introduced into the solution through a combination nozzle, foaming agents in the concentrate allow the combination of water and surfactant molecules to form bubbles, increasing the surface area of the water available to react to heat. A mass of Class A bubbles contains water to cool the fire, a surfactant to allow the water to penetrate burning and unburned material, and air, which makes the mass lighter in weight and provides an effective heat insulating medium.

In an earlier chapter, I reported on research conducted by the Osaka Fire Department in Japan to determine the most efficient water droplet size for firefighting. Testing by the Osaka Fire Department as well as the National Institute of Standards and Technology determined that droplets in the 250 micron–350 micron size range were more efficient in absorbing heat than larger, heavier droplets. Unfortunately, the tiny particles are so light that they are almost impossible to apply safely using manual firefighting methods. Research has determined that the stream of Class A agent, when directed into a fire, will begin to penetrate through the fire's heat. Then, as the air within the bubble becomes heated, the bubble will pop, fracturing the water solution into extremely tiny particles. The actual size of these particles is almost impossible to measure; however, it is felt that they are so tiny that they convert immediately to steam and will absorb heat at the same or a more rapid rate than the 250-micron ideal droplet size.

Researchers believe that Class A agents can allow the vehicle to deliver a more efficient droplet size into the flame/fuel interface area without having the droplet evaporate en route, as would be the case with plain water, small droplet application.

It can be seen that Class A agents are not one-trick ponies. They increase the fire-killing power of water by providing a number of extinguishing advantages:

- The surfactants reduce water's surface tension, providing immediate penetration of burning materials.
- The same surfactants emulsify grease, petrochemicals, paints, and other barriers to water penetration. Class B agents do not contain emulsifiers. If they did, the foam blanket would sink to the bottom of the flammable liquid instead of floating on the surface.
- Class A agents can generate a lightweight foam which, by defying gravity, allows water to stick to three-dimensional surfaces for a longer period of time, giving the solution more time to penetrate and cool the surface.
- The foam provides a thick insulating barrier of air and water that can help protect exposures from radiated heat.
- The white foam provides a reflective surface, increasing radiant heat protection.
- Since each bubble has a hollow center, it absorbs heat faster than a water droplet with a solid center.
- The combination of lightweight foam, water, and surfactants form a barrier at the flame/fuel interface that suppresses the liberation of flammable vapors being distilled from burning materials.

In practical firefighting terms, these figures can conservatively translate, for example, into a mini pumper with a 250 gal. booster tank having an estimated effectiveness potential of a pumper carrying 750 gal. of water. If a fully involved room could be knocked down in 45 sec. using plain water, the use of Class A solution will black out the fire in about one-third of the time. In terms of increasing firefighter safety, there will be much less exposure of suppression personnel to the effects of heat and combustion products by using Class A agents. If an interior attack is made with only treated tank water, the additional knockdown power of agent solution will mean more backup water will be available for handling undiscovered additional fire and for overhaul.

Putting Class A Agents to Work

Class A agents are provided in a concentrate form that is mixed with water to form a working solution. There are three basic ways of mixing or proportioning Class A agent solutions:

- Premix or batch mixing

- Discharge-side eductor or proportioner

- Positive-pressure injection metering system

Each method has its advantages and disadvantages, but all serve the purpose of increasing the firefighting potential of water by the addition of Class A agents. A department wanting to explore the benefits of Class A additive use initially does not have to invest in anything other than the concentrate.

One positive characteristic of Class A agents is that they are added to the water stream at extremely low percentage rates. These rates begin at 0.2% and can go as high as 1% depending on the intended use of the finished product, water hardness, and type of application equipment. Any higher percentage rate does not make the agent more effective; it just wastes concentrate.

At 0.2%–0.6%, treated water will exhibit extremely high penetration qualities, much more than old-style wet-water additive. If used with air induction nozzles or compressed air injection systems, the finished product will leave the nozzle as a snowy-looking foam. The finished foam can either be runny (similar to AFFF attack solutions), thick and creamy (similar to shaving cream), or extremely dry and fluffy (similar to the highly whipped cream on the top of a pie), depending on the nozzle and if compressed air is injected. The solution's penetration qualities can differ depending on how much air is entrained in the finished product. One advantage of entraining air is to increase the foam's insulating qualities and extend the drain time (how long water will be retained in the foam blanket), providing increased surface wetting and reducing the risk of ignition or reignition.

Class A agent solution can be mixed with water in four basic ways:

- Premix or batch mixing

- In-line eductor

- Balanced-pressure (bladder type) injection system

- Pump panel mounted positive-pressure injection system

Premix or Batch Mixing

The simplest way to use Class A agents is simply to pour them into the onboard water tank to premix them with initial attack water. This method has the advantage of being simple, easy, and accurate because the exact amount of concentrate needed to form an effective solution is known beforehand. Since the only cost involved is the price of the foam concentrate, premixing is the least expensive method of introducing the use of Class A agents into most departments' operations.

After adding the concentrate, it is wise to use the fire pump to circulate the tank water to ensure that the agent is properly distributed. A word of caution: leave the pump idling when circulating the tank water. If higher pump pressures are used, the agent will do what it was meant to do and produce foam which will blow out the tank fill tower and overflow pipe in great quantities. In actual use, it is simply an annoyance to wash away as the residue will continue to generate more bubbles when hit with a hoseline.

Because the concentrate is completely synthetic and contains little solid matter, it will remain in solution without dropping out over time. One drawback to premixing is when the tank water is depleted, there is no way of continuing the supply of foam solution unless additional concentrate is poured into the tank before refilling. If water is being supplied from a secured source such as a hydrant, tank-mixing may not be practical for extended operations. For structural use, the treated water is discharged through standard hoselines and nozzles, which usually generate a frothy, wet foam. This foam will coat and surround fuels with a layer of water, decrease knockdown times, and allow for the foam solution to be quickly distributed around the fire area. As with plain water use, both smoothbore and combination nozzles will work well and no change in tactics are needed. Standard water distribution techniques of bouncing the stream off walls and ceilings will help mix air with the stream and generate a limited amount of actual expanded foam. Standard combination nozzles will provide about a 2:1 foam to air expansion ratio. Smoothbore nozzles are excellent for use with Class A agents for interior attack operations without limiting their effectiveness for penetrating areas of intense fires. The addition of Class A agents to smoothbore attack will provide extremely quick knockdown of hot fires and will provide the added advantage of helping prevent reignition of extinguished solids.

The use of combination nozzle when producing Class A foam will produce a foam having anywhere from a 5:1 to 10:1 ratio (the amount of air to foam solution), depending on design. A thicker foam is extremely effective for coating exposures or covering a three-dimensional fire, such as a pile of tires, allowing the water more time to penetrate. An added advantage is that blowholes will form in the foam blanket after the complete foam coverage of an area, making it easier to locate hot spots.

Rather than premixing solution in the onboard water tank, many rural departments carry containers of Class A agent on their tankers for on-scene mixing. The concentrate is poured into portable drop tanks before the tanker discharges its load of water. The agent mixes into solution by the action of the offloading water, allowing the pumper to draft the treated water and deliver it to the fire. This method will also cause some frothing but, because it is restricted to the area surrounding the drop tanks, the froth will have little effect on water delivery operations. Care should be taken to prevent the froth from being suctioned into the pump as it could cause loss of prime.

The surfactants contained in Class A concentrates are highly effective degreasers, and there has been some concern about premixed solutions attacking pump packings. Hundreds of departments have been using premixed solutions for years without any abnormal deterioration of packings in large fire pumps; however, more frequent inspection of pump packings and valve operations would be a common-sense method of preventing maintenance problems. Many newer fire pump designs use mechanical seals or Teflon-impregnated packing rather than old-style graphite-impregnated asbestos packing. These seals are not seriously affected by the detergent action of Class A agents. While older methods of pump shaft sealing may lose some efficiency when foam solutions are run through the pump, normal packing adjustments should handle the problem. Because the foam solutions are effective in dissolving grease and oil, some increased maintenance can be expected in valves and fittings as lubricants are broken down. This lubricant dilution is not a major problem, but if foam use becomes frequent, one of the other two mixing methods should be considered.

Discharge-Side Eductors and Proportioners

The next step up from premixing foam concentrate with water is to educt the solution into the water stream by using a discharge-side foam eductor or proportioner, similar to those used for Class B foams.

The major difference between a Class A and Class B eductor is that the metering valve on the Class A unit allows for lower proportion rates. While eductors are simple and easy to use, they may have a problem providing the proper proportioning rate at lower temperatures. This is because low temperatures increase the viscosity of the concentrate, making it difficult to pass through small metering valve openings. In addition, all in-line proportioners must have nozzles matched to their flow rates or excessive backpressure prevents the pickup of the concentrate. There is also a limit to the amount of hose that can be stretched on the discharge side of the eductor. Too much hose will again result in increasing the backpressure, not allowing the eductor to pick up the concentrate.

Since most attack work around the country is done with preconnected lines, the limitations of hose length and nozzle size should not cause a problem. The eductor can be left attached to a discharge and agent concentrate containers stored in a compartment nearby. If it is decided to use Class A agents for attack, it is an easy procedure to place the pickup tube into the concentrate container to provide treated solution as soon as water begins to flow. Class B eductors can be used for Class A agents if the proportioning meter can be dialed to 1% or below. While eduction at 1% will provide more concentrate than is actually needed, it will do little harm other than increase concentrate costs. Some eductor manufacturers offer replacement metering valves that are calibrated to proportion at low percentages and that can be easily retrofitted into Class B eductors presently in service.

Balanced- and Positive-Pressure Injection Metering Systems

An accurate method of proportioning Class A concentrates is by using discharge-side, balanced-pressure, or positive-pressure injection metering systems. Introduced to the fire service around 1988, these devices utilize a foam-filled bladder inside a cylinder. When charged, the water pressure inside the cylinder squeezes the foam through metering devices to inject the proper amount of foam agent into the discharge stream. Because the concentrate solution is delivered under pressure, the device will function through a wide range of operating pressures and flow rates. This system is simple in design yet is extremely accurate in supplying concentrate to the water stream in the desired percentages. The bladder system is available in various sizes; these are usually refilled with a small hand pump, delivering concentrate into the device after the water pressure from around the bladder is relieved. Although water flow is not affected if the bladder empties, the foam flow is stopped until the bladder is refilled and repressurized. Bladder systems are the simplest balanced pressure injection systems presently available, require no electrical power for operation, and are widely used in wildland operations.

The fourth style of solution metering is by the use of a positive-pressure, discharge-side injection system, which is permanently mounted on the pump and uses an electric pump to meter the foam concentrate into the water discharge system. While these units are more expensive than other systems, they are extremely convenient to use and are very accurate. Another advantage is that they can deliver foam concentrates in different proportioning percentages. If an engine is equipped with both Class A and Class B tanks, the pump operator can easily change the injection percentage to deliver accurate flow.

In actual practice, positive-pressure proportioning systems require little attention other than turning them on by activating an electrical switch and opening the tank supply valve. The desired proportioning percentage can be preset before use. Normally, these devices are permanently installed on the apparatus and are fed directly from a tank of concentrate, though some smaller portable units are available. If a department plans to extensively use Class A agents for fire attack and exposure protection, a positive-pressure proportioning system can be a worthwhile investment since it makes agent management rapid, accurate, and cost effective.

Compressed Air Injection

The original Class A agent equipment developed in Texas used compressed air to cause soap-skim concentrates to foam. During later testing and operational use, two schools of thought developed; one preferred the application of Class A solution through standard hoselines and nozzles, while the other preferred the injection of compressed air to the solution before application. While both air injection and noninjection delivery systems offer more knockdown and sealing power than plain water, the use of compressed air has proven most effective in wildfire applications.

In practice, when using a compressed air foam system (CAFS) for air injection, concentrate is proportioned into the water stream by a positive-pressure proportioner. Downstream from the concentrate entry point, compressed air is injected at a rate of approximately 1 cu. ft. air per 1 gal. water. When the air and foam solution is mixed, it immediately forms an extremely light and fluffy product that is propelled through piping and hoselines by a combination of water and pneumatic energy. This action causes radical differences between plain water/agent solution and water/agent/air foams.

Compressed air foam displays the following qualities:

- Since the finished foam consists of over 50% to 67% air, it is extremely light and normally has the consistency of shaving cream. Hoselines are literally filled with air and have little weight other than the hose itself. Advocates say that this translates, for example, into a 1¾" CAFS line having a 70 GPM/80 cu. ft. per minute (CFM) flow rate, generating about the same amount of agent as a 1¾" plain water line flowing 150 GPM.

- Air horsepower, along with water pressure, comprise the energy that moves the product through hoselines, so the product's light weight, coupled with the high horsepower supplied by the pneumatic system, will propel the foam long distances with little effect from friction loss. The water/foam/air mass does have a boundary layer and center stream turbulence; however, overall friction loss is less because of the stream's lightweight pneumatic component. Because of the CAFS stream's lessened weight and density, there is little pressure loss due to head weight when the nozzle is elevated above the pump/compressor unit.

- The nozzle needed on the discharge end consists simply of a ball valve shutoff with a large tip. Common on 1¾" CAFS lines are 1¼" tips which deliver about a 70 GPM/80CFM of solution/air mixture. Depending on air pressure, air quantity, and size of hose, the stream will have a throw of between 80ft. and 180 ft. Flowing CAFS foam through a combination spray nozzle will strip a large portion of air from the stream, breaking down the foam's bubble structure, and is not recommended. CAFS application must be gentle and is best made with smoothbore nozzle equipment. CAFS can be applied with great effectiveness through a rotating cellar nozzle which will distribute large foam flakes in all directions. This makes it easy to apply Class A agents into attics, blind spaces, or basements that cannot be entered directly.

- The fluffy foam provided by CAFS is excellent for exposure protection and should give effective protection to treated surfaces for a minimum of 1 hour–2 hours in hot weather when properly mixed and applied.

- The bubble structure of CAFS is smaller and more uniform than bubbles formed by aspirating nozzles and is considered more effective by some users in separating the flame/fuel interface during extinguishment.

There are also some disadvantages to using compressed air foam systems:

- The air compressor must be able to supply 1 cu. ft. of air per 1 gal. of water. If a 20 CFM compressor is used, then the water flow should not exceed 20 GPM. While Class A foams increase the effectiveness of water, water flow rates such as those described in chapter 2 may still be needed to safely control interior structural fires. If an area calls for a flow of 95 GPM, then 95 GPM plus 95 CFM of air is needed to properly control that fire with CAFS Class A foam.

- Large-capacity air compressors supplied with CAFS systems usually powered off the fire pump, and its injection hardware are expensive and can be a drawback to CAFS use on general structural firefighting. Costs of a CAFS installation on a structural pumper can easily add $70,000 or more to its total cost.

- The large air compressor will need a certain number of revolutions per minute to operate properly. When supplied from tank water, this is usually not a problem. But when a hydrant supply line or relay hoseline from another pumper is connected directly to the pump suction, the incoming pressure will increase the pump pressure supplying the line. The pumper's pressure controller, if properly set, will reduce the engine's revolutions per minute to keep the line pressure at the required setting. This action may reduce the compressor's CFM outlet whereby less air is supplied, reducing the effectiveness of the CAFS stream. To remedy this, proper operation of a

CAFS system calls for all supply lines to be run through the tank. There are valves available that will shut off the supply line flow when the tank is filled and open when it reaches a predetermined level. This will have a material effect on operations if both CAFS and high-volume streams are wanted to be supplied from the pumper.

- The operation of compressed air units in conjunction with fire pumps will double the amount of operator calculations necessary to produce effective fire streams. If your operators have trouble understanding hydraulics, they will certainly be baffled understanding the principles needed to operate a combination of hydraulic and pneumatic devices. Automatic foam and air injection control systems are now available which help reduce this problem.

- When the ball valve at the end of the hoseline is shut off, pressure builds in the hoseline as the air/foam mixture is compressed. If care is not taken when the nozzle valve is opened, the high-pressure air slug, resulting when this compressed energy is suddenly released, can tear the hoseline from the operator's grip and whip it dangerously about. It is fairly easy to handle an air slug by slowly opening the nozzle to gradually vent off the energy contained in the line by the compressed foam/air product. If a hoseline bursts or a coupling blows off, the increased pressure of the moving force will cause the broken ends to whip about in a much more dangerous manner than with a standard hoseline.

- Foam produced by the CAFS can be too dry if too much air is injected into the water stream. Even though the stream appears effective, it cannot release enough water to wet the material it needs to cover. When flames reach the foam blanket, the dry material underneath will ignite. Operator experience is necessary to determine the proper water/air/concentrate mixture to provide an effective blanket that will stay on the material but deliver enough water to prevent ignition.

- At a ratio of 1 gal. water:1 cu. ft. air, there will be seven times more volume of air than water in the finished product. This air actually blows up water droplets, making more surface area available to react to heat. Some claim that a 30 GPM waterflow rate when used with CAFS can be as effective as 200 GPM of plain water applied by conventional means. Similar claims were made 70 years ago for high-pressure fog. Experience later proved that flow rate, not pressure, is what extinguished the fire. High-pressure delivery may have increased distribution effectiveness but put out little more fire than the same gallonage delivered at normal pressures. While Class A agents increase knockdown times and help seal burning surfaces more efficiently than plain water, exaggerated claims for the foam's efficiency should be investigated closely (fig. 10–3).

Despite CAFS foam's drawbacks, most of which are centered on efficient operation of its generating equipment, its advantages can outweigh its disadvantages, especially in wildland firefighting. It has enjoyed limited success in preventing the spread of wildfires, especially to structures adjacent to burning wood and brush lands. It also has many supporters, especially for some in western states, who advocate its use for interior and exterior structural fire attack (fig. 10–4).

Chapter 10 **Class A Foam** 247

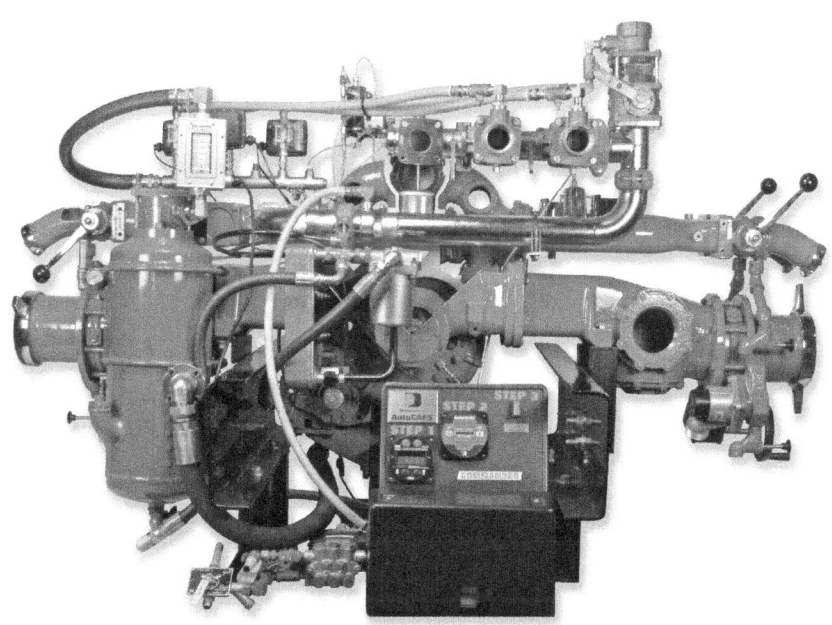

Figure 10–3. A CAFS system adds an air compressor with most installations driven off the pump driveline system, quite a bit of plumbing, and additional valves and sensors to allow the air to be injected into the discharge plumbing. (Phto courtesy of W. S. Darley)

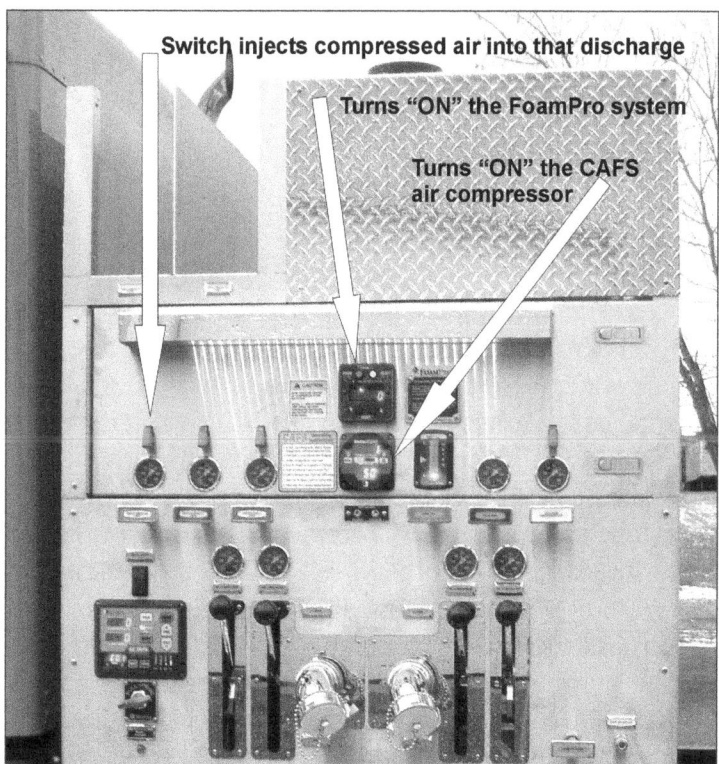

Figure 10–4. A pump panel on a typical CAFS-equipped pumper includes a foam proportioning system, air compressor control system, and switches that allow compressed air to be injected into individual discharges. (Photo courtesy of W. S. Darley)

Using Class A Extinguishing Agents for Structural Firefighting

Once it is decided how Class A concentrate will be mixed with the water stream, it is a simple matter to begin its deployment.

It is not necessary to generate true foam for Class A agents to be effective for structural firefighting. Most manufacturers and users agree that standard smoothbore or combination nozzles will provide effective application. It is important to remember that Class A agents make plain water more effective by increasing its penetration rate and providing more surface area to react to heat. However, while research has indicated that Class A foam can provide spectacular knockdown rates, it has not been proven that it will compensate for inadequate flow rate. If the area involved in fire requires 185 GPM to accomplish extinguishment, a line flowing Class A agent at 95 GPM will have a difficult, if not impossible, time putting the fire out. For example, if the area involved requires 95 GPM for extinguishment and a line delivering Class A solution at 95 GPM is used, the fire will be knocked down faster than with plain water, and the area will be more secure from reignition when the nozzle is shut off.

Increasing Firefighter Safety by Using Class A Agents

One major advantage to using Class A agent rather than plain water is that it provides a dramatic increase in firefighter safety. Let's assume fire is involving a couple of rooms in an apartment building. Entry is made by advancing a line down a long hallway. At the end of the hallway, flames are lapping out of the apartment and beginning to mushroom on the hallway ceiling. If the nozzle crew gives the compartment burst of agent, whipping the nozzle across the ceiling, down the walls, and across the floor, the agent will immediately begin to penetrate the surface material. If plain water were to hit the surface instead, much would immediately boil off. With Class A agents, less steam is created initially as the water soaks in; however, if flames continue to hit the ceiling and walls, the water will slowly steam off, protecting the surface and slowing the distillation of flammable gases.

Once the crew reaches the doorway and sweeps a stream around the room, a number of changes will be noticed over the use of direct attack with plain water. Wall and ceiling panels will immediately turn dark as the solution penetrates the panels through paint and wall coverings. It will also be noticed that almost no water runs off the walls and onto the floor. Burning ceases once the solution contacts burning material. There will also be less steam generated because not all the applied water is being boiled off by the fire's heat. In fact, most water will be absorbed by the material it contacts.

We know that ventilation should be accomplished before the crew advances into the fire area, but this is not the case in many situations. Many times, after the crew has knocked down the fire with plain water, the area will light up again as soon as the space is ventilated. If ventilation is somehow delayed, the use of Class A solutions can decrease the chances of reignition, helping prevent flashover within a fire area because of water and agent penetration into burned and unburned material. Penetration allows water to cool heated material more effectively, preventing the further distillation of combustion gases.

It will be noticed that the immediate suppression of the combustion process will also reduce the amount of smoke present within an involved structure. Ventilation will still be needed; however, firefighters will notice a pronounced increase in visibility in fire areas. Less water will be needed during overhaul to douse smoldering materials, and treated water applied to areas such as partitions and shake roofs will spread rapidly into void areas.

A painting technique can be used in certain circumstances to reduce the fire's intensity and accomplish extinguishment of large areas. The idea is to tackle one panel at a time. Since it is located in the hottest area of the fire, the ceiling should be taken care of first. This is done by working a straight stream back and forth rapidly until the entire panel is coated. This action soaks the panel and helps prevent the release of flammable gases. One wall panel at a time can then be soaked. As each panel is thoroughly wetted, it reduces not only the fuel load but also its radiation ability. A panel penetrated by Class A agent radiates almost no heat and can no longer contribute to the total heat load, helping reduce the chances of flashover.

While not intended to compensate for an inadequate flow rate, removing the fire's fuel by panel soaking does have a cumulative heat-reducing effect. By eliminating heat and fuel piece by piece, large fires can sometimes be successfully extinguished piece by piece, provided the remaining fire does not drive the crew from the area.

After the fire is knocked down, many times crews will notice a radical reduction in the atmospheric humidity normally experienced when plain water is used for attack. Most of the water contained in Class A solutions will have penetrated solid material and not boiled off into the atmosphere. It will also be noticed that the ambient temperature of the fire area will tend to be lower than with plain water attack because solid material has cooled much faster as it is thoroughly wet.

Quick knockdown, securing heated surfaces, reduction in steam generation, and lower temperatures after the fire is blackened all contribute to greatly reducing the amount of time personnel have to spend operating within hostile, dangerous atmospheres.

External exposures will rapidly absorb water when wet with a stream of Class A agent solution, making the surface harder to ignite from radiant heat. Especially in rural areas where water is scarce, it makes more sense to coat an exposure with a blanket of foam than to continue to wet the surface with a continuous stream of water, most of which runs off without effectively penetrating the material.

Another ideal device in which to use Class A agent is in the can. Adding Class A agent to a 2½ gallon water extinguisher will increase its effectiveness and provide a quicker, safer knockdown.

Class A Agents on Class B Fires

It must be remembered that Class A agents work differently from Class B foams. While Class A agents could perhaps extinguish some flammable liquid fires, they have limitations and should not be considered for primary Class B protection.

If the flammable liquid has a relatively high flash point, such as kerosene, its vapor pressure is rather low. That means the rate of flammable vapor production produces low vapor pressure when burning. This low vapor production can be limited, or stopped altogether in most cases, by the application of water spray in such a volume that the burning material is cooled below its flash point. With these types of liquids, Class A agents certainly help in extinguishment since they hold water in place longer, helping it perform its cooling function. They also provide water with more surface area to more quickly absorb the fire's heat.

When a material having a relatively low flash point, such as gasoline, is burning, the fire's heat is generating a much greater amount of vapor at a greater pressure. Water itself has a poor chance of cooling burning gasoline below its flash point because of the large amount of vapor being generated and burned. Class B foams form a barrier over the surface of the burning liquid that can resist high vapor pressures and prevent further production and escape of flammable vapors into the flame/fuel interface.

Class A agents are not designed to resist high vapor pressures. They are designed to amplify water's cooling and penetrating qualities when applied to burning solids. It can be generally assumed that if water by itself can extinguish a fire in a certain type of flammable liquid, then Class A agents will assist in the extinguishment. If water by itself would not be successful in extinguishment of fires in low flash point liquids, the addition of Class A agents will not be of much help and should not be depended upon for extinguishment.

Since automobile fires are basically Class A fires that could involve a limited amount of Class B material (until the fuel tank splits open), Class A agents are extremely effective for extinguishment of vehicle interior fires.

There have been cases where Class A agents have successfully extinguished fires in low flash point hydrocarbon fuels. However, it must be remembered that Class A agents will not provide a suitable seal against vapor escape through the foam blanket. Hydrocarbon fires may be extinguished by Class A agents but the danger of reflash is great. If a large amount of Class B material is involved, switching to Class B agents is the correct course of action.

The Future of Class A Agents

The introduction of Class A extinguishing agents into attack operations should begin immediately if for no other reason than because of the increased safety they provide for suppression personnel. No amount of protective clothing or breathing apparatus worn during extinguishing operations can equal the protection effectiveness of an extinguished fire. Certainly, we must provide firefighters with improved protective gear; that issue is not in dispute. However, $30 of Class A concentrate can provide quite a bit of additional protection for trained suppression personnel during attack operations.

Class A extinguishing agents alone will not answer all the problems of providing effective structural fire protection, but they will certainly help stack the deck in favor of the firefighters where most of the firefighting is accomplished—at the end of a handline. Paul Blankenship, formerly of the California Department of Forestry and Fire Protection and an enthusiastic and experienced Class A user, summed it up when he said, "There is no realistic way to say it's not worth doing."

Already proven by hundreds of departments are the facts that Class A agents make the firefighter's job much safer, knocking down flames more quickly, and providing a more secure fire area after knockdown. Safety in and of itself is reason enough to immediately put the agents to work everywhere.

Chapter 10 Review Questions

1. What is the primary ingredient in Class-A foam?
2. What is the primary extinguishing agent in Class-A foam?
3. What would be the agent of choice for an automotive junk yard fire?
4. What are the main disadvantages of a compressed air foam system?
5. Can standard spray nozzles be used on compressed air foam delivery systems?

11

Tactical Fire Attack

After the completion of response district strategic research and evaluation of water delivery equipment, the design and implementation of the resulting strategic fire stream management plan lays the foundation for the tactical utilization of water on the fireground. The strategic plan will help provide water application tools and strategy for the incident commander to tactically utilize in extinguishment operations. The strategic plan by itself cannot guarantee the water will be utilized in the quickest, most efficient manner. Successful suppression operations will depend on the level of firefighters' training and experience with using the plan, their ability to carry out the plan, and the incident commander's ability to make correct tactical deployment decisions, including making plan modifications, on the fireground.

Applying water to a fire is not a science but a fine art. Combustion engineering is the exact science of fire. Firefighting is not. While there are many scientific principles on which tactics are based, the decision of when to rapidly advance down a burning hallway, deciding when to open or close the nozzle, properly directing a stream, and knowing when it's time to get out are all decisions based not on scientific fact, but on a great amount of a firefighter's gut-level response to the situation that is influenced by actual experience and training.

When an engine company is first to arrive at the scene of a fire, decisions must be made efficiently and safely to combat the blaze. The tactical firefighting decision process encompasses these three basic steps:

1. Size up the situation
2. Form an attack plan
3. Remove threatened or trapped occupants while simultaneously fighting the fire

Engine Company Size-Up

The officer of the first-arriving engine should attempt to quickly but accurately evaluate the situation and then form a plan of action based on as many facts as possible. What looks like just light smoke from the front could be a fire blowing out three windows in the back.

We have all observed situations where lines were stretched too quickly and ended up in the wrong building because time was not taken to pinpoint the fire's exact location (fig. 11–1).

Formulation of tactical attack strategy begins with the arrival of the first company on the fire scene. The officer or firefighter in charge of the first-arriving apparatus should size-up the situation taking into consideration

- Location of fire within structure
- Amount of fire
- Life hazard
- Occupancy
- Tactics available to limit the spread of or to extinguish the fire
- Available water for initial attack
- Additional water sources for sustaining attack
- Personnel available to carry out operations
- Building construction

Figure 11–1. First-arriving units must stretch lines and perform ventilation to locate the seat of the fire. It is important that the size of lines be selected to cope with a fire's potential. In a building of this size with this much smoke present, it is a sure bet the fire cannot be handled with anything smaller than 200 gallons per minute (GPM) flows. The department handling this fire wisely stretched and charged 2½" handlines.

Location of Fire Within a Structure

The location of most fires will be readily apparent by the amount of flames or smoke exiting from windows or doorways. Locating a fire hidden behind clouds of smoke and heat is more difficult. If there is no doubt a fire is in progress but it cannot be located, hoselines should be positioned and charged and the building must be thoroughly ventilated in order to find the fire. A few broken windows could make the difference between a successful save or the creation of another parking lot (fig. 11–2).

Many times, officers are reluctant to take out a few panes of glass or cut holes in roofs because they think those actions cause unnecessary damage. Meanwhile, companies are inside the building, groping around in the smoke while the fire continues to consume the building. If suppression personnel are committed to interior search and attack, the cost of replacing window glass or even repairing a hole cut in the roof is minimal when balanced against the chances of the interior crew being injured or killed. Ventilating well and ventilating early is the interior firefighter's greatest ally. If the fire cannot be located, it cannot be extinguished. Conversely excessive ventilation or ventilation in the wrong place can certainly contribute to increasing fire intensity and to its spread. You must control the flow path in, around, and above the fire.

Figure 11–2. Proper ventilation, such as removing windows in this situation, will make hoseline advancement safer and provide improved visibility to help locate victims as well as the source of the fire. Lack of visibility makes it impossible to locate the source of the fire and unsafe for nozzle crews to advance. Sometimes, lines must be operated into areas of probable involvement before ventilation operations are complete to help prevent possible fire spread.

When evaluating conditions in an attempt to locate a fire, it is important that an officer be able to *read* a fire building. Unfortunately, the best way to learn how to read a building is through experience. Some items to consider are

- Color of smoke and the amount of pressure pushing it out of the building
- Location of flames or excessive heat
- Level of fire above or below point of entry
- Direction of probable fire travel
- Type of construction that may contribute to fire spread

Amount of Fire

The amount of fire present will dictate the flow required for extinguishment. If the fire is in a bedroom of a single-story dwelling, 150 GPM should be able to easily handle the fire. If the fire is in the attached garage of the same house and flames are showing from all windows, the amount of fire will require more flow for extinguishment, possibly supplied by a larger line. If fire is showing from a number of second-floor windows in an industrial building, the amount of fire would demand that master streams be used (fig. 11–3).

It takes experience to judge the amount and type of fire present on arrival if it has not already vented itself. Some visual and physical indicators are

- If the building is shrouded in smoke from ground to roof level, or if there is a considerable amount of smoke and heat present inside but no flames can be seen, there is a good chance the fire is located in the basement. If entry is made into an interior area and considerable smoke and heat are present but no fire is found, check areas below.
- Black smoke, characteristic of heavy burning of solids or flammable liquids, would indicate a need for large lines.
- Fire showing from multiple windows should be closely evaluated to determine if different rooms or different sections of the building are involved.
- Smoke showing from the complete eave line of connecting occupancies, such as a row of stores, could indicate the fire is traveling horizontally across the individual units within a concealed space.

Life Hazard

When occupants are threatened, the first lines stretched must be positioned and operated to help ensure life safety. By initially evaluating fire location and amount of fire, the level of threat to trapped occupants can be determined. Consider for a moment the practical amount of actual rescue work that a two- or three-person crew can actually accomplish. Raising ladders may be the only method of quickly removing endangered persons. Raising ladders and bringing down occupants, however, is a very time-consuming operation. If the entire first-alarm

Figure 11–3. Facing heavy fire that started in the garage, spread to the rest of the house, and threatened the house next door, the Delavan, WI, fire department initially hit the fire with their monitor while stretching a 2" 250 GPM preconnect. This department's frontline engines have 2,000 gal. tanks. The civilian at the left rear of the engine is an off-duty firefighter of a neighboring fire department who lives nearby. The two departments train together often, so he knew the department's procedure for securing a hydrant and where the necessary equipment was located. Before the second engine arrived, he stretched a 5" supply line and made the hydrant. Nothing can beat proper planning and superb execution through training. (Photo courtesy of Chief Tim O'Neill)

crew is engaged in rescue work, the fire can continue to expand, threatening even more victims and their rescuers.

In 1958, the Our Lady of Angels school fire in Chicago killed 95 students and teachers. The fire had started at the base of an open stairwell, and smoke and gases quickly spread upward. The first engine to arrive was confronted with a massive number of threatened victims showing from the second-floor windows. Making rescue a first priority, they raised both the extension and roof ladder carried on the engine and began to evacuate trapped children. As other companies arrived, they also raised ladders and removed trapped occupants. It was not until an extra alarm engine arrived that the first line was operated into the stairwell area.

By this time, the fire was roaring up to the second floor where it spread rapidly and claimed its victims. Many officers who were present on the scene felt that if water had been applied sooner to the fire in the stairwell, stopping its upward spread, more lives might have been saved than were rescued with ladders alone. Faced with a massive number of trapped children and teachers showing at the classroom windows, it is not fair to second-guess the rescue efforts of the first-arriving companies. They did an excellent job. The suggestion of stretching a line surfaced during postfire discussions, and its lesson should be considered by every engine company officer when faced with a similar situation.

Early in my career, the engine on which I was riding arrived on the scene of a fire in a four-story flophouse hotel. About 12 persons were at upper floor windows, screaming for help. The engine officer ordered us to lead out a line to the fire, which had involved a room on the second floor. The fire was quickly extinguished. As we were picking up, I asked the officer why we made no attempt at rescue. He said we must never lose sight of the fact that the primary duty of an engine company is to extinguish the fire. Certainly, if flames were seriously threatening a victim, which would have been indicated by fire or heavy smoke pushing out the same window, we would have raised a ladder and made the removal. He said that even though the victims were yelling loudly, there was only a small amount smoke coming out the windows around them, so we could assume they were not in immediate danger. There may have been other victims trapped within the building who were not immediately visible but possibly in greater danger. Other arriving companies were being assigned to primary search duties. Our duty was to protect those personnel and trapped victims by suppressing the fire. The officer also said the ladder truck was following right behind us and they carried more and longer ground ladders as well as the aerial with which they could make a greater number of removals more efficiently. Most important, if we had not stopped the fire when and where we did, it could have easily involved the hallway, seriously threatening tenants on the second floor as well as trapping those still evacuating the floors above the fire.

All firefighters should consider life safety, both their own and the victims', as a first responsibility. Engine company firefighters have the means to prevent the fire from spreading and reaching victims trapped within a fire building. It does not make much sense to run into a burning building to locate a victim and then not be able to remove the person because the fire has spread and blocked the exit path.

If the fire is threatening parts of the building that may be occupied, the best tactic for the engine company is usually to stop the fire's spread with hoselines, providing more protection for trapped victims and allowing more time for their evacuation. A limited number of engine company personnel might be better utilized and save more lives in this manner than if used to immediately raise ladders (fig. 11–4).

Figure 11–4. The first-arriving engine team is going to work on a fully involved vacant dwelling, first knocking down a large portion of the fire with its monitor while other members stretch handlines for further knockdown and exposure protection. When the tank is empty, the driver will proceed to the hydrant and secure a water supply.

The decision of how to best assign firefighters on the fireground must, unfortunately, be made on the spot, usually under extreme pressure. Of course, if someone is showing from a second-floor window with heavy smoke, gases, or flames coming out of the same area, a ladder must be raised to make the removal. Quick deployment of a preconnected line could make the difference between success or failure of the removal operation if the victim falls back into the room or if large amounts of fire are present. At least one firefighter, possibly the pump operator, should be stretching a line for immediate use during rescue operation.

Occupancy

Obviously, if the fire were in a nursing home rather than in a vacant garage, when the threat of life safety becomes more important than saving property, operating tactics could differ. When evaluating the building's use or its contents, some indicators are the following:

- Life hazard probability varies considerably. Hospitals, nursing homes, schools, occupied theatres, a parish hall on bingo night, and apartment buildings at 4:00 a.m. are all examples of structures with serious life hazard problems. Chances are the first lines laid into these types of structures will be used to cut off fire spread to help aid evacuation.

- High-hazard contents can be found in almost any industrial or commercial occupancy. If the building is being used as a factory, or if storage tanks are in the yard outside, chances are high that combustibles will be far from ordinary. Hay in a barn, because it is fast-burning and can become deep seated, could easily be considered a high-hazard material.

- Type of construction influences rapidity of fire spread. Balloon-frame buildings, common in older sections of the country, allow fire to spread upward rapidly. A building of mill construction is sturdy and will withstand a great amount of fire exposure before failure. Newer-type construction with laminated floor and roof joists as well as lightweight wood roof and floor trusses will fail quickly under fire conditions.

- Any occupancy with sprinklers or other type of fixed extinguishing system may be considered as a moderately high hazard. These systems cost quite a bit of money and their installation is mandatory in certain occupancies to protect an insurance company's investment. If the insurance company made the owner install a system, you can bet it was concerned about paying out a large loss in case of fire. More recently, sprinkler installations have been made in private dwellings and other occupancies not deemed hazardous. This makes determination of hazard severity difficult if based simply on the presence of a sprinkler installation.

- A good way of estimating the amount of water needed to fight a fire in a commercial or industrial occupancy is to determine the amount of storage present within a structure. Storage has greater density and presents a heavier fire loading than areas containing office equipment or production machinery. The more storage a building contains, the more water will be needed for extinguishment.

- Give consideration during evaluation to determine if any hazardous materials are in the involved building. Many materials react violently to the application of water and some compounds form hazardous gases when exposed to heat, water, or both. If there is any doubt as to the amount or location of hazardous material within a fire building, it may be best to use defensive tactics until more facts are available on which to base a safe firefighting strategy. Detailed pre-fire planning of each hazard can help supply the incident commander with facts needed to make intelligent fireground decisions.

Line Placement

Every department should determine an order in which firefighting lines are placed for attack, depending on type of occupancies encountered within the department's own district. Some suggestions include the following:

- First line rescue: If it is determined that there may be people trapped in the structure, the first line should become the life safety line. It should go through the doorway being used for evacuation to secure the area beyond. If fire is found that prevents further searching, the line can be used to fight the fire, allowing the search to continue. If the personnel must immediately evacuate the building, they can easily follow the line out. It is not good practice to put firefighters in a burning structure for search operations unless lines are being laid to fight the fire. Having the lines available can lessen the chances of the rescue crews being trapped if the fire spreads. Engine companies operating these lines should take care not to push fire and steam into the rescue crews operating far ahead of the line or above the fire.

- First line attack: If not needed to cover life-threatening situations, the first line stretched into a building should be positioned to cut off the fire's internal travel and rapidly advanced to the seat of the fire. Consideration should be given to the area opposite the direction of stream travel where steam, combustion gases, and heat may be driven. Ideally, the stream should hit the fire so the flames and heat can easily vent to the outside atmosphere. If the fire has already vented itself through a window, door, or other opening, the fire will be much easier to fight from the inside since the fire is accomplishing its own ventilation. Engine companies should take steps to immediately secure interior stairways and hallways to permit safe movement of evacuating occupants and incoming firefighters. Holding the interior stairways is extremely important, even in single-family dwellings, as the stairway will provide the fire with the easiest and quickest channel for upward extension.

- Second line placement: If the first line is having trouble advancing, the second line should back up the first line to help ensure the crew's safety. It can be taken over fire to cut off vertical fire extension. It can be taken to locations to the left or right of the first line to head off fire extension on flanks of the main body of fire. Care must be taken if a second line is positioned to operate opposite the first line, as crews can drive heat and combustion products back on one another.

- Third line placement: The third line stretched should be of greater volume than the first two unless it is needed in an area removed from where the first and second lines are operating. It should back up the first or second line if needed to ensure crew safety. It can be used to protect exposures from heavy flames, and care must be taken to not position it opposite other lines as heat and smoke can be driven at the opposing crew (fig. 11–5).

Additional lines should be deployed tactically to head off probable areas of fire extension. Extra alarms are sometimes sounded for spreading fires, not necessarily to develop volume streams but to provide personnel to operate a number of smaller streams in an attempt at containment. For example, extinguishing a basement fire in a balloon-frame apartment building usually requires a number of mobile, medium-diameter handlines placed upward on all floors, along with personnel equipped with proper tools to open up hidden spaces.

If the fire is in the basement, the first line should cover the interior stairs to limit the upward fire spread. This line may not necessarily be the direct attack line. If the fire area can be more easily entered from the outside, a second line should be advanced from there while the inside line prevents upward extension. If it is decided to advance the first line down the stairs to the fire, a second line should be positioned at the top of the stairs to prevent upward fire spread.

The direction of the interior stream in relation to the uninvolved sections of the building is important. If the fire is blowing out the front windows, the entry is normally made from

Figure 11–5. Depending on arrival conditions, exposures become a priority if threatened by a large amount of fire. Here, firefighters deploy a 250 GPM line to keep the church fire from spreading to a private dwelling. (Photo courtesy of Gordy Nord)

a doorway located in the rear or side of the structure. There is a great tendency to go in the front door on every fire, but proper size-up should determine the best entrance in which to advance an attack line to keep the fire from spreading inside the building and to help avoid the possibility of the line pushing fire into unwanted areas.

If the initial attack line has made entry from a position other than the front door, there is usually good reason to enter the front door with another line, especially when searching residential occupancies, because the staircase leading to other floors is usually located near the front entrance. This second or backup line may be needed to secure the stairs to protect occupants or search crews.

Water for Fire Attack

The importance of flowing the proper amount of water in the proper place cannot be overemphasized. When selecting a line at the fire scene, the officer should remember these strategic facts:

- The stream must have minimum or greater rate of flow to overcome the heat of the fire at hand.
- The stream must be positioned properly to penetrate into the fire.
- The stream must be distributed in an efficient manner around the area of burning.

The attack line's proper size and length should be selected by considering the required flow both for the size of the present fire and also the size of its reasonably anticipated growth. Be sure the lengths stretched will reach both the fire and areas of probable fire extension. Nothing is more frustrating than to run out of line when attempting to head off a spreading fire (fig. 11–6).

Number of Available Personnel

It would be foolish for an incident commander to devise a tactical firefighting plan that utilizes a number of high-volume lines and a great amount of entry and ventilation work if only a small number of firefighters are available to put the plan into action. Help should be called as soon as it is determined that the first-arriving forces cannot hold or extinguish the fire or provide for life safety. In many cases, a commander will have to make compromises in order to save lives or stop the fire. For example, salvage has a low priority when fire threatens to spread to an exposure. Conversely, salvage may have a high priority if fire threatens priceless historical material or artworks. New hardware, medium-diameter handlines, and high-flow, low-pressure nozzles allow fewer personnel to fight more fire than ever before, but those few personnel can perform initial attack, rescue, and ventilation operations for only about the time it takes to use one air cylinder. If the firefighters are in good physical condition, and if the weather conditions are not too hot, too cold, or too humid, many of them may last through a second change of breathing apparatus cylinders. After that time, personnel should be given a rest before they can safely perform additional firefighting duty. Help in the form of relief personnel should be called in as soon as possible if it appears the structural firefighting operations will be extended over one hour in moderate test penetration into the fire area.

Figure 11–6. Deploying a line for initial attack that has an insufficient flow rate for the fire at hand wastes time and destroys property. Rather than attempt defensive operations with a 95 GPM handline, this crew could have knocked down the source of their problem with a 250 GPM stream.

If the fire is confined to a residential kitchen, a combination nozzle adjusted to flow a straight stream for attack and overhaul is quite adequate. If long apartment hallways or a great amount of area in a warehouse occupancy is involved, a smoothbore nozzle on a large line will work most efficiently.

It is generally desired that an attack line flow as much water as can be easily handled by the nozzle crew. Surveys have shown that the most popular attack stream flows 150 GPM through a $7/8$" smoothbore nozzle. The second most popular is 170 GPM flowed through a $15/16$" smoothbore nozzle, both lines being supplied by 1¾" hose. There is a tendency to stretch a medium-sized line on all fires. In many cases, however, use of a larger-flow line is definitely indicated. If there is any doubt as to flow selection, the larger line should be used.

The Art of Applying the Water

Now comes the hardest part.

To achieve penetration and to properly distribute the water, the interior suppression crew must take the stream to the area of operation, sometimes through punishing heat and smoke. If transitional attack is decided upon, the line should be positioned to knock down the most amount of fire before it has to be moved. When the nozzle is opened, the water must be distributed to the burning material so the fire will be extinguished. At the same time, care must be taken to prevent an imbalance in atmospheric conditions that could injure the crew and jeopardize maintaining the line's position. The nozzle must be shut off immediately after blacking out the fire to prevent both excessive water damage and excessive flow from turning into unwanted steam, which will create a saturated atmosphere which hinders ventilation efforts (fig. 11–7).

There will be many times when a complete evaluation of the fire involvement cannot be made because of fire conditions, weather, location, and other factors. In this case, you enter the gray area of success probability called luck. Luck can come in good and bad forms, but relying on luck should never take the place of sound tactics and training.

Bad luck is not being able to locate the fire until it vents itself out the roof, or stretching a line only to find out you are on the third floor of the wrong building. There can always be a measure of bad luck waiting to happen at every fire. Its effects can be countered by training and experience.

Of course, there can be a stroke of good luck, such as the time a truckman climbed the aerial to the roof to ventilate, only to find the roof completely covered with thick brown smoke

Figure 11–7. A combination of ineffective flow rate, improper stream placement, and lack of opening up the fire building are working to make this fire a total loss. To accomplish extinguishment, the stream must reach the burning material, not tickle tongues of flames. In this case, rapid truck work to force the front door, open windows, and pull ceiling will be needed to stop this blaze. If a truss or steel bar joist roof is discovered when the ceiling is pulled, all personnel should be removed from the building showing this much fire because of the danger of collapse.

behind the parapet. Trying to dismount the ladder without injuring himself, he threw his axe and Halligan bar on the roof. They hit an unseen skylight, fractured the glass, and fell inside, releasing great billows of smoke and heat. He finished the job with a pike pole, and the engine company was on top of the fire by the time he reached the ground.

I remember responding to a fire on the top floor of a two-block, square, eight-story building that housed a candy manufacturing company. We could see a great amount of smoke as we responded, and the chief rapidly called for a second alarm. Our engine was ordered to lead out and charge lines into the sprinkler connection. We did this, and the company entered the building as they were being charged. Somehow, the company got separated as we looked for a way to the roof. I saw an elevator door open and the operator step out, quite relieved to see a firefighter. He quickly took me to the roof, where employees had stretched a house line to a fully involved cooling tower. The line was not long enough to reach the fire, so I directed one employee to remove the hose from another rack, and we connected it to the first line. The chief arrived via the inside stairs and the two of us proceeded to extinguish the fire.

Luck in the form of the elevator got me to the roof, and luckily the flow from the 1½" line was adequate for the size of fire. Some talent was needed to extend the line and extinguish the fire, but luck played a large part in getting the chief and me to the fire area. Not relying on luck alone, the first-alarm truck raised its aerial to the roof and the second-due engine brought its line over it to the fire.

Moving In

Before masking up and making entry, establish a working plan with the crew outlining the line's basic goals. Where do you expect to cut off the fire extension? Where will you be pushing steam, gases, and smoke? How will you handle removing a fire victim? These items need to be briefly planned before charging into a fire building.

As you move in, continually assess the conditions. An experienced firefighter will always be reading the fire. Is there fire overhead? Is the heat dramatically increasing, indicating a fire is burning elsewhere? Is the smoke lifting as extinguishment takes place or is it beginning to get dark and starting to bank down, indicating a potential rollover or backdraft? Everyone inside a fire building must constantly remain aware of conditions in and around the fire atmosphere to help ensure the crew's safety (fig. 11–8).

Suppression crews will experience discomfort during interior operations. It is an inescapable fact of firefighting life. Protective clothing and breathing equipment must be used to provide a certain level of protection to the firefighter. The best protection, however, is provided by the firefighters themselves. The faster the line is moved to the seat of the fire and the faster the fire is extinguished determines the length of the firefighter's exposure to a hostile environment. Proper ventilation also helps reduce the amount of heat and combustion products threatening the firefighter's well-being.

When moving in on an interior fire, movement should be rapid but thorough. One big mistake made by inexperienced firefighters is the tendency to remain in a ventilation flow path.

If a line is being advanced into a rear bedroom, for example, the fire can be knocked down from across the hallway, allowing the crew to advance into the room for further extinguishment. Too often, personnel make their attack from the doorway. Since this is a primary ventilation channel, they are subjecting themselves to much more heat and smoke than they

Figure 11–8. To keep the interior crew safe, the officer must know the whereabouts of all members at all times and must be in a position to observe the objective and the effect of the stream upon it. Here, the officer is positioned above and to the side of the nozzle operator. Directly behind the nozzle is not the place for the officer to be.

would if they had entered the room or stayed on the opposite side of the hallway when making their attack (fig. 11–9).

There were numerous occasions in the days before the widespread use of breathing equipment where a member of an engine company, crawling on the floor while attempting to make their way to the seat of a smoky fire, would stand up and punch the face of a hapless firefighter standing idly in a doorway, blocking the incoming flow of fresh air. This action would certainly make that person think twice about blocking a doorway in the future.

When advancing on a fire in a cellar when no outside entrance is available, the more quickly the crew can pass through the upper level of the fire area on their way down the stairs, the more quickly the relative safety of the cellar floor can be reached. If the group's leaders hesitate, you can be assured they will find themselves rapidly helped down the stairs, usually propelled by the tip of a boot. It may be felt that conditions may call for the nozzle to be operated for a few seconds to cool down the atmosphere around the stairwell before entry is made. This may not be a good idea. It must be remembered that this action could create a cloud of steam which can obscure vision and increase humidity to uncomfortable levels at the cellar's ceiling.

The whole idea is to advance the line as quickly as possible while combining ventilation and common sense to help reduce the crew's exposure to heat and combustion products.

If the flames are concentrated on the floor or are extending up a wall, a straight stream should be directed low, at the base of the flames, after quickly sweeping the stream in a rectangle to cool the compartment. This is the quickest, most efficient method of extinguishing interior fires without causing thermal injuries to the suppression crew by the creation of unwanted steam. If the fire is more advanced and flames have traveled up the interior walls

Figure 11–9. For this crew to advance safely and effectively, ventilation needs to be performed, preferably on the other side of the building. While thermal imaging cameras can be a big help in locating the fire in low-visibility conditions, lifting the smoke a bit through control of the flow path helps as well. Also remember, when advancing into low-visibility conditions, if you can't see the floor when you bend over, you need to be on your knees.

and are crossing the ceiling, again, a direct attack using a straight stream on the base of the burning materials will eliminate the heat source and stop flame movement. If applied properly, the water will cause the flames to retreat, looking like a movie of the fire being run backward. It is important not to overapply water during this operation. If more water than necessary is used, excess steam will be generated, expanding in the overhead and causing imbalance in the atmosphere. The nozzle must be shut down as soon as the flames cease.

As discussed in chapter 3, using a spray or fog pattern to attack a fire while inside the building is no longer considered good practice. A wide fog pattern will not reach the burning material until the crew is extremely close to the fire, and a narrow fog pattern will concentrate the effects of entrained airflow and will fan the flames, in many cases causing them to intensify and spread with a blowtorch-like effect far beyond the immediate fire area. To prevent thermal injuries, put out the fire, not the smoke.

Applying Water as an Art

As discussed in earlier chapters, a handline should be capable of flowing as much water as the nozzle crew can safely handle. This is a reflection of the Big Brick theory of hitting a fire with all the water you can until it is knocked down. The more water you can apply, the faster the fire is darkened.

It must also be remembered that water does not have to be dumped on every fire just because it is available. The late George Hughes, former fire marshal of North Richland Hills, Texas, and prominent fire service educator, equated applying water with a delicate surgical operation. He stated, "Years ago, we used to be proud to see how little water we could use to put out a fire, not how much. Inside a fire building, during offensive attack, the work is finite, very delicate. The nozzle stream should be used like a surgeon's scalpel, not like a machete."

The decision of when to open a nozzle is easy to determine. As soon as the stream has the reach to accomplish extinguishment, as soon as it is determined where the resulting steam will vent, and as soon as it is determined that the crew is ready to safely handle the line, the nozzle can be opened.

The line should flow only enough water to extinguish visible flames, then should be shut down immediately. In some cases, the crew might feel that more fire lurks behind the smoke. This is where the art part comes into play. Should the crew continue to apply water from the same position? Should the crew move in immediately? Should the crew wait until the smoke, heat, and steam have a chance to lift? Should the line possibly be repositioned to get a better shot at the blaze? These are questions that can only be answered by assimilating facts and feelings present during the time of attack.

If it is suspected that victims might be trapped, the best decision may be to quickly advance into the smoke and heat and continue the search. If there is a great amount of shelving and boxes in a storage area, and if it has been determined that no one is trapped, it might be wise to let the smoke lift and reposition the stream to make water application more effective and less damaging.

To help make water use more effective during attack, aggressive *hit-and-move* or *flow-and-go* tactics should be utilized. Hit-and-move means that water is flowed into a compartment, sweeping the walls, ceiling, and floor. When the fire is knocked back, the nozzle is closed, and the crew moves to the next operating position. With the flow-and-go tactic, the nozzle is left open as the crew moves in, again sweeping the compartments as they go. This tactic is useful when advancing down a long hallway in an apartment building or large house.

If it is delivering the proper flow to the fire, a stream should extinguish all the fire it can within about 30 seconds. Knockdown occurs much more quickly in individual rooms of residential occupancies. If the area is wide, knockdown may take a little longer as the nozzle is swept up and down and from side to side. If the flames do not subside within 30 seconds, the nozzle may not be delivering the needed flow rate or the stream may not be hitting the burning material. Water may not be the most effective extinguishing agent in some cases. It could be that the flow rate is adequate but, because of the mass of burning material or the way in which the material is arranged, it is taking longer for the water to be distributed and do its job. Deflecting the stream off of side walls and the ceiling will help distribute the water in total flame involvement situations.

If the flames subside and then pop back, chances are the fire is being pushed into an unventilated area, such as an enclosed overhead or unventilated room. Rapid ventilation work will be needed to open up concealed spaces to avoid a backdraft.

Three firefighters were killed in Oklahoma City by a rollover caused by this situation. They knocked down a fire in a bedroom, then experienced the fire popping back at them. It had burned into a blind area between an old and new roof. Investigation determined the stream kept the fire inside the area boiling around until it pushed out with such force that it enveloped the crew.

On some very hot, confined fires, it may be necessary for the crew to hit the fire briefly, then back down the stairs or out into a hallway to give the products of combustion and heat a chance to clear. The crew can then quickly move in again and continue to repeat the process until the fire is extinguished. It is not necessary for a nozzle crew to remain within a high-heat situation just to see if they can take it while extinguishment and ventilation are taking place. There may be a great deal more fire in other rooms that the crew will have to extinguish. It does not make much sense to take a beating if it can be tactically avoided.

A nozzle crew may find themselves entering an area filled with thick smoke under pressure and high heat, yet no flames are visible. In these cases, it may be necessary for the crew to discharge water into the area in an attempt to hit fire they cannot see but can reasonably assume is burning in the area. As the stream is deflected around the area, the atmosphere should be monitored to detect changes in heat level, smoke thickness and pressure, and the presence of steam.

Many times, after lines are worked in this manner and after ventilation has removed combustion products, it has been revealed that the stream had extinguished a fire which could be sensed but not seen. If a fire is threatening a structure, streams should be applied immediately into all areas of suspected fire involvement even if flames cannot be seen. There is a possibility that the water will have no effect on the fire, but some water damage to the contents is certainly more acceptable than the unseen fire damaging the structure to the point of collapse.

Good fire stream practice dictates that the fire should be located before discharging water. In most cases this is done visually. In some cases where smoke prevents flames from being seen, other indicators such as high heat, thick smoke, and smoke pressures can indicate areas of probable involvement. A thermal imaging camera should be carried by every officer operating inside a fire building to help locate areas of high heat and possible victims.

From time to time as the line is flowing, the officer or nozzle operator should shine a light in the direction of the stream's target to monitor any buildup of condensing steam. Using this technique, the stream can be shut down and the crew moved to a safe position before the condensing steam blanket overruns them. Remember, true steam is not visible. Spotting an oncoming steam balloon with a light is no guarantee that the crew will not be burned by rapidly expanding steam; however, if water is being applied properly (i.e., straight stream directly into the heated compartment), this technique can provide reasonable warning of deteriorating atmospheric conditions.

At least one person not involved in the direct physical operation of the line, preferably the line officer, should be alongside the nozzle crew to continually evaluate the results of the stream and to stay in touch with the fire environment. When directly behind a nozzle flowing water, it is almost impossible to observe danger signs such as darkening of smoke, intensifying flames, and flames retreating and then popping back.

Commanders should provide for extra personnel to relieve interior attack crews if it is evident the fire will not be knocked down in a minute or two. For example, the New York City Fire Department (FDNY) operating procedure calls for personnel from two engine companies to advance each single line into a fire. By having a number of firefighters immediately available, rotation can be frequent, helping prevent thermal injuries to attack crews. If a single firefighter suffers heat exhaustion or burns, then that firefighter's removal to a medical care area requires two additional firefighters to be taken away from the attack. Having interior personnel overexert themselves in high-heat conditions creates a domino effect in which even more personnel will be involved in removal and care of injured firefighters, reducing the number of available suppression personnel even further.

It is wise to frequently relieve interior crews to prevent heat-related injuries and exhaustion from adversely affecting search and attack operations. This rotation, along with a positive accounting system to track all personnel involved in interior operations, must be put into place as soon as possible during attack operations.

Fighting a fire, especially an interior blaze, is a dynamic process; conditions are changing by the second. Ongoing evaluation of operations must be constantly made by both interior and exterior commanders.

Engine company officers should constantly evaluate the effects of their water delivery (fig. 11–10). If the stream is having no appreciable effect on the fire, immediately assess

- Flow rate—are you applying enough water?
- Line operating position—are you hitting the base of the fire or simply pushing flames to a different location?
- Ventilation—where are the smoke and heat going?
- The fire's fuel—is it a chemical or other material that cannot be extinguished with water?

Figure 11–10. The 1¾" attack line is not an all-purpose firefighting tool. It is excellent for interior attack and exposure protection; however, if the department is serious about extinguishing this fire, one or more 250 GPM lines will be needed.

- Is the fire burning elsewhere and supplying the fire you can see?

Getting water to the seat of the fire might be adversely affected by the following conditions:

- Heat and smoke preventing the crew from rapidly advancing.
- The area of the building may be so large that a stream may not have enough reach to hit all parts desired.
- Fire may be concealed within walls, ceilings, or other spaces.
- Structural obstructions such as walls, dividers, desks, or machinery may be deflecting streams.

Ventilation

Once all visible fire has been knocked down, the nozzle should be immediately shut off to let natural ventilation remove smoke and steam so the fire situation can be better evaluated. Crews should let the smoke lift, when possible, before pouring on more water. If water has been applied in the proper amount, the steam in the overhead will be buoyant and should quickly move out of the building through flow path channels (fig. 11–11).

Figure 11–11. Proper ventilation is one of the most essential, yet one of the most poorly practiced firefighting tactics. Because of large holes cut in this apartment building roof, firefighters on the third floor can advance on the fire in relative safety.

Lack of visibility due to unventilated steam and smoke can lead to a dangerous situation if the integrity of the floor and building structure cannot be quickly assessed by the interior crews before further firefighting operations take place. Smoke and steam can easily obscure holes in floors or other signs of a weakened structure. If the integrity of the structure cannot be continuously assessed as the attack advances, the suppression crews could be moving deeper into areas without knowing the danger that surrounds them. It makes much more sense to safely advance on a fire in a properly ventilated building than to grope around in clouds of smoke and steam, thereby inviting serious injury. There was a saying among Chicago politicians back in the 1950s urging people to "vote early and vote often." It is much the same with fire building ventilation—vent early and vent often.

Remember, the most expensive and insulative structural fire clothing cannot provide visibility nor can it relieve the effects of continued 1,200°F heat common in interior attack operations (fig. 11–12).

Overhaul

Many buildings initially saved using sound interior firefighting tactics were later destroyed due to rekindled fires caused by inadequate overhaul procedures.

Use water wisely during overhaul. Many times, a nozzle operator attempts to soak a pile of debris with a spray stream only to have the water not penetrate and run off around the edges, soaking everything on the floor but the pile. A better tactic is to stick the nozzle right in the material and use short bursts of water to drive the liquid into the pile. Put the nozzle parallel to the floor and drive a forward stream horizontally between the debris and floor to thoroughly soak the material. Remember, you don't have to open the nozzle all the way during overhaul if a full stream is not needed. The use of Class A agents during overhaul operations will dramatically increase the certainty of complete extinguishment (fig. 11–13).

Overhaul operations are often delayed so that investigators can examine burned areas to determine origin and to gather needed evidence. Overhaul should not be considered complete unless all burned material is examined after investigators are finished with their work.

During after-fire operations, all burned or suspect material should be separated using proper tools and all hidden areas opened up for inspection. A minority of fire officers feel that opening up walls, ceilings, and floors or the removal of window or door casings during overhaul operations creates unnecessary damage. If not done wisely, it will.

Opening up should be considered a form of controlled damage that must be accomplished to make sure the fire is completely extinguished. Contrast a few stripped window casings with the additional damage caused by a fire raging within outside walls for an hour or so before someone sees the flames exiting the roof—all this after the fire department has left the scene. In any rekindles, the second fire usually causes much more damage than the first.

Allowing a fire to rekindle is not just embarrassing, it is unprofessional. Remember, thousands of fires each day are successfully overhauled and do not reignite because proper tactics have been employed, not only for attack, but also for overhaul.

Chapter 11 **Tactical Fire Attack** 273

Figure 11–12. Some indicators of a backdraft condition can be seen here. Melted curtains indicate high heat; brown-stained windows indicate heated smoke. Smoke pushing out from around window frame and through cracks in mortar joints indicate high smoke pressure. In this situation, rapid and thorough ventilation from the top down is necessary after hoselines are positioned and outside the fire area and ready to immediately move in once the vent takes place.

Figure 11–13. Proper overhaul calls for effective coordination between engine and ladder company personnel to assure all hidden or suspect areas are inspected to make sure that all fire is extinguished.

Water Damage

Water damage to a building and its contents is directly related to the nozzle operator shutting down the nozzle as soon as it has completed its work. Besides making a mess of the atmosphere within a fire building, excess water oversoaks the area, causing damage along the way as it is absorbed by the contents of the building and its structure. Remember, water damage occurs after the fire is out. A high-flow handline will cause no more damage than a booster line if it is closed immediately after flame production ceases. Master streams used on large fires can cause a great amount of damage if they are not shut down when they have extinguished all the fire they can reach. If a building is still burning while tons of water run down the stairs and out into the street, feel the water with a bare hand to see if it is warm. If it is, the presence of heat would indicate the water is cooling at least some material. If the water feels cold, the streams supplying that water are not extinguishing any fire and should be repositioned or shut down. A few officers are so overly concerned about water damage that they make salvage, not fire extinguishment, a primary priority. I remember a chief in a neighboring community who was so concerned about water damage that he wanted his men to spread salvage covers before flowing water. Needless to say, his department had an extraordinarily high fire loss rate, commensurate with its extremely high salvage cover loss rate.

Sometimes it seems that the only people who complain about water damage are the ones whose house you just saved. The fear of water damage should never enter into a fire flow decision when beginning attack. Excessive water damage is a result of poor nozzle control discipline after the fire is extinguished. It is not a result of handline flow capabilities or nozzle selection.

There are times when excessive water damage cannot be avoided, such as when a distributor nozzle is used on a basement fire. Of course, here you can ask the question: What would you rather have in the basement: (a) 12" of water, or (b) the first and second floors?

What Happens When Unwanted Steam Is Generated?

The action of a fire's heat converting water into steam is instantaneous. The resulting burst of high atmospheric pressure will cause the steam to penetrate turnout gear inward beginning from the neck area, then upward from beneath the coat and pants hems, and will push through woven hoods and wristlets. Hoods are good defensive protection from the effects of direct flame contact, but because they are constructed of open-weave material and do not have moisture barriers, steam can quickly and easily penetrate to burn the neck and face areas (fig. 11–14).

Ventilation sometimes cannot vent all of this unwanted steam fast enough to prevent thermal injuries and visibility problems, even if it is properly performed before water is applied. The best way to handle unwanted steam is not to generate it in the first place.

Figure 11–14. Actual steam is colorless. Steam exiting this vent stack on a New York street is invisible until it begins to condense as it contacts the cooler atmosphere. What is thought of as steam is actually a cloud of water vapor. Condensing water vapor clouds can still cause burns; however, it is the superheated invisible steam which will do the most thermal damage because it is hotter.

Pure steam is colorless and odorless. When a white cloud exits a building, the heat has already begun to leave the pure steam. What you are seeing is steam that has condensed into water vapor. You cannot readily see actual steam.

As a fire burns in a room, its heat will rise. Cool air will be drawn in by the fire, usually entering at the bottom or floor level. Tests have shown the heat in the overhead can reach over 1,800°F while the floor remains relatively cool at 100°F–250°F, easily tolerable by a firefighter in protective clothing and breathing equipment. As long as the fire remains in the free-burning stage, a type of balance will be maintained, with the heat and combustion gases collecting at the top of the structure and cool air entering below. This thermal stratification must remain in place if a fire is to be successfully fought from close range.

The fire service has been conditioned beyond all practical reason to fear the heat in the overhead. Even some seasoned officers and firefighters say they feel the fire will race across the ceiling at every interior attack, and most advocate performing steps to cool it down or control the overhead by applying a spray pattern of water into it.

In a way, they are right. A fog stream applied in the wrong place from inside a fire building will cause injury by driving excessive heat and steam in the overhead down on the crew. The rapid onrush of air supplied by a fog stream can pump the superheated flammable gases lurking in the overhead with needed oxygen, causing the gases to ignite. Unfortunately, the fastest way to cause the overhead to erupt is to spray it with water—exactly what many firefighters think is protecting them.

To help define the extent of overhead heat danger, the University of Illinois Fire Service Institute performed a number of high-intensity fire experiments in the late 1980s. It was found that the massive air delivery to the superheated gases by directing fog streams upward could cause ignition of fire gases in the overhead as well as the violent expansion of water into steam. These actions drove the flames and heat down on the operating forces. If flames were present in the overhead, applying water to the base of the fire caused the flames to retreat backward along the ceiling, all the way down to the level of the fire's base where they were extinguished. This effect is compared to a movie of a fire running backward.

The overreaction to the perception of the ceiling immediately flashing over upon entry into the fire area is usually based on experience. In most cases, firefighters actually created their own problems after the employment of outmoded fire attack techniques, filling the overhead with oxygen generated by improper application of water spray.

When explaining the theories of horizontal, direct fire attack to groups of firefighters, I have found many skeptics who, even after listening to a detailed explanation, still felt the overhead needed to be cooled. In many cases, this perception exists because the questioning firefighter was burned by steam or flames from the overhead while making an attack in the past.

If this feeling still exists, one question should be answered: How many times have you been burned before you opened the nozzle? If thermal injuries occur to nozzle crew members during or after water application, it may be that improper tactics are to blame. A firefighter does not have to be burned in order to indicate that an aggressive interior attack was made. A safe, aggressive interior attack calls for sensible water application and ventilation tactics. The ceiling needs to be evaluated along with other fire atmosphere indicators. However, with proper tactics, it does not need to be a matter of overriding concern. It certainly should not be the target of an automatic discharge of water. If, after careful evaluation, the officer feels the overhead needs to be cooled because it is preventing the crew from moving in, a quick 2 or 3 second dash with a straight stream moved side to side will provide temperature reduction without generating a massive amount of steam.

When used within a fire building, fog streams should be considered to be non-standard specialty streams best suited for rare situations encountered in fire extension protection or ventilation. Consider them unsafe for general firefighting when applied inside structures by firefighters within the fire environment. For outside fires, where containment of unwanted steam is not a factor, any pattern that penetrates the burning material can be safely used.

Distance also offers protection. Nozzles should be opened up as soon as their reach will bring the stream to the base of the burning material. When combating a fire, remember that the firefighter's protection problems are being caused by the fire itself. Heat, smoke, and the possibility of a weakened structure—all threats to the firefighter's well-being—are only symptoms. The disease is the fire itself and it can only be cured by extinguishment. It serves no useful purpose to use the means of extinguishment in the form of a wide spray to get firefighters physically closer to danger if that same means can be effectively utilized from a distance to extinguish the fire.

One technique that should never be used on an actual fire is pencilling (where the nozzle is opened and closed a number of times before the fire is fully extinguished). A firefighter was killed in a Chicago suburb and another seriously burned a few years ago while attempting to use this technique inside a well-involved dwelling. Origins come from training fires where the instructor wants the fire knocked down but not completely extinguished so the next crew in can have something to hit. Once a line is flowing on a fire, it should not be shut down until the fire is extinguished. Your life depends on it.

At a vehicle fire training session in the eastern part of the country, the instructor told a nozzle crew to advance on a burning car behind a wide fog pattern. For a good minute, as the crew moved toward the vehicle, none of the water was hitting the fire. A tire popped and startled the crew, making them back up. That action demonstrated the amount of confidence they had in the stream's protection. The pattern was changed to a medium-wide fog and the crew continued to advance, but the choice of pattern still permitted no water to hit the flames. As the group got closer, the concentrated onrush of air produced by the fog nozzle began to fan the flames, increasing their intensity, but water was still not hitting the fire because the pattern did not fit the job. After an agonizingly long time, some of the water started reaching the flames and began to reduce their intensity, but the engine's water tank ran dry before extinguishment was complete. When questioned later about his tactics, the instructor said, "I had the men advance behind the wide fog pattern for their protection."

Protection from what? The heat was being carried upward by thermal currents. If the gas tank had ruptured violently, the thin sheet of water would have offered no protection from flying debris and flames (fig. 11–15).

The next fire was handled by an experienced company officer. He set the nozzle on straight stream and hit the blazing car from about 30' away. The flames immediately darkened. He rapidly advanced to the car with the nozzle open, sweeping the stream underneath the car, then inside the passenger compartment. Extinguishment was complete within 30 seconds.

To repeat, putting out the fire solves a multitude of protection problems. All a wide fog pattern accomplishes is to postpone extinguishment while pumping more oxygen into the fire area. Remember, the most efficient extinguishment is obtained by water reaching the burning material. Extinguishment can be performed at any distance as long as the stream is hitting the fire.

Figure 11–15. Every step forward taken by these firefighters brings them closer to danger. If someone were to slip and fall, the fire would envelop the entire crew. It makes absolutely no sense to advance into danger without some extinguishing action taking place. Many schools use this technique to get students used to operating close to live fires but neglect to tell them never to use wide-angle streams on actual fires. Many firefighters have been burned as a result. (Photo courtesy of Becky Gerard)

Preventing Fire Spread by Stream Placement

Fire in an internal exposure can be quickly spread down hallways or up staircases by convected or radiated heat. The most efficient tactic to help stop a fire from spreading internally is to cool the exposed compartment before it ignites. In some cases, a stream will have to be directed down a hallway, across a stairwell opening, or across the ceiling of a large area in an attempt to reduce the traveling heat. Flames may or may not be visible during these operations; however, smoke may make it difficult to tell if actual flame is present. If a great amount of smoke and heat are moving through an area, it's a safe bet that the nozzle should be opened.

This tactic should not be confused with an indirect attack. For example, to protect a warehouse ceiling, the straight stream of water is played on the ceiling material to cool the material itself, not to create a volume of steam below the ceiling in hopes of extinguishing the fire.

After knocking down the heat and flames in a stairwell or hallway being held for defensive purposes, the nozzle crew may decide on its own to complete extinguishment by advancing on the fire. There could be cases where the line must remain in place to keep the exposure secure, even though the crew feels it is accomplishing no useful purpose by remaining stationary. If a stairway must remain secured for evacuation, advancing on the fire could have adverse results in the event the nozzle crew fails to extinguish the fire or if the fire travels around or over the nozzle crew and reaches the exit area. Proper supervision may be needed to keep the crew in its assigned location.

Ventilation in Conjunction with Engine Work

Other than flowing water, *vertical ventilation* (opening up the building structure over the fire) is probably the single best tactic available to prevent fire spread and allow for rapid extinguishment. As the overhead spaces are opened, the nozzle crew must be ready to go into immediate action with a charged line to extinguish the fire. Horizontal ventilation, while not quite as efficient as vertical ventilation, is much easier to accomplish and in most cases provides a great amount of relief for inside crews.

Vertical ventilation is a tactic that might not be needed at every structure fire, especially those where the interior crews are knocking down the fire and the atmosphere is relatively clear and tenable. The incident commander reading the building is in the best position to decide.

In my early days as a firefighter, before the widespread use of breathing apparatus, I can remember the relief felt by our crew as plaster and lath rained down from above, caused by the truck crew punching down the ceiling over our heads through a hole they cut in the roof. The instant lifting and removal of smoke, heat, and combustion gases allowed us to rapidly advance the line further into the fire area, making short work of the fire with much less discomfort (fig. 11–16).

Figure 11–16. One member assigned to outside ventilation duties can make a drastic difference in the safety and comfort of crews working inside the fire building. Working from the roof, a member with an 8' pole can rapidly vent top floor windows.

When former colleagues and I discussed our earlier days of firefighting, many remarked how rarely someone was burned during attack operations. While our interior tactics were quite aggressive, moving in without the protection of breathing apparatus required the prompt and effective application of ventilation techniques before and during advancement to the fire. All of us felt (and it is now universally agreed within the fire service) that prompt ventilation of the fire area is the key to successful and safe attack operations.

In these days of limited attack personnel, the tendency is for the crew on the line to bull its way through unventilated smoke and heat in their attempt to find the fire source. Some recommend ventilation by placing a high-volume fan behind the attack crew to help push combustion products away. This action may move the heat and products of combustion away from the crew but most certainly will push them into uninvolved areas or on top of trapped victims. My rule of thumb is that the positive-pressure fan is best turned on when the interior attack crew tells the incident commander it is safe to do so.

There is no substitute for immediate and thorough ventilation of the structure when interior crews are making an aggressive attack. One firefighter taking out windows from the outside with a pole, ahead of the nozzle crew, can affect the success of the operation more than a dozen hoselines in the front door. In order to prevent unwanted fire spread, coordination between the exterior and interior crews is a must.

It is simply a matter of physics. The products of combustion and steam must occupy a certain amount of space. If they are pushed into an area, or if an attempt is made to force them out a single small opening, they will compress until they reach a pressure that will force them back on the interior crews. If flames are present, they, too, will be picked up by this moving mass of atmosphere and will travel with it.

It is generally agreed that rapid, high-volume ventilation, which was then, and still is, common practice in large-city departments, vented almost all of the hostile fire atmosphere, preventing many thermal injuries to firefighters and victims.

Vertical ventilation helps remove smoke, improving visibility and firefighters' safety. The officer should be positioned away from the nozzle to better observe the effects of the stream on the fire and to continually assess fire conditions.

If no one is assigned to force entry and ventilate along with the nozzle crew, the officer directing stream operations should carry a tool with which to force doors and take out windows. The ideal situation is to have the fire ventilated ahead of the nozzle. If not, someone will have to push into the fire area after the fire is knocked down to get the windows open and begin ventilation (fig. 11–17).

The correct sequence is vent, then enter for extinguishment while controlling the flow path. Venting should start as soon as nozzle crews are ready to immediately advance with charged lines. In some special cases (for example, a ladder company arriving on the scene before the pumper and finding a number of victims endangered by spreading flames), ventilation can be started immediately to help direct flames and combustion products away from those trapped. It must be remembered in these cases that fire will be much more difficult to control if it becomes free burning before lines can be stretched.

Figure 11–17. Firefighters prepare to mount a heavy stream attack on a fourth-floor fire. Unfortunately, the attack began about an hour earlier at the point of origin in the basement. Failure to thoroughly ventilate the upper floor and roof allowed smoke, combustion gases, and fire, running the walls upward from the basement, to accumulate and ignite with explosive force, driving interior crews from the building. (Photo courtesy of Alan Chaniewski)

If ventilation is not accomplished before the attack crew moves in, it could happen that the crew extinguishes the fire only to have it reflash when ventilation channels are opened after initial knockdown. What happens is that oxygen supplying the fire is replaced by a volume of contained steam that dilutes the atmosphere to a level that cannot support combustion. When the nozzle is shut down, the fuel and heat needed for combustion remains. When air is admitted to the area by opening ventilation channels, enough oxygen is supplied to cause the extinguished area to light up. Adequate ventilation before attack helps prevent area relighting.

Consider Water Supply

Many departments have the attack pumper go to work on the fire scene unsupported by a supply line or other water source. If a large volume of fire is suspected when the call is dispatched or is evident when nearing the fire's location, the attack pumper should lay supply lines on the way in or use reverse lays in order to provide a constant supply of water.

If the first-in crew operates streams using only tank water, action must be taken to immediately secure additional water supply. If this is not done, one of three things could happen:

- The fire will be extinguished with tank water.
- Another company will arrive in time to supply the first with water.
- The first company will run out of water, will have to abandon the attack, and may lose the building.

If an aggressive attack is mounted using only tank water from the attack engine to sustain the flow, there is a great risk of having the crew trapped deep inside a fire building without any water. If the tank runs dry before the fire is stopped, from the standpoint of reducing injury exposure to firefighters, it probably would have been better not starting the interior attack in the first place. Running out of water before the fire is extinguished is embarrassing as well as dangerous. Doing the right thing is more important than merely doing the fast thing.

Pretty much standard practice in the fire service today is to have the first engine mount an attack using tank water, and the probability of knocking down the fire is quite high. My feeling is that the majority of progressive departments are flowing 150 GPM or more on the initial line, and this flow rate, applied externally or internally to the base of the fire, has materially increased our successful batting average.

On the way in, when fire or smoke is showing, dry lines can be laid from a hydrant to be charged by the next arriving firefighter or company. In rural operations, lines can be laid from the location where you intend to place a portable drop tank or drafting pumper. This action is far preferable to running out of water and having to reassign attack personnel to hand-lay supply lines or redirecting another engine (which may be needed elsewhere on the fireground) to provide the hose for a continuous water supply. If the fire is well involved, a sustained water supply will eventually be needed. It is more efficient to lay the lines at the outset and have a constant water supply ready for connection when attack is begun (fig. 11–18).

Figure 11–18. More than one pumper should be in operation at multiple line fires. If this department depended solely on this pumper for attack, the broken radiator hose and subsequent shut down could have spelled disaster for interior crews.

If operating procedure calls for engine companies to make the initial attack using only tank water, it is good practice for the pump operator to announce over the fireground radio when a secured water source has been established. Knowing that more water is available will certainly have a positive effect on operational decisions made by the incident commander and officers inside the building.

Review

While not intended to cover every aspect of tactical hoseline use, these suggestions should form the basis for discussion among your department's operations personnel. A technique which may work extremely well with five-story tenements in New York City may not be as acceptable in Chicago's three-flats or in a barn loaded with hay in Ohio. Many tactics will remain the same no matter what area of the country the fire is located in.

There are no single quick fixes when it comes to fire attack tactics. All too often, members of the fire service have picked a single tool, such as fog streams, 1¾" hose, or positive-pressure ventilation, and have attempted to make the tool work for all fires. There is nothing wrong with these items if they are employed to the best advantage after evaluation of the problem. It must be kept in mind that there is no single tactic or tool which can be used on all fire-extinguishing problems. Answers and hardware choice can only be decided after careful consideration of a number of factors.

Successful fire stream management consists of efficient accomplishment of the following tactics:

- Water supply
- Flow rate
- Stream penetration
- Ventilation
- Safety during all of the aforementioned operations

The tools and tactics presented in this book should not be considered as absolute answers. They should be used as starting points for development of your department's individual, custom-designed fire management plan (fig. 11–19).

Figure 11–19. When the fire is of such intensity, such as this fully involved second-floor apartment, knocking the fire down from the outside is now a proven technique to make interior advancement safer and to reduce the interior temperatures to make it more tenable and survivable for any victims trapped inside. When teaching, I use the phrase *give it a toot* until the fire has materially reduced in intensity. (Photo courtesy of Jim Regan)

Chapter 11 Review Questions

1. How does the location of the fire within a structure impact engine company operations?
2. Where should the first line be placed at a working fire?
3. What are some of the primary considerations when estimating the volume of water that will be needed in a commercial occupancy?
4. What is the best position for the hose line officer when advancing on a fire?
5. When do you need to be on your knees during fire attack?
6. Describe a flow-and-go operation.
7. What are a couple ways to handle unwanted volumes of steam?
8. What are the possible outcomes of using only the tank water for an attack?

Managing Hose

Unless some advance consideration and planning is given to how hose will be carried, stretched, and controlled at the fire scene, a needless amount of time and energy will be expended in the effort to bring effective streams to bear on the fire.

There are a number of different ways to pack hose in a bed. The best method of loading hose for a particular operation can only be determined by repeated evaluation and practice stretching of each line by the individual department within its own operating environment. For example, taking a line to the top floor of one of Boston's frame apartment buildings will be quite different from stretching a line to protect an exposure during a Los Angeles County brush fire.

Deciding how to pack and deploy attack lines on the first-arriving apparatus is determined by the line's intended flow, length, and number of personnel responding on the unit who will be available to stretch lines. Required flow and length should be determined by evaluating the results of response area planning.

To be able to use a line at a fire, some means must be provided to get it to the scene. Almost all attack hose carried by fire departments is packed in working lengths inside apparatus hosebeds. How quickly and easily these lines can be stretched and operated begins with how the hose is loaded.

Attack Line Hose Loads

An attack line can be defined as hose packed in a bed with a nozzle attached, arranged for quick deployment. There are three basic types of beds used for attack line storage:

- Preconnected beds
- Bulk hosebeds
- Reverse lay or day line beds

Preconnected Beds

Most of the fires in this country are extinguished with preconnected hoselines. If their length, flow, and operation are planned in advance, preconnected lines can provide the hose lengths the pumper needs for average fire distances and supply streams of adequate flow capacity to extinguish most common fires while having the decided advantage of being rapidly deployed.

There are two operational problems which can sometimes hinder efficient operation of preconnected lines. The first is lack of adequate hose length. Unless a department conducts response district research and measures distances between probable pumper positions and various points in a number of target hazard buildings, preconnected lines will many times come up short when stretched into a fire building. This is not the fault of the preconnect but a result of improper or nonexistent planning. Too many departments merely fill up the standard size bed supplied with the new pumper without evaluating their own particular needs. When the heat is on, it pays to have measured to assure that preconnected attack lines will be able to reach the fire (fig. 12–1).

The second most frequent preconnect line operational problem is lack of sufficient training and experience in sizing up a fire situation. If the size-up is not accurate or complete, it can have a negative effect on proper hoseline deployment. Since small- or medium-diameter preconnects are the most frequently used hoselines and are easy to get into position, they are too often stretched to fires obviously beyond their extinguishing capacity. The *theory of*

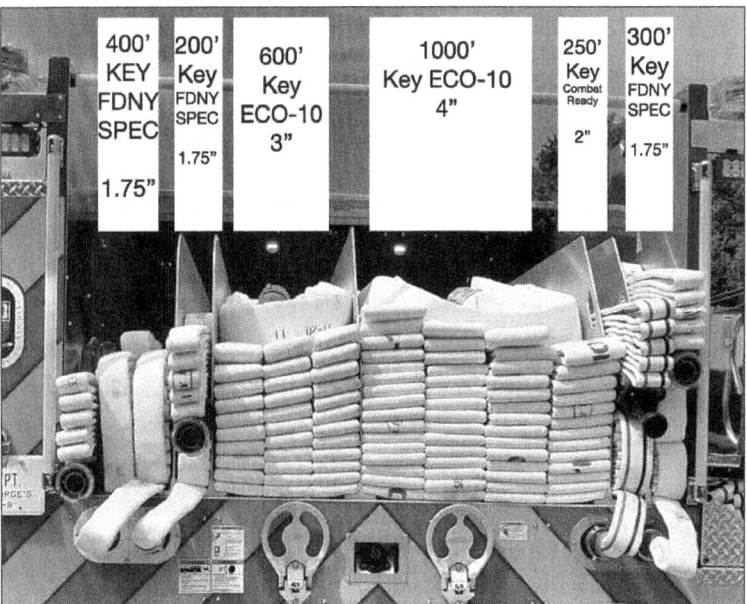

Figure 12–1. The Prince Georges County, VA, fire department, after much research, developed this hose load to provide proper hoseline reach and flow for their diverse response district. All 1¾" lines flow 150 gallons per minute (GPM) through a combination low-pressure nozzle attached to a shutoff with a built-in $^{15}/_{16}$" smoothbore. The 2" bed is equipped with a 1$^{1}/_{16}$" smoothbore flowing 240 GPM. The need for longer preconnects is predicated on long lays into garden apartment complexes and commercial buildings. After quite a bit of thought, they developed a hybrid hose load that delivers the required flow exactly where it is needed. (Photo courtesy of Prince Georges County Fire Department)

recency was discussed earlier, describing situations in which a firefighter, unless they have used something recently, may have forgotten its use or availability. Unfortunately, this theory frequently rears its ugly head when an inexperienced or poorly trained officer must rapidly make a size-up and decide what firefighting tactics are most appropriate. One way of limiting the theory's effect is frequent training in the use of larger flow streams. Training in evaluating burning buildings and potential fire spread hazards is extremely important if increased survivability for trapped victims and limited fire damage are desired. All first-arriving suppression personnel should be capable of making correct decisions as to line flow and placement when an attack is made. In many cases, the fire's size-up is correctly made but effective attack suffers because personnel were never adequately taught how to quickly deploy large, high-volume lines.

New Hardware to Increase Efficiency

Along with training in decision-making, departments should also evaluate high-flow equipment to make sure large-stream deployment is not hindered by heavy, out-of-date appliances.

Advantages of 2" and 2½" Preconnected Lines

One of the most effective tools departments can add to their firefighting arsenal is a preconnected attack line flowing more than 250 GPM. If the large preconnected line is as readily available as smaller attack lines, the decision to immediately deploy the large line becomes an easy one to make.

If a smoothbore or low-pressure combination nozzle is used in place of a standard-pressure combination nozzle, the line will become much easier to control, a plus if it must be operated by a limited number of personnel (fig. 12–2).

I'm surprised at how many departments have never explored the possibilities of deploying a high-flow preconnect line immediately upon arrival at a building showing a heavy volume of fire. Attempting to make up a large offensive line from hose contained in a supply

Figure 12–2. This department has a selection of 1¾", 2", and 2½" preconnected attack handlines, all accessed from the rear, and all packed in a low hosebed due to the rig's L-shaped tank. Two beds of 3" are carried as supply line.

bed is a throwback to the old days when streams were supplied directly from a hydrant without going through the apparatus pump. Hunting around in compartments for a nozzle and needed fittings can seriously delay getting a large line into service at a time when rapid deployment may mean the difference between a good stop or a large loss. In too many instances, company officers not wanting to go through the time and effort of making up a larger line will frequently attempt to make do with a smaller line, hoping that an increase in aggressiveness will make up for the reduced flow. No matter how much effort is expended in moving in the line, it cannot accomplish extinguishment if the flow is not sufficient to overcome the heat being produced by the fire. If a line that cannot provide the required flow is mistakenly selected, the fire will continue to burn.

Another use for a preconnected 2½" or 3" line is to quickly supply a portable ground monitor. The short hose length has little friction loss and is able to deliver 400 GPM–500 GPM to get water on the fire or exposure quickly. If the fire is of such volume on arrival that defensive tactics are used immediately, it makes much more sense to apply a heavy stream with a monitor than have a couple of firefighters wrestle a high-volume handline for what could be an extended period of time. In many cases, the tendency to pull a too-small line can be corrected with training in both making decisions concerning proper deployment of high-flow lines and the actual hands-on use of larger lines and master stream devices (fig. 12–3).

Locating Preconnect Hosebeds

Preconnected lines can be located practically anywhere on a pumper as long as they are able to be easily stretched and the supply piping is of adequate size to efficiently handle the desired flows.

Figure 12–3. This department carries an attack ground monitor attached to 3" supply line along with a 2½" preconnect, which is quite a bit of firepower, in case heavy fire is encountered.

Rapid deployment of attack lines, and supply hose for that matter, can be greatly aided by proper design and placement of hosebeds. Formerly, all hose was laid from the rear of the unit, but today, hosebeds can be found almost anywhere from the front bumper to the back step. The most popular location for 1¾" attack lines is probably across the body, from side to side, over, or ahead of the pump area. Hose packed in *cross lay* or *mattydale* (named after the town in western NY which pioneered their use) beds is usually connected to the pump by the use of a swivel connection. This allows the line to be pulled off either side without kinking. The location of these beds permits attack lines to be quickly stretched from either side of the unit, and, because attack lines are carried forward, more space is made available in the main bed for carrying more supply line or additional attack lines.

When specifying beds to be installed on new apparatus, their width should be determined by measuring the actual flat width of the hose which will be carried. For example, 1¾" hose is not 1¾" wide when laid flat. It measures approximately 3¼". If not specified by the purchasing department, many apparatus manufacturers fabricate standard beds based on the flat width of 1½" line. If the width is not wide enough to accommodate the hose the department is using, the hose will have to be jammed in the bed, making it hard to deploy. Packing attack hose on edge, rather than flat, can sometimes help solve the problem of insufficient bed width if an older rig must be pressed into service.

It is wise to specify a minimum piping and valve size of 2" on all new apparatus if 1¾" hose is to be carried. The increased piping size will allow higher flows while keeping friction loss at a minimum. Beds for 2" or 2½" hose should be piped with 2½" piping and valves. It is also important to specify that a minimum of elbows be used in the plumbing of attack line beds. I've seen as many as five 90° bends used in the pipe run to a cross lay bed. No more than two should be allowed. Another way of assuring that the piping is sufficient is to specify a maximum flow rate desired from each outlet. Cross lay beds should be capable of flowing 250 GPM with 2" piping and 400 GPM with 2½" piping.

A variation of the cross lay is the speed lay, in which hosebeds are built into the front of the pump compartment from the step area upward. These beds, also fed by swivel discharges, offer the location advantages of the cross lay, but can be more easily deployed by firefighters without having to climb up on a step because they are mounted ahead of the pump rather than over the pump (fig. 12–4).

Preconnected beds exiting at the rear of the pumper offer the advantages of increased flexibility in hose loading and increased hose capacity over what is available in most cross lay hosebeds. In addition, after the lines are stretched and charged, they do not hinder the movement of the pump operator, as happens when cross lay lines are deployed over or around the pump operator's panel.

Engines can be specified with L-shaped water tanks which materially lower the rear hosebed height. Lines off the rear of a well-designed engine should be able to be stretched with the firefighter standing on the ground and not having to mount any steps (fig. 12–5).

Many departments are now locating preconnect hosebeds carrying 100'–200' of hose ahead of the cab, recessed into an extension built behind the front bumper. This location is becoming more and more popular for trash or nuisance lines because the hose can be easily deployed and can be repacked with minimum effort (fig. 12–6).

Figure 12–4. Speed lays are similar to cross lays except that they are mounted vertically and are lower than hosebeds over the top of the pump area. It is important to specify speed lay bed dimensions and capacities carefully. It is practically impossible to add more hose, if desired, after the beds are built.

Figure 12–5. This department runs all its working line off the back with 2½", 2", and 1¾" attack lines, 3" supply hose, and 5" supply hose coupled to a hydrant relay valve. The low height hosebed allows for line removal without having to mount the rear step.

Figure 12–6. Storage wells and prepiped discharges can be easily provided by apparatus manufacturers, built into the front bumper extension of new apparatus. This location makes deployment of short lines rapid and convenient for trash, grass, and vehicle fires.

Packing Preconnect Hosebeds

One of the most common loads used for packing preconnected lines is the flat pack, laying the hose on its flat side rather than on its edge, about two or three hose widths wide, and stacked as high as needed. Usually, two or three loops are provided halfway up the bed, forming handles of sorts, or bunny loops, that provide a grip for ease of pulling. Unless sufficient hose is provided below the loops, between them and the hose connection, the pack will only come out of the bed as far as the hose connection will allow. In most cases, this results in dumping the hose in a pile on the ground next to the engine. Evaluations of the effort involved in stretching this piled line have shown that finding the nozzle and advancing it toward the fire usually causes the hose to become twisted as it feeds from the jumble of hose. In addition, as the amount of hose from the engine to the nozzle becomes longer, the friction caused by the hose contacting the ground becomes greater, requiring more effort on the part of the firefighter to pull the desired amount of line to the fire (fig. 12–7).

What sometimes happens on the fireground is that the firefighter will pull just enough line to be able to make the first shot, usually with the excuse of saving time. The tendency is that, if the firefighter does not take enough hose, extra hose will end up at the pumper, not

Figure 12–7. Loops inserted in a hosebed will make it much easier for a firefighter to pull and shoulder-load a preconnected line.

at the fire. Hose can be easily stretched when dry but becomes more difficult to move when filled with water under pressure. It is certainly more energy efficient to pull the needed line to the point of operation before charging the line, even if it takes a few more seconds. Remember, almost every attack line will have to be moved forward from the point at which water is first applied. It simply makes sense to have extra hose available near the nozzle, rather than 200' back at the engine.

Shoulder-Loading Hose

Always remember to work smart when stretching hose; expend the least amount of effort to accomplish the most amount of work.

One example of working smart is to pack lines so that part of the hose can be shoulder-loaded. When the line is deployed, if the hose load has been packed so that at least half a standard-length, 150'–300' hosebed is packed underneath the loops, the nozzle operator can shoulder-load the top part of the load and easily transport that hose to the area of operation. The hose remaining in the lower portion of the bed will feed out as the hose is advanced. If another firefighter is available, assistance can be provided to clear the remaining line from the bed by grabbing a few folds and pulling them toward the fire as far as the line allows. Only half the effort will be required by the nozzle operator to overcome the friction created when the hose slides along the ground (fig. 12–8).

Figure 12–8. Carrying the weight of the hose on the firefighter's shoulder is a great way of working smart, allowing the firefighter to safely carry a number of lengths of hose a required distance. When moving hose, reducing friction between the hose and the ground makes life much easier for the firefighter.

If the bed contains more than 200' of hose, it may be necessary to provide a second set of loops near the bottom of the load. For best operating efficiency, the maximum amount of hose shoulder carried by each firefighter should be no more than 150'.

The wisdom of shoulder-loading hose can easily be demonstrated on the training ground by having the nozzle operator stretch a line to a third-story landing using first the shoulder-load, and then simply pulling the line up the stairs. It will be much easier to carry the hose to upper floors by having both hands free to assist in climbing than to drag it up along the stair treads. When stretching hose above street level, the legs, not arms, should be used whenever possible to help reduce fatigue.

When firefighters carry the hoseline into a building instead of dragging it, extra hose will usually be available near the nozzle, not strewn around the pumper. This makes advancement into the fire area a much smoother operation than having to yell downstairs or outside for someone to feed in more line.

It makes little sense for a firefighter to expend great amounts of energy stretching hose if the exertion can be safely and efficiently minimized. Energy should be conserved as much as possible for nozzle and search operations inside the fire building. Practicing energy conservation can become second nature if the hose loading and hose laying operations are adequately planned and practiced on the training ground before being deployed at a fire.

Some preconnect line hose loads are designed with the nozzle located in the middle of the pack. In this way, the shoulder-loaded hose can be fed off the top of the load when the building is reached. This pack, called the *minuteman load* in some areas, is a variation of the flat pack and is only one of a number of time and energy-saving hose loads, each of which should be evaluated to find which combination works best for a particular situation.

It might be worthwhile to visit other departments to better evaluate different ideas in hose loading techniques. Sometimes ideas from two or three different departments can be combined into a single operation to help increase speed and efficiency when stretching and flowing lines on the fireground. The variations available in hose loading and deployment are limited only by imagination and the amount of study and planning completed before placing new-style lines into service.

Bulk Hosebeds

Attempting to determine the length of a standard preconnect line is difficult for many departments because widely varying distances are normally encountered between the pumper and the fire. In these departments, if a certain length of hose was placed in a preconnected bed, it might frequently prove too short or too long for the job at hand.

One way of addressing this situation is with bulk beds—simply a large bed of attack hose with a nozzle attached and the end left unconnected to the engine (fig. 12–9).

In normal use, the nozzle and the needed amount of hose are removed by the attack crew and stretched to the fire. When enough hose reaches the objective, the pump operator disconnects the hose at the next coupling, connects the line end to a discharge, and charges the line when the attack crew calls for water. With practice, this operation can be accomplished in almost the same amount of time as stretching a line from a preconnect bed and offers the advantage of not having a large amount of extra hose strung around the fire scene.

Figure 12–9. Because its hose is stretched different distances at practically every fire, the New York City Fire Department (FDNY) runs bulk hosebeds, each bed with all hose coupled together and topped with a nozzle. Typical hose loads are 500' of 3½", two beds of 500' of 2½" with 300' of 1¾" on top, and 600' of 2½" with nozzle attached. No preconnects are provided off the main hosebed. (Photo courtesy of Joe Pinto)

Storing such a large amount of hose in the attack line bed also allows for ease in reaching distant objectives, such as a long line stretched to protect an exposure supplied from a pumper on a hydrant the next street over. Use of bulk beds is ideal when a line must be stretched to the rear upper floors of a multiple-unit fire building to cut off a spreading fire.

When rear-facing beds are used as bulk beds for carrying attack hose, they appear similar to reverse lay beds, except when stretching from a bulk bed, the engine remains in place and the lines connected to it when the stretch is complete. In a reverse lay operation, the line is removed from the engine which then proceeds to the water supply source laying in effect a supply and attack line.

When stretching hose from a reverse lay or bulk-load bed, place an amount of hose on the shoulder when advancing the line rather than just grabbing the nozzle and moving in the direction of the building. Taking only the nozzle while trailing hose behind when stretching hose from any type of bed is an energy-expending practice that can be made more efficient by simply grabbing a few folds of hose with the free hand and carrying them along to the fire.

A bulk bed line can greatly increase an engine company's efficiency if varying distances between the pumper and the fire are frequently encountered. One length as well as twenty can be easily stretched and quickly charged, depending on the situation at hand.

Reverse Lays

Reverse lay hosebeds are similar to bulk beds except that they are stretched by removing the needed amount of working line and the engine, continuing to lay hose to the water source. Reverse lays offer the distinct advantage of being a combination of both supply and working line together, allowing one company to quickly drop attack lines at the scene and then proceed to the water source with the pumper to establish a continuous supply operation. Since the pumper is always located at the water source when using reverse lays, proper nozzle operating pressures can be more effectively supplied, because the pump operator, located at the water source, can exercise greater control over incoming and outgoing flows and pressures. Widely used in larger cities, reverse lays are now being increasingly used by smaller departments, which, because of limited personnel availability, must use a single pumper to accomplish both attack and water supply operations (fig. 12–10).

Figure 12–10. This reverse lay hosebed set up to lay a 2½" working line supplied by 700' of hose will prove useful if the 200', 2½" preconnect bed at left is too short to reach the fire.

In areas where hydrants are available, there is really no reason for an attack to be stopped because the pumper runs out of water before the fire is knocked down or fully extinguished. The main reason attack companies frequently run dry is because a supply line was not laid. It may be that the officer did not order a supply line stretched on the way in because the fire could not be seen before arrival, or perhaps they didn't want to commit one third of a three-person pumper crew to hydrant duties, preferring to have that person assist with actual attack.

Efficient and safe firefighting calls for both an aggressive attack and an aggressive supply operation. If personnel are committed inside fire buildings, everything possible must be done to keep water supplied to the attack crew. Most of the time, the tank water carried by the attack engine will be sufficient to handle the fire. However, chances are someone is going to get hurt at those fires where the tank runs dry without a secured source of supply being established (fig. 12–11).

Some departments rely on a second engine company to establish the secured water supply. If the arrival of the second company is somehow delayed, the results of running out of water could also prove disastrous to the crews and victims inside the fire building. In an area served by hydrants, there is no good reason why three firefighters cannot make an attack supplied with a secured water source. Granted, the reverse lay is not as convenient or as easy to deploy as simply stretching a preconnect, but it will ensure a constant supply of water once

Figure 12–11. A simple reverse lay hosebed consisting of a number of feet of supply line, 50' or more of working line (loaded horseshoe style for ease of stretching) coupled to a 1¼" smoothbore nozzle. This set-up can easily supply over 300 GPM to a distance of 600 or more feet from the engine.

a working hydrant is found. When using reverse lays, the pump operator becomes more productive, sending water as well as making the hydrant hookup. It's not as if the pump operator is unproductive when a forward lay is used; it is just more desirable that a higher priority be placed on securing a constant water source than on having the operator assume less important duties, such as traffic control, during initial attack.

It is unfortunate that, due to the lack of personnel, the fire service has to modify otherwise sound and workable tactics with sometimes less desirable compromises. While some compromises, such as two-person companies, should be unacceptable from a safety standpoint, the reverse lay can actually be a positive compromise, especially during daytime hours in volunteer departments where a large number of firefighting personnel may not be quickly available.

The main advantage of the reverse lay is that it allows a limited number of personnel to secure a constant water source. It also allows a limited number of personnel to flow large quantities of water if portable master stream devices are used. In rural areas, the first-arriving pumper can reverse lay lines from the fire to a drafting source or portable tank location to help assure a constant water supply (fig. 12–12).

If a large number of personnel are available (nowadays a five-person engine can be considered large), the reverse lay has the advantage of supplying as much water as can be handled, provided the water source is adequate.

Packing Reverse Lay Beds

Reverse (or fire-to-water-source) lays differ from forward (or water-source-to-fire) lays basically by the direction in which the hose is packed. The male end is at the end of the line for reverse lays and the female end is at the end of the line for forward lays.

Figure 12–12. Two Chicago engines with reverse lay hosebeds await orders to stretch a line or two on an industrial building. The reverse lay allows for the size-up to be determined, and then the number of lines and length of the stretch needed can be immediately deployed. The engine proceeds to a hydrant away from the area of action to leave room for additional apparatus. (Photo courtesy of Steve Redick)

For example, many departments carry a bed of 2½" or 3" hose, approximately 500'–800' in length, with a smoothbore nozzle attached. On the threaded end of the large nozzle, 100' of 1¾" line with an attack nozzle is attached. The smaller line provides around 150 GPM of flow and is easily maneuvered inside a fire building. If the fire overwhelms the smaller line, it is detached and the larger line with its higher flow capability is immediately available. This use of two different sizes of hose, coupled together in a reverse bed, is commonly called a *leader line bed* or *day line* (fig. 12–13).

Usually, a second bed of 2½" or 3" line with a nozzle attached is packed alongside the first. Special folds, sometimes called *skids*, *horseshoes*, or *flaps*, are packed at the top of the load so a predetermined amount of working line can be easily shoulder-loaded when being pulled out of the bed.

In use, the day line is easily shoulder-loaded for rapid advancement, in this case, up an outside fire escape. As much line as needed can be pulled from the bed before either connecting to the pump or laying to a hydrant. A day line bed overcomes the length limitations sometimes found when using preconnected beds.

Operating with Reverse Lays

When using reverse lays, the pumper stops near the fire building, the nozzle crew removes enough hose to reach the fire or its estimated point of extension, and the signal is given for the apparatus to proceed to the water source, laying the supply hose behind. When the pumper arrives at the water source (a hydrant, for example), the operator puts the pump in gear, goes to the rear of the pumper and disconnects the bed at the first fitting, connects it to the pump, and, depending on department policy, sends tank water. Some departments have the pump operator flush the hydrant first to make sure it is operational. If it is found inoperable, the pump operator simply drives to the next hydrant (fig. 12–14).

In city or suburban areas where hydrants are regularly inspected and the water system is in good repair, waiting to charge the line until the hydrant is inspected is usually not delayed, and the tank water is sent immediately.

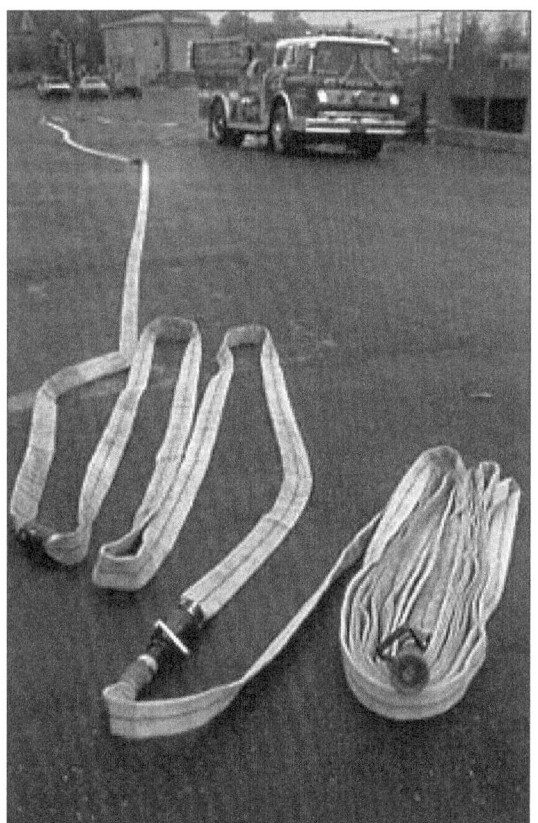

Figure 12–13. A day line is simply a preconnected hoseline that combines an interior working handline along with a high-volume back up line. Depending on local conditions, there should be at least 100' of 2½" line to insure that, if needed, it can reach an effective point of operation.

Figure 12–14. The engine shown here is located at a safe distance in case of building collapse. A reverse lay puts the engines at the water source where they can better control water delivery, keeping the engines away from the working area around the fire building. (Photo courtesy of Dennis Walus)

In inner-city areas where vandalism is high, not charging the line with tank water is certainly good practice. In Chicago, a hydrant person dismounts the pumper when it stops at the fire and proceeds to the hydrant before the pumper crew finishes unloading attack line. They then attach a gate valve to the second port and flush the hydrant, blowing out any debris that may be present and, in winter months, making sure the hydrant is not frozen. The flushing is done before the pumper's soft suction is attached. If the hydrant is found to be inoperative, the engine picks up the hydrant person and fittings and proceeds on to the next hydrant to try again.

Reverse Lays

As with all tools and tactics, reverse lays have their good points and bad points as listed here:

Advantages
- It places the pumper at the water source, allowing it to better utilize available water resources.
- It clears the front of the fire building for placement of aerial or squad apparatus.
- The apparatus becomes self-sufficient and not dependent on another pumper for an adequate water supply.
- A more constant water supply is assured, which is important in providing protection for interior attack forces.
- Fewer personnel are involved in the water supply operation, freeing more firefighters for search, attack, and ventilation operations.
- The fire can be sized up before apparatus are committed to operating positions or lines are laid. This can be important if the fire is showing from one position but can only be approached from another. If an engine lays a line on the way in and the fire cannot be accessed from that location, the effort expended to lay the line will have been wasted. Larger cities use reverse lays in part because the officer can see exactly what is needed and where it is needed before making any apparatus assignments.

Disadvantages
- Unless equipment, such as forcible entry tools, ladders, or poles, is removed at the fire scene, it might end up being located some distance from the fire.
- The reverse lay operation can be somewhat slower in getting water on the fire than when deploying preconnected lines. In actual practice, the only delay is the travel and hookup time before charging the line with tank water.
- If additional lines are needed, hose usually must be stretched from another unit because of the distances involved.

The third disadvantage can be an advantage in some cases, because after the lines are laid to a pumper operating at a water source, the second pumper then becomes immediately available. It can then be used to lay additional lines to other pumpers or lay to another water source to supply its own lines.

In years past, if the hydrant was found frozen, some rags, stored in a coffee can, soaked in kerosene or diesel fuel, and hung from a coat hanger, were ignited and applied to the operating stem from the inside of the barrel through the large hydrant port. In many cases, the stem is frozen at the top where it passes through the cap. The heat from the burning rags, applied as high as possible inside the hydrant, was usually enough to melt the ice and allow the stem to be broken loose by pressure from the hydrant wrench. If this did not work, the freezing problem was at the bottom of the barrel. Dropping the burning rags to the bottom of the hydrant many times provided enough heat to allow wrenching of the operating nut to break the valve loose.

Using Large-Diameter Hose for Reverse Lays

When large-diameter hose is available, it can be an excellent tool for supplying multiple attack lines by using the reverse lay. Because of its low friction loss, large-diameter hose has the ability to move great quantities of water long distances, especially when a pumper is utilized and is pumping on the supply end.

Most large-diameter hose is equipped with sexless Storz couplings. Since the couplings on both ends of the hose are the same, no double male or female adapter fittings are needed to make either a forward or reverse lay. Reverse lays using large-diameter hose can be made to supply another pumper, an aerial apparatus, or distributed by using a manifold or portable hydrant device.

Since movement of great quantities of water through large-diameter hose creates high kinetic or moving energy, attempting to stop this flow quickly has the same effect as running a car into a concrete retaining wall. To help relieve this excess kinetic force, relief valves should be provided on all intake and manifold fittings. Water hammer in large-diameter hose can be extremely damaging to equipment and to any nearby personnel if a hose bursts.

The use of large-diameter hose and multi-port distribution devices can make operations at a large fire more efficient than having a number of companies lay multiple smaller lines in an attempt to supply the same flow.

In addition to moving great quantities of water, large-diameter hose can help develop tactics that place pumpers at high-flow hydrants or drafting sites, help keep the fire scene and access roads clear of unnecessary apparatus and make picking up after the fire is out easier and quicker than if multiple smaller lines are used.

Forward Lays

The forward water-source-to-fire lay is probably the most frequently used method of supplying water to a pumper. It is simple and can be quickly deployed. If large-diameter hose or multiple lines are utilized, the forward lay can supply a sufficient amount of water to handle most fires.

In practice, when using a forward lay, the pumper stops at a hydrant located close to the fire building. The designated hydrant person removes about 25' of hose from the rear of the pumper and wraps it around the hydrant to prevent its being pulled away from the water source once the pumper moves toward the fire. As the pumper is traveling to the fire, the hydrant person attaches the line to the hydrant, gating other ports for later use if necessary, and awaits the order to charge the line. At the pumper, the pump operator charges the attack line and then pulls line from the supply bed until a coupling is reached. This connection is uncoupled and the line attached to a gated inlet (fig. 12–15).

If only three members are assigned to the pumper, the remaining firefighter or officer has the sole responsibility of size-up, planning, stretching, and operating the attack line if necessary. If more than three members are riding the pumper, the extra firefighters should assist with stretching and operating the attack lines, search operations, and ventilating.

Remember when using a forward lay for water supply that its waterflow capability will be determined by two factors:

- The pressure and volume available at the supply source
- The friction loss within the supply line

The water flow can be increased when using a forward lay by taking the following steps:

- Laying multiple lines
- Laying large-diameter hose
- Having a second pumper supply the lines from the water source
- Laying additional lines from a different hydrant
- Any combination of these steps

Figure 12–15. This engine has two forward lay supply beds, a 3" on the left and 5" on the right. Once the supply line is laid, preconnects are deployed to attack the fire.

Forward Lay Hosebeds

A forward lay hosebed is usually ended by having a female fitting attached to the line and a hydrant wrench and hydrant gates either attached to the line or available in the bed. It is important to have the hydrant wrench and necessary adapters secured to the line. If it is not, there is too great a chance of having this necessary tool and fittings travel on to the fire, with the hydrant person running after the pumper attempting to retrieve it. One way of carrying the wrench is to stuff it in the open butt of the hose. Another is to clip it to the snap of a hose tool. The idea behind fastening the wrench to the hose is to make sure it stays at the hydrant when the pumper leaves (fig. 12–16).

It is a good idea to gate an additional port, even if a large-diameter hose is being laid, so it can be used as a bleeder to relieve pressure when the line is shut down (figs. 12–17 and 12–18). This is especially important if the hydrant is located below the pumper or delivery point. Head or back pressure is best bled off using a 2½" gate after the hydrant is shut down rather than attempting to wrench the couplings apart. The couplings could be damaged and the resulting blast of water exiting the couplings could injure firefighters (fig. 12–19).

Figure 12–16. If more water is desired than the initial lay can deliver due to a combination of hydrant pressure and friction loss in the hose, a second pumper attaches suction and discharge lines to the valve and then using the valve to redirect the flows, it pumps down the first line without interrupting the attack pumper's supply.

Figure 12–17. It's always a good idea to gate the opposite port of a hydrant when laying a supply line, so additional water can be obtained if conditions warrant. On this older hydrant that has only two 2½" outlets, it is rarely a problem supplying large-diameter supply hose when you consider that a 2½" hydrant outlet can flow 1,040 GPM at 40 pounds per square inch (PSI) pressure. With the second port gated, depending on main size, this hydrant has the potential of flowing over 2,000 GPM.

Chapter 12 **Managing Hose** 303

Figure 12–18. A lineman's operators tool bag makes an ideal container in which to carry hydrant fittings, valves, and wrenches. The West Carrollton, OH, Fire Department keeps their bag on the rear step for easy removal when making a supply line lay.

Figure 12–19. One effective method of increasing fireground water flow is by the use of a hydrant relay valve. In use, the valve is carried attached to the end of a forward lay hose bed and attached to the hydrant when the initial lay is made.

Relay Valves to Increase Flow

When making a forward lay, the use of a relay valve on the hydrant allows another pumper to make suction connections to the same hydrant through the valve, then pump into the relay valve and down the initially laid line (fig. 12–20). This will increase the flow without having to interrupt the supply flow to the attack engine (fig. 12–21).

Figure 12–20. If more water is desired than the initial lay can deliver due to a combination of hydrant pressure and friction loss in the hose, a second pumper attaches suction and discharge lines to the valve and then using the valve to redirect the flows, it pumps down the first line without interrupting the attack pumper's supply.

Figure 12–21. In areas where older hydrants with just two 2½" ports are common, the hydrant relay valve can be attached to a 2½" port with an adapter. It is good practice to gate the other port so the engine can be supplied with additional water.

Because of variations in main size, age, and water distribution design, a water system will rarely supply the same amount of water and pressure throughout an entire fire operation. The use of a relay valve will help cope with the system's extremes by allowing the department to easily insert a pumper into the supply line to boost delivery rates (fig. 12–22).

Figure 12–22. In this case, the supply engine connects to the relay valve and then stretches an additional supply line to provide increased water supply.

Flying Lay

A variation of the forward lay is the *flying* or *engineer's lay*. It is used when the pumper goes to work at the fire scene, supplying attack lines from the tank without laying a supply line (fig. 12–23).

After charging the attack lines, the pump operator will stretch the supply line to a nearby hydrant. If the hydrant is located within 300' of the pumper, this operation can rapidly establish a constant supply source without having to utilize additional personnel.

To make a flying lay, the pump operator or hydrant firefighter removes the supply line from the bed and advances to the hydrant. If more than 50' of hose is needed, the pump operator should shoulder-load as much as possible before proceeding to the hydrant. Dragging hose to a hydrant from a bed becomes extremely difficult if over 50' is needed. The pump operator may have to make two trips to obtain enough hose for the lay. It is wise to take the necessary tools and gates along with the first line of hose to eliminate multiple trips to and from the apparatus.

The flying lay takes more energy and time than using either forward or reverse lays. It should be considered more of a crisis situation evolution rather than a standard operating tactic.

To review, using the forward lay for supply lines can cut the time necessary to establish a constant water source. Its efficiency depends on how many lines are laid, the line's length, and the line's diameter. It is possible to increase the flow by using certain hardware, inserting

Figure 12–23. A Chicago engine company carrying the standard 150' of 4" supply line for a flying lay on the front bumper extension. In practice, the engine pulls past the fire to allow room for aerial equipment and lines are led out to the fire area. The engineer or hydrant person performs a flying lay to the closest hydrant either from the front or rear of the unit. The pump has 4" intakes at both the front and the rear. Since Chicago hydrants are spaced no more than 300' apart, the pumper is usually within 150' of a secured water supply. A 25' length of 5" soft suction is carried on top of the 4" hose. (Photo courtesy of Jim Regan)

another pumper in the initial layout to boost the flow, or using another pumper to lay lines to another hydrant.

As with all other tactics and equipment described in this book, the best way to properly design and evaluate the most efficient tactics for a particular firefighting operation is by actual testing, utilizing the personnel who will actually place the equipment in service at a fire.

Handling the Line

Nozzle reaction, or *back pressure*, is the hoseline's way of reacting to the energy of water exiting the tip. Newton's third law of physics states that for every action, there is an equal reaction in the opposite direction. The combination of the weight of the water flowing and the pressure pushing it out the nozzle determines the amount of nozzle reaction. To reduce it, it is necessary to reduce the flow, the nozzle pressure, or both.

Contrary to some texts, a combination nozzle does not have less reaction than a smoothbore nozzle at the same flow rate. In normal operation, a standard combination nozzle flows its rated capacity at 75 PSI–100 PSI nozzle pressure, and a smoothbore nozzle is most effective if it flows its rated capacity at 50 PSI nozzle pressure. At the same flows, the smoothbore nozzle will cause less nozzle reaction because it is delivering the same flow at half the nozzle pressure. By consulting the nozzle reaction charts or by using the formulas listed in other chapters to compute reaction force, it will be generally shown that smoothbore or low-pressure combination nozzles are less stressful to operate at the same flows than standard combination nozzles. If combination nozzles are preferred, low-pressure combination nozzles can help address the high reaction problem by delivering their flow at lower pressures than standard combination nozzles. In actual use, low-pressure combination nozzles exhibit reaction forces equal to smoothbore nozzles at the same flow and pressure (fig. 12–24).

When comparing reaction forces of smoothbore and combination nozzles, the combination nozzle should be set on the straight stream position to provide an equal basis for evaluation. As the combination nozzle's stream shaper is adjusted to provide a wider spray pattern, reaction energy will be spread out in the opposite direction of the pattern. If a wide pattern is used, the sides of the stream are drawn back to the nozzle, and it will be found that the reaction will be directed toward the centerline of the nozzle instead of being directed backward as with the straight stream, causing less backward force. A combination nozzle set at its widest spray position exhibits almost no backward reaction because the force is directed at the center of the nozzle and is balanced by flowing water on its opposite side.

Although setting the combination nozzle to a wider pattern will reduce backward reaction, the energy is now being expended in oblique directions instead of one, and this spread will cause the effective reach of the stream to decrease. Since the water must hit the fire in order to perform extinguishment, the crew must utilize ways to counter the nozzle reaction while it is operating at the maximum reach, straight stream position (fig. 12–24).

Figure 12–24. Probably the dumbest way for a firefighter to hold a hoseline is between their legs. In this case, there are 126 lb. of force of nozzle reaction only inches and two slippery gloves away from a painful injury that is certain to take you out of the game. If one person must operate a hoseline alone, the best method is to put a knee on top of the line and create as much friction with the ground as possible.

Countering Nozzle Reaction

While it is desirable to reduce friction between the hose and the ground when stretching hose, the opposite is true when operating hoselines. Utilized properly, friction is one of the nozzle crew's best assets in helping counter the effects of nozzle reaction.

When holding a line, as much friction as possible should be created by wrapping arms around the line and holding it close to the body. One good method of increasing the second or third firefighter's amount of control on the line is to use the Seattle Grip (fig. 12–25). Its use makes holding the line more secure than using hand pressure alone. It is simply a method of locking the arms around the line, providing both a hard-to-slip grip and allowing more hose surface to come in contact with the firefighter's body. The more line coming into contact with the body, the more nozzle reaction force can be absorbed by friction. If operating close to the floor inside a building, keep as much line on the floor as possible and create friction by kneeling or lying on the line during operation. Another technique is for the nozzle operator to back up against a wall or door frame, pinning the line between the operator and the wall or door. This allows the building structure to help absorb some reaction energy (fig. 12–26).

If the nozzle crew has the luxury of a second member, that person, using a combination of body weight and friction, should be capable of absorbing all of the nozzle reaction force. It is difficult for the operator to direct the stream as well as using their body weight to anchor the line. Unless the stream is to remain motionless, the operator's body must flex in many directions when distributing water. They can be easily thrown off balance if care is not taken when directing the stream.

If the line starts to slip from the operator's grasp, it should be quickly shut down until control can be regained. For safety reasons, the nozzle operator's hand should never leave the shutoff while water is flowing. If the line manages to get away without being shut off, firefighters should immediately run forward to prevent injury from the whipping nozzle. Because reaction will force the nozzle backward, common sense dictates that running forward will put as much distance as possible between the whipping hose and the firefighters. If the hose cannot be immediately shut off by the pump operator, lying spread-eagled on the

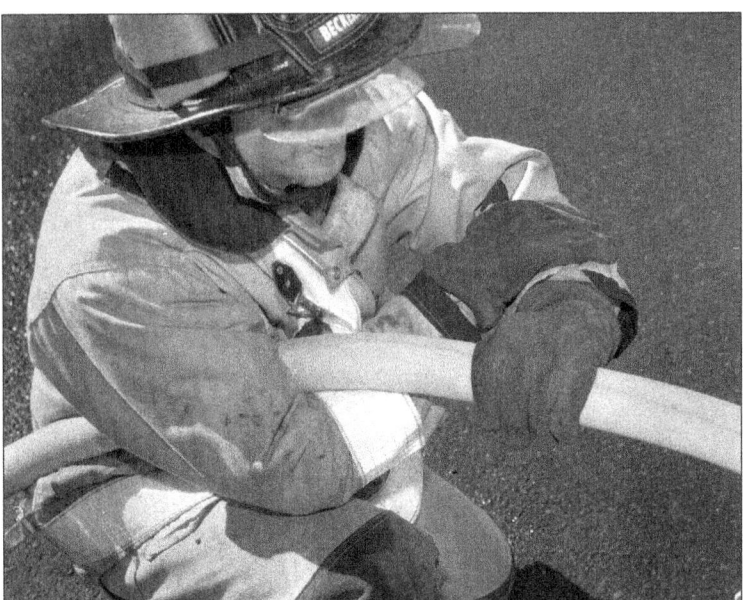

Figure 12–25. The Seattle Grip. By bringing one arm under the hose and then using that hand to grip the forearm of the other arm, the line can be contained and gripped securely. By hugging the line in this manner close to the body, more friction is created than if hands alone were used.

line and moving forward will gain control of the hose until the nozzle can be reached. In all cases, if over 50' of line is involved, it is much safer to have the line shut down at the pump or to have personnel kink the line shut a few lengths back from the nozzle to slow the flow before control is attempted (fig. 12–27).

Figure 12–26. Hoselines are best controlled when personnel are properly spaced behind the nozzle and grip the line in such a way as to create as much friction as possible. The third man on this line is attempting to keep the second person in position by holding on to his self-contained breathing apparatus cylinder. If the line were to get away, this pressure would be of no use except to place the second person in danger. For positive control, hoseline crews must grip the line, not each other. Two members on a hoseline should never position themselves back-to-back. The reverse person cannot see what is going on with the nozzle operator and cannot do an adequate job of being another set of eyes and ears. (Photo courtesy of Jo. L. Keener)

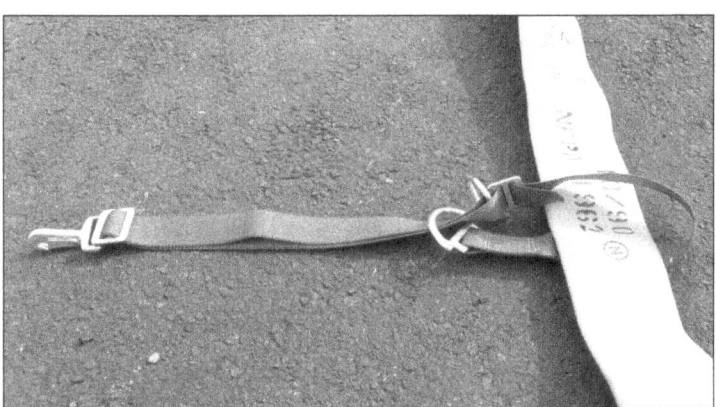

Figure 12–27. Some departments issue special belts or straps which are designed to help control hoselines. The Chicago belt allows for firefighters to carry tools as its basic function; however, when looped around a hoseline and the opposite arm placed inside the double webbing section, more control can be gained over nozzle reaction.

Use of Tools to Help Control the Line

One simple item of equipment, under-utilized for many years, is the rope hose tool. Formerly made of manila rope, it was stiff, especially when wet, and prone to mildew. Today's versions are fabricated of nylon rope or nylon flat strapping, making them light, strong, and flexible. The tool can be equipped with a snap hook in place of the traditional open hook, so the hook cannot be lifted off the support without opening the snap-latch when lines are secured to ladders or fire escapes.

Some cities use a variety of belts as well as rope hose tools to help keep lines under control. The whole idea of using belts, straps, or ropes is to secure the line to the crew or structure by using a means other than a hand grip, which can slip.

If a high-flow handline is to be used for defensive purposes, the operators can form a full 50' length into a loop and secure it with the tool (fig. 12–28).

In this operation, the nozzle end goes on top. It will then be in a position for easy operation without having to be pulled upward. A greater amount of hose will then be in contact with the ground, increasing friction that will, in turn, increase security and safety. If the nozzle end goes underneath, the small surface area parallel with the line of reaction force will act as a skid, offering little resistance to back pressure (fig. 12–29).

Pistol Grips

Pistol grips can make the handling of nozzles on small- and medium-diameter handlines safer. Originally designed for high-pressure booster line operation, pistol grips have proven effective in providing a secure gripping surface when attempting to control the reaction from high-flow, medium-diameter handlines. Keep in mind that at higher flows with increased

Figure 12–28. When securing hose into a loop, the nozzle end should be on top. This puts the nozzle in a natural position for stream operation, and having at least 50' of hose looped under the nozzle provides more ground contact, creating more friction. For defensive operations, this method provides more operator safety that merely sitting on the line and pulling the nozzle up between the operator's legs.

nozzle reaction forces, the pistol grip may not provide enough surface area in which to safely control the line. When experiencing flows above 150 GPM, it is safer to firmly control the hoseline behind the nozzle rather than depend on the grip itself.

The removable grip is the most convenient since it can be easily replaced if damaged or removed entirely if the shutoff is to be used on larger lines where other means of control are more effective.

When in use, remember to keep the pistol grip at least 12" forward of the body (fig. 12–30). Inexperienced firefighters have a tendency to keep the grip even with the front of the body. If the arm holding the grip is not extended forward in an almost straight out and locked position, it cannot bring as much strength to bear in the control of the reaction force. The folding arm tends to permit the nozzle to slide backward as soon as it is opened or when the operator begins movement to distribute water (fig. 12–31).

Figure 12–29. When looping a hoseline for defensive operations, a full 50' length should be allowed for the loop to properly counteract the nozzle reaction. It is also important to provide as much hose as possible at right angles under the nozzle as more hose creates more friction and makes the line safer to handle.

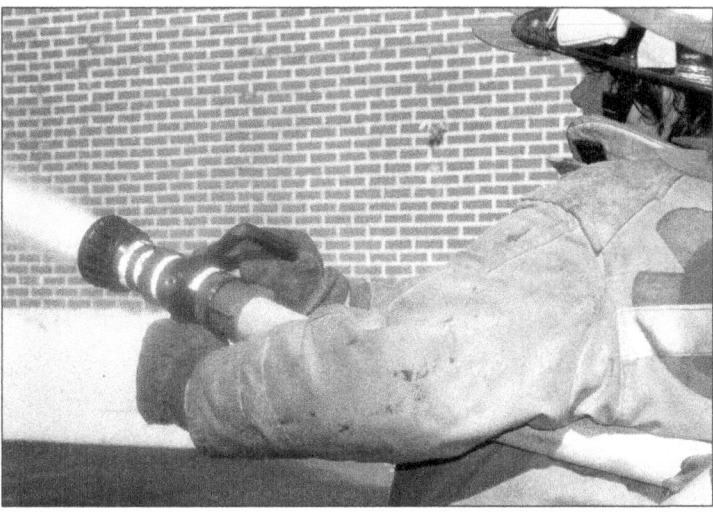

Figure 12–30. For effective control when using pistol grips, the arm holding the grip should be as straight as possible ahead of the body.

Figure 12–31. If the operator loses control, the line slips backward, and the arm becomes bent, diminished strength will be available to counteract nozzle reaction. When the nozzle becomes buried in the firefighter's gut, it's time to shut down and reposition.

Once nozzle operators get used to pistol grips, they sometimes find it difficult to control lines without them. A firefighter holding hose without the use of a pistol grip will tend to wrap an arm around the line in an attempt to increase contact area between the hose and body, creating friction to help in controlling the line. There is less tendency to wrap the arm around the hose when a pistol grip is in use, so the primary source of control friction is from the hand wrapped around the grip. If the hose slips backward and the arm bends, the small amount of control friction will many times not be enough to overcome the nozzle reaction force. Then, the line will have to be shut down or it will whip away from the operator.

While some officials continue to debate the effectiveness of pistol grips, the majority agree their use will make for safer operation.

Review

As with all other operations described in this book, handling hose and nozzles should be practiced regularly on the training ground in order to increase operating effectiveness on the fireground. To efficiently deploy hoselines, two important operations must be mastered by all fire fighters. The first is laying lines—transporting hose from the engine to the fire using as little energy as possible. The second is being able to handle the line once it's charged. Firefighters should know the amount of reaction generated by each size and flow line so they will not be surprised by high forces when operating on an actual fire. If hose loads and methods of advancement are planned and practiced, great amounts of energy will be saved on the fireground—energy better put to use fighting the actual fire rather than the hose.

A chief told me once that the best way to see how sharp a department operates is to pick up the tarp and take a look at the hosebeds (fig. 12–32). It doesn't take any more effort to pack hose neatly to make sure it can be properly deployed without getting tangled (fig. 12–33).

As in any mechanical field, competency and professional attitude are reflected in how much care you take with your equipment.

Chapter 12 **Managing Hose** 313

Figure 12–32. Imagine trying to deploy hose from this bed without it getting tangled. It takes no more effort to pack hose properly than it does to throw it in the bed in a careless manner.

Figure 12–33. If hose is packed and deployed properly, it is hard to imagine how this could happen. Whoever was responsible had to work really hard to cause this. (Photo courtesy of Jim Regan)

Chapter 12 Review Questions

1. What are the basic criteria for designing an attack line?
2. What are the three basic attack line loads?
3. What are some examples of working smart when stretching a hose line?
4. Define a reverse lay attack line.
5. Describe a day line.
6. Define a forward lay.

13

Evaluation and Maintenance

The nozzle is the business end of the entire firefighting system. No other device has more control over the outcome of the fire-extinguishing process than a nozzle. Not only does it deliver the water to extinguish a fire, it needs to be considered as important a safety tool as breathing apparatus or a helmet.

A nozzle is not merely a piece of aluminum, brass, and rubber that happens to be needed at the end of a hoseline to shape the water stream. It is a carefully engineered piece of firefighting equipment needing trained operators, proper water supply, and meticulous maintenance to function as designed.

Selecting the proper nozzle for handline or master stream use involves a good bit of research and evaluation. However, this effort is time well spent when tactics, firefighting efficiency, and safety are improved by the addition of high-performance water delivery equipment.

Nozzle selection by individual departments can be determined by examining a few basic factors. Among them are

- Type of tactics employed or desired to be employed by the department
- Required flow performance
- Personnel resources available on the scene
- Typical hazards within the response area
- Purchase cost and service support

A department utilizing the recommended guidelines in chapter 2, applying the research contained in the strategic water management plan, and combining that research with the answers to the aforementioned points, should have a basic idea of nozzle performance requirements before requesting samples for evaluation testing and purchase.

Much of the data presented in this chapter reviews material presented in previous chapters, highlighting important items to be considered during equipment evaluations.

Nozzle Operation Review

To evaluate a nozzle properly, its operating functions and the manner in which it achieves those functions should be thoroughly understood. A short review can be helpful in understanding nozzle operation.

To effectively form a useful stream, a firefighting nozzle performs four main functions:

- Controls flow: The size of the orifice in the outlet of the nozzle, along with the nozzle's construction geometry and inlet pressure, controls the flow. The nozzle may be equipped with a single orifice (fixed-gallonage combination or smoothbore), multiple fixed orifices (adjustable-gallonage), or a self-adjusting variable orifice (automatic).

- Provides reach: By creating a restriction that changes pressure to velocity at the end of the waterway, the nozzle causes the water exiting the nozzle to obtain distance, necessary if water is to hit the intended target.

- Creates shape: Selection of a nozzle determines if it will discharge a long cylinder of water, such as in a smoothbore, or a spray pattern provided by a combination nozzle. Certain models can give the spray pattern a distinct droplet size that may increase or decrease firefighting effectiveness, depending on tactics and composition of the burning material.

- Determines firefighting action: For example, a distributor nozzle will form the discharge into a different style of stream than that formed by a foam playpipe. A low-pressure combination nozzle provides a stream that exits a nozzle differently from that of a standard-pressure fog nozzle.

When determining the desired performance of firefighting handlines, it is important to consider the maximum flow rate desired, length of hoseline, and number of personnel available to operate the nozzle.

Performing Evaluations

A salesperson places a new nozzle on the chief's desk and, other than its blue bumper and gold-colored plating, there isn't much to differentiate it at first glance from nozzles the department presently has in use. Its exit opening is round, and it has teeth, a swivel inlet, and a shutoff handle. Should it be purchased? Will it have a positive effect on the department's firefighting efficiency? A reasonable decision can only be made by carefully evaluating the device's flow characteristics and handling qualities along with price, technical support, and maintenance considerations. Actually, it might be found after evaluation that the nozzles that have been in use for years are still the most effective for the department's particular manner of operation.

When making a nozzle evaluation, answers to the following points can help a department organize its data:

- What flow range is desired?
- How much flow can be controlled by available personnel?
- Are different discharge patterns required?
- How will the flow be controlled by the pump operator?
- How durable must the nozzle be and who will perform preventative maintenance and servicing?
- What level of training is available?

Flow Range

As was pointed out in chapter 2, the department should develop a strategic plan to help determine the type of equipment needed to implement that plan. A nozzle's desired flow rate should be based on the amount of water needed to control a baseline fire within the response district, as well as having that flow controlled by the actual number of personnel who will be using it. If all the department has available are the minimum two firefighters for initial handline attack, one on the line and one officer alongside, attempting to flow 300 gallons per minute (GPM) on the first line will not be a practical or safe operation.

Handline nozzles are available with advertised flows of up to 375 GPM but are generally utilized on medium-diameter (1½", 1¾", 2") attack lines. Most of these devices are supplied with 1¼" threads, causing many departments that have purchased them to believe they will have that top flow available on their medium size preconnected lines. These nozzles are certainly capable of flowing their top-rated GPM, but these flows are available only when the nozzle is supplied by 2½" or larger lines unless unrealistic engine pressures are employed. It must be remembered that because of high friction loss at high flows, medium-diameter attack lines cannot flow much over 180 GPM unless extremely high engine pressures are used and unless there is more than one person available to help control the line. Beware of promised unrealistic flow rates, especially when using 1¾" or 2" hose. Be reasonable and practical when determining desired handline flow requirements. If an advertised claim sounds too good to be true, it probably is.

It is important that equipment be evaluated using a calibrated flowmeter to accurately determine the true amount of water being delivered. I call it the water flow lie detector. Many departments are surprised to learn that their 350 GPM nozzle is only flowing 100 GPM when used on their own pumper and hoselines at their standard operating pressure (fig. 13–1).

If the nozzle is intended for use on a line that will always have the same basic flow requirements, for example, always to be used at the end of a 200' preconnect, a single fixed-gallonage nozzle, smoothbore, or low-pressure combination might be the proper choice because of increased durability, ease of operation, and lower purchase cost.

Just because a salesperson says a certain nozzle is the latest firefighting innovation should not automatically justify the nozzle's purchase without extensive testing. Any new nozzle should materially add to the department's firefighting needs and effectiveness before being purchased.

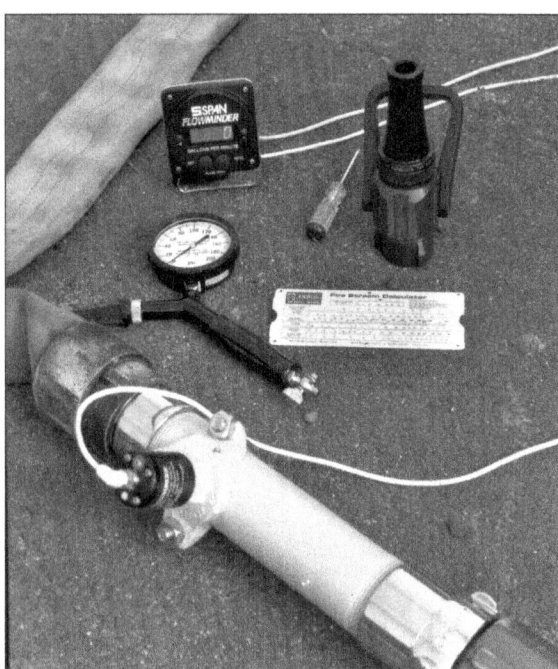

Figure 13–1. A portable flowmeter is the best device for quickly and accurately determining flows and is especially useful when evaluating automatic or conventional combination nozzles. It must be calibrated by using a pitot gauge and smoothbore tip to insure accurate readings.

How Much Flow Can Be Controlled by Available Personnel?

The amount of both physical and mental firefighter stress experienced during fire attack can be an extremely important consideration when evaluating stream equipment. Can you imagine how efficient a present-day firefighter would be advancing old-time 2½" all-cotton hose with brass couplings and 30 lb. brass playpipe while weighted down with present-day protective clothing, breathing apparatus, radio, and tools? As the weight of protective equipment increases, there should be a corresponding reduction in the weight of water flow equipment to keep firefighter efficiency at least at the same level.

To review, stress evaluation of equipment should begin with the hose load. Preconnected lines in lengths up to 300' should be packed so that the firefighter can advance the hose from the bed with at least 100' resting on the shoulder. When advancing lines from reverse lay or bulk beds, the same 100' should also be pulled from the bed and placed on the shoulder for carrying. This action reduces dragging effort and makes it easier to stretch up stairs or around corners. Dragging creates friction and the firefighter must overcome not only the hose weight but also the increased effort needed to counteract the effects of friction. It is much safer and more efficient to have the firefighter carry as much hose as possible on the shoulder to the point of operation while leading out attack lines.

As we already discussed, nozzle reaction is force created by a combination of the weight of flowing water and the pressure that pushes it out the nozzle tip. Newton's third law of physics states that for every action there is a reaction of equal force in the opposite direction. A good example of this effect is when a handgun is fired. The force of the bullet exiting the barrel causes the weapon to kick back. Water flowing through a fire nozzle causes the same type of kickback force. An operator must be able to counteract this in order to safely operate the stream.

A previously discussed rule of thumb stated that a firefighter can usually control their body weight in GPM flow. That rule applies only to being able to counteract the nozzle reaction.

How long the firefighter can continue to effectively operate the line after stretching to an upper floor, for example, depends on their physical condition and fire conditions. A single firefighter can control a line flowing more water by kneeling or sitting on the hose, but it then ceases to be an offensive attack line, becoming instead a stationary defensive line. Tests have shown that one firefighter of average size can safely and effectively operate an attack line, using sustained hit-and-move or flow-and-go tactics, with the nozzle flowing 125 GPM at 100 pounds per square inch (PSI), 150 GPM at 75 PSI, or 185 GPM at 50 PSI, for at least 10 minutes.

For example, if three-person engine companies are the norm for your area, the officer and one firefighter will have to make the attack while the pump operator charges the lines and attempts to establish a continuous water supply. The nozzle operator may have some help from time to time from the officer. For all practical purposes, however, the officer will have their hands full performing size-up, searching, and directing the line. It is extremely difficult for a single firefighter to safely direct a high-flow line operating at high nozzle pressures inside a fire building. This critical fact must be taken into consideration during evaluation. There is a difference between flowing water on the training ground and actual use on the fire scene. There should be an attempt to simulate realistic tactics when determining nozzle selection utilizing the same personnel who will be using the equipment if it goes into service. Evaluating firefighter stress and the effects of high flows as they relate to fatigue and safety should be primary considerations.

The late George Hughes, the Texas fire educator, recommended evaluating the handling requirements of high-flow nozzles by performing a simple test using regular suppression personnel. After determining flows and discharge pressures, he recommended stretching the same size and length line as will be used on an actual fire and flowing the entire contents of the booster tank without reducing the nozzle pressure and with the nozzle wide open. The nozzle crew should flow water, work forward on their knees using flow-and-go tactics, and after advancing about 75', make an abrupt 90° turn and continue advancing, all without throttling the nozzle's flow with the shutoff. In many cases in the past when this test was performed, he said a good share of the firefighting companies found they could not safely or effectively control the line for the entire period of time it took to use up the tank water. If this evaluation test proves that the flow is too high for the operating crew to handle, adjustments in pump pressure, hose diameter, training, and nozzle equipment might have to be made before placing the equipment into frontline service.

How Will the Flow Be Controlled?

Evaluation of tactics should determine whether it will be necessary to change a line's flow rate during firefighting operations and, if so, what nozzle hardware is needed to handle this change. Both the pump operator and nozzle operator are responsible for delivering an effective firefighting stream. If tactics are planned in advance, all the pump operator has to do when using preconnected lines and predetermined pressures is pump to that pressure to provide the desired water flow. On other lines, if nozzle flow capacities and pressures are standardized, the pump operator can more easily compute discharge pressures by counting the number of hose lengths out of the beds, then consulting a chart or calculator to provide the figures needed to supply proper pump pressures.

If it is desired to vary the flow rate during firefighting operations, coordination between the nozzle crew and pump operator is critical. The pump operator should understand what effect raising and lowering the pressure has on both the stream and the crew. If the flow rate is controlled by the nozzle operator using the gallonage selector ring on an adjustable-gallonage nozzle, both the nozzle and pump operators should understand the effects this action has on pressure and flow.

Durability and Service

If an engine company has less than 10 working structure fires a year, durability may not be as important a factor during nozzle evaluations as it would be in a city the size of San Francisco, Boston, or Detroit. Frequency of nozzle repair is determined by the amount of use, amount and quality of preventative maintenance, and the firefighters' attitude toward care of equipment.

Generally, the more complex a nozzle's construction, the more moving parts it contains and the more chance for damage and the less durable it becomes. The more a nozzle weighs, the more destructive force is transmitted when striking the ground or an object. Years ago, heavy duty nozzles were constructed of massive amounts of brass to provide strength. An old-time heavy brass nozzle, if dropped on a curb, landed with such force that its weight could increase the damage, eventually causing it to self-destruct.

Brass, the traditional nozzle construction material, is a soft metal with a lower yield strength than anodized aluminum. When dropped, it may bend or become misshapen. Depending on the nozzle part, this may be major damage if it involves the pattern selector, handle, or inlet swivel. Practically no brass nozzles, or couplings for that matter, are used any longer in the fire service (fig. 13–2).

Newer nozzles constructed of anodized aluminum are much lighter and stronger than the older brass nozzles. Aluminum has a rather porous surface, but the anodization process

Figure 13–2. Nozzles should be accorded the same care given to other life-saving devices such as breathing apparatus and turnout clothing. This nozzle was in frontline service in a large city. The gallonage ring was jammed in the flush position, the stream shaper could not be adjusted, and the shutoff leaked. It was finally removed from service after a nozzle crew was burned.

applied to firefighting hardware causes a durable coating to fuse itself to the aluminum's surface, sealing the pores and making the surface extremely hard. The anodizing process is easily recognized by its dark gray or black color (although dyes can be added to color it red or green). Since nozzles constructed of aluminum are lightweight, impact damage caused by the device striking a hard surface is minimized, helping increase service life even in busy departments. Aluminum used in fire service water delivery devices is half the weight of brass yet has twice the strength.

The firefighters' attitude toward equipment will have a great effect on the length of a nozzle's service life. If personnel continually drag nozzles along the ground, bounce them upstairs, or use them for wheel chocks, they will fail rather quickly. All firefighters must be made aware of the reduced safety and effectiveness of damaged or malfunctioning nozzles and should be required to inspect them after each use. I've seen nozzles in use in large cities that look like a dog's chew toy simply because of the firefighters' lack of interest in properly caring for the piece. A nozzle's service life can be considerably lengthened by regular preventative maintenance. Periodic inspection, cleaning, and lubrication will also ensure that the device will perform properly under fire conditions.

Because they are in constant use and generally taken for granted, maintenance is usually not performed unless something breaks. The number of working parts a nozzle contains and ease of repair are important considerations for busy departments to evaluate.

The following list shows the relative durability of various nozzle designs, the most durable being listed from the top:

- Smoothbore nozzles
- Fixed, single-gallonage combination nozzles
- Automatic combination nozzles
- Adjustable-gallonage combination nozzles
- All types of plastic nozzles

What Level of Training Is Available?

To get the most performance out of any fire stream equipment, pump and nozzle operators must thoroughly understand the operation of the equipment they are using, must be properly trained, and must continue to train to remain proficient. They should think and act as a team because they are dependent on each other for efficient and safe fireground operation. Nothing is more disheartening on the fire scene than to see poor, ineffective streams in use, caused not by inadequate water supply but by pump operators who do not understand what they are doing. Inadequate water supply creates tactical problems that must be solved on the fire scene by laying more lines or providing more tankers. Operator errors must be remedied by additional and continuous training in nozzle and pump operations.

After initial familiarization, the almost-daily use of hoseline equipment in a busy department helps keep firefighters familiar with its operation. In departments with less fire call frequency, more formal and frequent training will be required to put the equipment in service and to maintain a high level of proficiency.

Once the engine pressures and flows are determined and the nozzles go into service, all the pump operator has to do is charge the line to a predetermined discharge pressure. In the case of preconnects, a stripe marked on the pressure gauge with tape to mark proper pressure can be a helpful indicator (fig. 13–3).

The use of smoothbore tips, while posing no unusual problems for the pump operator, requires somewhat more training at the nozzle end to be operated safely and effectively within a building. To have maximum effect, smoothbore nozzles must be kept in motion to properly distribute their water and, because they generally flow more than a comparable combination nozzle, the use of the shutoff to stop the flow is critical when the fire has darkened down if water damage is to be kept to a minimum. Remember that 3' of hose behind the nozzle is also the nozzle. If this area of hose is kinked or severely bent, it will cause turbulence that will severely degrade the stream. The use of a stream shaper materially helps in maintaining stream quality in these situations.

The use of calibrated flowmeters during evaluation and training will simplify the determination process over the use of other measuring methods. Once the results of flow evaluations are compiled and operating flows and pressures determined, the operation at the pump is relatively simple—the operator supplies the lines at the preselected pressures.

Figure 13–3. After calculation with a portable flowmeter, these gauges were marked with the pressure needed to flow the calibrated amount of water. The markings help prevent mistakes under times of stress.

Equipment for Evaluation

As previously stated, the entire evaluation process is greatly simplified and made more accurate if a portable flowmeter is used during testing. This device eliminates having to constantly replace combination nozzles with smoothbore tips for purposes of measurement with a pitot gauge. Smoothbore tips and the pitot gauge are necessary to calibrate the flowmeter device initially. After this is done, however, all tests can be run under the same calibration.

Many departments are equipping new pumper discharges with flowmeter devices. These can make the evaluation process even simpler because the meters are calibrated at the factory for the flow range expected through the size piping in which they are installed. It may be that not all discharges have flowmeter devices. While they can be used to test hose and nozzles that will be placed in service on other discharges, be sure the discharges intended to supply the lines being evaluated are of proper size and have properly-plumbed piping with a minimum of elbows, so they be will capable of efficiently delivering the desired flow (fig. 13–4).

The following equipment should be available when performing evaluations of nozzles and hoselines:

- Portable flowmeter
- Pitot gauge or water flow test kit
- Line gauges
- Smoothbore nozzles
- Smoothbore nozzle flowchart

A recommended procedure is to stretch the line from the pumper using the size and length of hose intended for service. Place a line gauge on the discharge end and attach a shut-off with a smoothbore tip, preferably about ⅞" for smaller lines and 1⅛" or 1¼" for larger hose.

The flowmeter can be attached at any point in the line; however, some require external battery power for operation, so it usually ends up near the truck. Unlike a pressure gauge, the flowmeter will indicate the proper flow no matter where it is located in the line. The flow reading will be the same at the pump, at the nozzle, or anywhere in between. If evaluating a number of discharges, it can be located in a supply line between the hydrant and gated intake. The tank-to-pump line must be closed so the flowmeter can indicate proper flow. In this manner one or a number of discharges can be metered without having to connect the device to the desired discharge each time.

Water should then be flowed at a pump pressure that will deliver 50 PSI at the tip. This pressure is measured with a pitot gauge. Note the engine pressure and the pressure on the line gauge. The line gauge reading will vary somewhat from the pitot gauge reading. This is

Figure 13–4. The basis for determining flow rates can be made by measuring nozzle pressure from a smoothbore tip with a pitot gauge, then consulting a flowchart which shows how much water is being delivered based on the size of the tip and the nozzle pressure. A simple alternative is to use a calibrated flowmeter.

because the line gauge is taking its reading from the edge of the water flow and the pitot is taking its reading from the center. Of the two, the pitot reading is the more accurate (fig. 13–5).

By noting the difference between the readings from the line gauge and the pitot gauge, a more accurate nozzle pressure reading can be calculated from the line when testing combination nozzles. Because of their design, combination nozzles cannot be successfully measured with a pitot gauge.

Setting up a line using a smoothbore tip is relatively easy. Simply charge the line and take pitot readings while increasing the engine pressure until the desired discharge and nozzle pressure are obtained. Readings should be taken with the tip of the pitot gauge blade positioned in the center of the stream. Flow is determined by matching tip size and nozzle pressure readings to columns on a smoothbore nozzle flowchart or by using a slide or electronic calculator. If a flowmeter is being used, the nozzle pressure is determined with a pitot gauge and the flow read directly from the flowmeter display.

When evaluating standard combination nozzles, the pump operator supplies increasing pressure until the desired nozzle pressure is reached after calibrating the line gauge installed behind the nozzle by using a smoothbore tip and pitot gauge. If a flowmeter is in use, the flow is read directly from the display. If a flowmeter is not available, the combination nozzle is removed, a smoothbore nozzle closely matching the desired flow is put in its place, and the line is charged to the indicated pressures. A pitot gauge reading is then taken and a chart consulted to determine flow.

If the high nozzle pressure proves too much to control, another procedure can be used. Using a smoothbore tip, flow the desired amount of water, taking note of the friction loss between the pumper and the nozzle. This friction loss is simply the difference between the two pressure readings. The proper engine pressure for this flow will be this friction loss added to the desired nozzle pressure.

For example, let us assume it is desired to evaluate a low-pressure combination nozzle rated at 150 GPM at 75 PSI attached to 200' of 1¾" line supplied from the swivel discharge connection in a cross-lay hosebed. By consulting a flowchart, it can be seen that a ⅞" smoothbore nozzle will flow approximately the same flow at 45 PSI. The ⅞" smoothbore nozzle is attached to the end of the hose and the line charged so that a pitot reading of 45 PSI

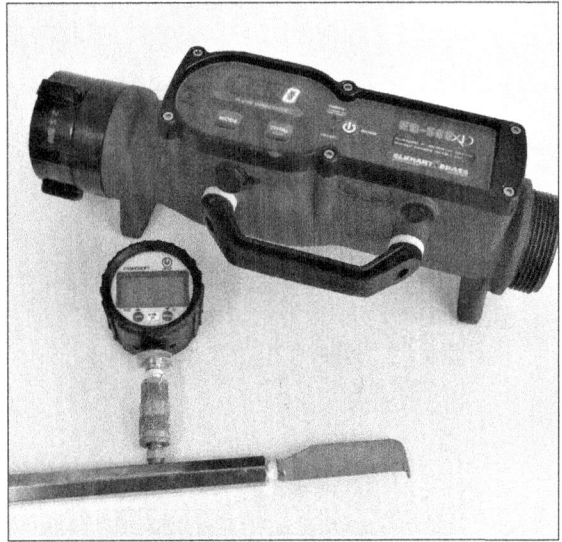

Figure 13–5. Newer flow measuring equipment includes internal battery-powered digital flowmeters and digital pitot gauges. While the older, straight mechanical devices can do the job, newer digital products produce accurate results with less effort.

is reached. This gives a flow of 153 GPM. Let's assume the difference between the nozzle pressure and engine pressure is 67 PSI. The friction loss added to the 75 PSI nozzle operating pressure is 142 PSI. That figure becomes the engine pressure necessary to supply the nozzle flowing 150 GPM at 75 PSI.

To quickly review this procedure, if a flowmeter is not available, the line should be flowed and measured using a smoothbore tip and a pitot gauge to determine the friction loss between the pump and nozzle—a measurement taken as the smoothbore flows the desired GPM rate. This friction loss is added to the combination nozzle's operating pressure to determine the total pressure needed at the engine to supply the line.

Master stream devices can be evaluated the same way. Using a smoothbore tip, determine the difference in pressure between the pitot reading taken at the tip and the pump pressure. The difference between the two readings becomes the friction loss. If the monitor is permanently mounted, this figure will take into account pressure loss through the device and piping.

Replace the smoothbore tip with the desired combination nozzle and supply the device with its operating pressure, determined by adding together the friction loss and nozzle pressure, which is usually 100 PSI. If an automatic nozzle is being used, it will be necessary to determine a number of operating pressures throughout its flow range.

A line gauge installed behind an automatic handline nozzle or master stream nozzle is not an effective indicator because it will always read the nozzle's operating pressure, normally around 100 PSI. Unlike other nozzles, the nozzle pressure remains constant as the flows change. Flow changes on automatic nozzles will have to be computed from the pump gauge rather than the nozzle gauge.

When operating pressures are determined, effects of reach, feathering, and handling qualities should be evaluated. Many departments have found videotaping these evaluations to be extremely helpful. The tapes can be reviewed later, at a much slower pace, away from the distractions that always seem to be present during field operations.

It is important to record pressures and other operating data for later evaluation and for future operational use. If not written down, many helpful variances in pressures and flows can be quickly forgotten. It is also important for firefighters to handle the lines during evaluations as if they are being used in a fire situation. They should be worked inside buildings and used for hit-and-move and flow-and-go operations, and the crews should be continually evaluated for signs of fatigue or difficulty in handling lines or flows.

If at all possible, live, controlled training fires make excellent evaluation conditions. However, determinations should be made later after reviewing all data, including videotapes, if available. These actions help to prevent subjective opinions, rendered after the heat of combat, from affecting evaluations of factual data. Subjective opinions are important. They should be recorded immediately after they are rendered and should be considered during evaluations, but the best time to weigh their worth is away from the fireground atmosphere.

Purchasing Equipment

Once the department determines which nozzle or device will best fit its needs through testing and evaluation, actual purchasing is a rather simple procedure.

All nozzle and most hose manufacturers employ sales representatives who travel extensively throughout their territories, calling on departments and dealers. They carry a full line of product samples and flow testing equipment. Obviously, the purpose of sales representatives is to promote their employer's products. Most all, however, will work side-by-side with the department and other manufacturers' representatives to help a department make unbiased equipment evaluations. Most larger dealers also have demonstration equipment from manufacturers they represent and can provide more frequent contact with a requesting department.

Pricing sometimes becomes an overriding consideration when departments purchase nozzles. While low price is important, a product should be selected based on performance.

A department should also consider what value is provided for the price and how the vendor responds to requests for evaluation, training, and service support. An item's low purchase price quickly becomes unimportant if repair parts are not rapidly available or if a warranty problem is not handled in a timely manner.

Pricing Factors

Discounts given to dealers from product list prices are set by the manufacturers on a volume basis. A new dealer wanting to handle a certain line will usually have to place an initial order of several thousands of dollars' worth of equipment, after credit approval. If someone just starting out in the business offers a department what seems like incredibly low pricing, they may be trying to sell off their initial order. There is nothing wrong with this, but delays in delivery could arise until they sell and receive their complete stocking order.

An established dealer has discounts set on a sliding scale. With some companies, the discount is based on the total yearly volume. With others, the discount is based on the order size. Some dealers try to concentrate their nozzle business with one certain company so their volume with the company becomes larger and the discounts greater. If a salesperson seems to be promoting a certain brand, it might be because their discounts are higher with that brand than with others. On the other hand, if a dealer's yearly volume with a certain brand is low, the manufacturer may be considering terminating the dealer. By offering low prices, the dealer may be trying to increase their volume so as not to lose the line.

It is important for the department to perform its own evaluations and determine through testing and evaluation of gathered information which nozzle, hose, or piece of equipment best fits its needs. A purchase based only on a salesperson's claims may not always be in the best interests of the department. A salesperson should be able to provide detailed information on the products they represent. The department should take that information into account during the evaluation process. It must be kept in mind that a salesperson's claims may be biased or slanted because of the seller's economic interests, not because of performance.

Factory representatives are salespeople in a sense because they represent a certain product. They do not sell directly to a department and are interested in presenting their products in positive fashion. It may be in the department's interest to have the factory representatives demonstrate their equipment separately from the salesperson or dealer. Dealers can arrange for manufacturers' representatives to call on the department or the department itself can contact the manufacturers directly. There may also be problems among certain dealers who sell equipment from manufacturers they are not authorized to represent. This practice is called bootlegging and could cause problems in delivery lead times, handling warranty claims, or after-sale service.

Bootlegging works like this: a dealer for a certain manufacturer makes a deal to sell equipment at a low markup to another dealer who is not an authorized representative of that manufacturer. The second dealer may handle a line of equipment needed by the first dealer, so equipment can be interchanged. If, for example, a nozzle fails because of a defective part and someone is injured, the manufacturer may later prove in court that they are not liable for damages because the unit was sold through an unauthorized dealer and the manufacturer had no control over training in the product's use. Warranty service is also slower because the selling dealer has to return the product to the supplying dealer, who then contacts the factory for service authorization. A call to the manufacturer can easily provide the names of authorized dealers and information on how the manufacturer handles after-sale product problems.

Specifying a nozzle or heavy-stream device is relatively easy. Rather than write a set of specifications, as would be done when purchasing a pumper, simply specify the device you wish to purchase by name, style number, or description of the device's features. Most dealers handle all lines, and it makes life a little simpler for them if they don't have to interpret your needs by wading through excess verbiage when preparing a quote. Nozzles, hose, and other stream appliances should be evaluated beforehand, and the department should know exactly what it requires when requests for pricing are mailed out. If your city or department requires detailed specifications, all manufacturers can quickly supply descriptions of their equipment for bid use.

Nozzle Maintenance

Retired Lieutenant Ted Aff of the Oakland, CA, Fire Department clearly explained some firefighters' attitudes toward equipment care when he stated, "If you give a fireman a 2" stainless steel ball bearing and put him in a bare, windowless room alone for an hour, when you open the door and ask about the ball bearing, he will have either bent it, broken it, or have lost it."

Manufacturers state that lack of preventative maintenance is the major reason that nozzles are returned to their service departments for repair or rebuilding. While most firefighters will carefully repack beds with dry hose, the nozzle sits on the running board, ready to be connected when the packing is over, without even a quick wash under running water. Nozzles are undoubtedly the most important pieces of firefighting hardware owned by the fire department. Without nozzles, fires would continue to burn—with malfunctioning nozzles, they often do. Winning or losing the fire attack game depends in large measure on the performance of nozzle and hoseline equipment. To provide optimum service during critical times, the equipment should be maintained as carefully as your breathing apparatus.

The Los Angeles Fire Department instituted a program to completely strip down, inspect, and repair an engine company's inventory of nozzles when the apparatus enters the shops for its annual service test. Preventative maintenance is the most cost-effective method of keeping nozzles and other equipment working properly and extending their useful service life.

Maintenance Procedures

To keep nozzles and other stream devices in top shape, they should be checked after each use as follows:

- Wash under running water, using soap if necessary to remove soil and grime. If soil proves hard to remove, soak the device in warm water and soap solution.
- Thoroughly check the nozzle interior, looking for pieces of trash or debris that may have become lodged inside.
- Check for proper operation of shutoff and pattern control and check teeth for damage or breakage. On smoothbore nozzles, check the waterway for dents, nicks, or burrs, especially on the exit hole edges, which could interfere with stream quality.
- Check all gaskets for cracks or cuts and make sure they do not protrude into the waterway causing excessive turbulence.
- Check all major parts—bumper, handle, stream control, gallonage control rings, and baffle—for looseness or damage. Tighten or replace as necessary.
- Lubricate according to manufacturer's recommendations, if necessary.

Broken or missing teeth, damaged or broken handles, or loose or cracked bumpers can cause serious problems during firefighting operations. Without thinking, many departments will put the nozzle back into service with these or other damaged items unrepaired. This certainly is not the wisest or safest thing to do. Remember, the nozzle is the last device between the firefighters and the fire. It provides not only the means of extinguishment but also protects the crew from the effects of heat and flames. Would a breathing apparatus be placed back on the rig if the shoulder strap were cut, the regulator providing erratic flow, or the warning bell inoperative? Nozzles must be considered an important piece of safety equipment and should be accorded the same care as helmets, coats, breathing apparatus, and personal alarm devices.

One large-city engine officer told how a malfunctioning nozzle almost seriously injured his crew operating on the third floor of a heavily involved apartment building. During the course of business, the center baffle somehow became loose and, as the company was preparing to make its attack on a heavy volume of fire, the water pressure caused the baffle to strip its last threads and blow out in the direction of the fire. The resulting stream was useless, and the fire would have rolled over the crew's heads if it had not been for the quick thinking of the officer. He jammed two wood wedges used for holding doors open into the throat of the nozzle. When the shutoff was opened, the resulting stream, while quite ragged, had reach and managed to extinguish the fire. Proper inspection after prior use would have discovered the faulty baffle before it caused grief on the fireground.

Combination nozzles should be carried with the pattern control bumper screwed forward in the forward-stream position to protect the center baffle from damage if dropped. If all nozzles are carried in this position, the nozzle operator will also know the pattern control is in the forward-stream position before beginning attack.

Repair Parts

Most manufacturers offer inexpensive repair and rebuild kits that departments can easily install to help put a damaged or leaking nozzle back into service. These kits contain seals, screws, and other high-wear items. Larger parts, such as bumpers, handles, and teeth sets, are available individually. It makes sense to have a complete set of spare parts available for each style of frontline nozzle in order to put a damaged device back into top shape immediately. Manufacturers will supply parts with drawings and repair instructions for the nozzles and other stream devices they make, and most provide repair classes to teach proper maintenance and repair techniques. It's a wise idea to contact the factory or local dealer to obtain technical manuals and parts before they are actually needed after a fire.

Some manufacturers prefer that the nozzle be returned to the factory for repair. While taking more time than repair in the department's shop, manufacturers take extra care to check the device over thoroughly and flow test it afterward. This could make the extra repair time it takes worthwhile if a department does not have flow equipment or trained repair personnel.

Depending on the product, certain tools may be necessary for disassembly and assembly. Before purchase, a determination should be made if these tools should be added to inventory along with spare parts. Manufacturers can supply the necessary information.

When any large department's annual budget is considered, nozzles cost only a small fraction of the total outlay. Given the increased safety provided and the possible increase in firefighting effectiveness, it makes sense to replace worn or outmoded nozzles on a consistent, planned basis.

I always wonder what disaster might befall a department when I see hand-me-down, older nozzles in service on spare or reserve pumpers. If the first-line piece is out of service, these elderly, worn devices can set the department's firefighting effectiveness back to the days of outside attack and low flows. Not too many years ago, departments considered 60 GPM–95 GPM flows from 1½" lines as standard fire streams. If these older, lower-flow nozzles are relegated to service on reserve pumpers, it is a good idea to remove frontline nozzles from the unit undergoing repair and place them in service on the reserve rig. Operating with damaged, worn, or low-flow nozzles not only reduces efficiency but it could also become dangerous to crews used to flowing more water from handlines.

At times, reserve pumpers are put into service in addition to firstline units to cover during large fires or multiple smaller alarms. It is important that the crew staffing these units understand the flows at which the nozzles operate and to check the nozzles, breathing apparatus, and other tools before the unit is placed in service.

In summary, not only will proper evaluation of nozzle and hoseline equipment help a department make a purchase decision, the time spent flowing water can also help accomplish the following operational and administrative procedures:

- Determine what hardware can increase efficiency and safety during fire attack operations.

- Plan tactics revisions and assist with developing training programs for the new equipment.

- Provide cost-effective analysis of various extinguishing devices to aid in budget planning.
- Improve fireground safety by determining in advance the effects of nozzle reaction and effective flow.
- Determine which suppliers and manufacturers are most responsive to a department's needs.
- Determine maintenance and repair requirements for the new equipment that will aid budgeting and help determine maintenance personnel requirements.

By actually handling the equipment before purchase, the positive improvements in operation that can result from evaluation will be worth much more than the few hours or days invested in testing.

In a Broadway stage production, what goes on behind the scenes is as necessary to the show's success as well as the actors' performance. While putting new equipment into service and waiting for a fire to see how it worked may have been an adequate practice in the past, it can be dangerous and actually hinder operations today. The fire service is being forced to produce more results with fewer resources than ever before. Inefficient equipment, be it a nozzle or a ladder truck, cannot be compensated for any longer by throwing more resources into the fight, as was done in the past. In many places, there just may not be any more nozzles, ladder trucks, hoselines, or personnel to operate them once initial attack is made.

Steps must be taken behind the scenes and before the show. Support the stars with the best equipment possible. Evaluation and maintenance can help accomplish this task before the curtain goes up on the next performance.

Chapter 13 Review Questions

1. List two points to consider when evaluating a nozzle.
2. Define nozzle reaction.
3. What is the most durable nozzle design?
4. What should be some determining factors for equipment purchases?
5. Describe your department's nozzle inspection and repair program.

Answers to Chapter Review Questions

Chapter 1

1. By creating a restriction that increases velocity.
2. Hose diameter, hose length, amount of water flowing, other factors such as kinks, rough liner.
3. On board tank, tank from another vehicle, supply line from hydrant, drafting.
4. A reverse lay is defined as the operating lines at the fire being supplied by the pumper positioned at the water supply.
5. 2.
6. Put the pump in gear, get water into the pump, get water out of the pump.

Chapter 2

1. A 2½" line flowing at least 250-GPM.
2. A strategic plan is based on evaluation of a number of buildings and water supply in a given area rather than on a single occupancy.
3. It may not provide enough flow rate to overcome the heat being produced by a large volume of fire.
4. Available personnel, available water supply, type and amount of equipment and apparatus, goals in performing size-up, search, attack, limiting fire spread, ventilation.
5. Flow Rate. Water must be applied in sufficient quantity to overcome the heat being produced.

Chapter 3

1. Utilizing a straight stream water pattern.
2. Knocking down the main body of fire from a safe, exterior defensive location then changing to offensive attack for final comprehensive extinguishment.
3. Rate of flow and proper distribution.
4. Flashover is a sudden, full-room involvement in flame, as being caused by thermal radiation feedback from a build-up of superheated gasses and heat. Flashover is best mitigated by flowing water into the superheated area.
5. The interior flowpath should be controlled to ensure that the fire is not unwittingly fed excessive amounts of outside air. In some cases, this can be accomplished by controlling the door through which the attack team enters.

Chapter 4

1. Controls flow, provides reach, creates stream shape, determines water direction and form.
2. It allows for quick attachment of additional hose line.
3. In operation, it allows water to be distributed effectively over a large area for fires located beneath the firefighters.
4. Low gallonage flows may not be able to overcome the heat being produced.
5. To stop fast moving fires in overhead cockloft spaces and ducting.

Chapter 5

1. A device to replace the baffle in master stream nozzles.
2. It maintains stream reach by varying the flow rate.
3. On high-flow master stream devices.
4. It provides maximum reach while developing additional water supply.
5. It is typically a good indication that the flow rate is being reduced.

Chapter 6

1. The size of the exit hole drilled in the tip and the geometry of the reducing area determine how much water will flow at a certain pressure.

2. A smoothbore nozzle reduces the generation of unwanted steam, provides greater stream penetration, generates less nozzle reaction, and produces less air retention in stream that can affect fire atmosphere.

3. As a measuring tool to determine the flow capacity of standpipe and sprinkler systems fed by fire pumps and gravity tanks.

4. The lower nozzle operating pressure makes lines easier to control.

5. An overpressurized nozzle will exhibit a stream that "rags" or "feathers" around the edges, caused by unequal velocity generated by water rubbing against the sides of the nozzle.

6. A problem, especially in older occupancies, is clogging of the nozzle due to rust, scale, and trash dislodged as engine companies begin pumping into the standpipe system. These debris will easily flow through a smoothbore without having to stop the attack to flush or shut down to clean the tip.

Chapter 7

1. Los Angeles and Chicago.
2. It reduces nozzle reaction while allowing for increased flow rate.
3. The lower nozzle operating pressure allows the decreased pressure (vacuum) ahead of the baffle to pull the hollow stream back into a more solid flow.
4. The pattern can cover a larger area for cut-off or holding operations, can absorb a wider range of nozzle operating pressures while still providing a stream with effective reach, a spray pattern delivers water more gently for close-in work, can provide limited heat protection for crew, useful for indirect attack in enclosed areas such as attics, can move a greater volume of air useful for forced ventilation.
5. The low-pressure nozzle's larger, heavier water droplets drop through the heat to the burning material with very little being converted into unwanted steam. This tends to keep the thermal layering intact which keeps fire atmosphere disruption to a minimum.

Chapter 8

1. Any stream flowing over 400-GPM.
2. Utilize an automatic nozzle.
3. Big water can be quickly and safely applied with a minimum of personnel; to protect exposures and to extinguish the fire.
4. Make sure everyone, including mutual aid companies know what it is.
5. Make sure no personnel are in or near the area where the stream will be operating.

Chapter 9

1. Inner tube, strength inner jacket, outer woven jacket, or protective rubber covering.
2. To provide strength.
3. To protect the filler (strength) yarns.
4. A 3' random section is tested by the manufacturer until it bursts.
5. Coating it with a liquid polymer or elastomer compound.
6. The hose is more prone to kinking as well as the rougher liner causing increased friction loss. Also, the thinner tube will provide decreased hose life with moderate usage.
7. Woven double jacket, plied rubber covered, and extruded rubber covered.
8. Standard 1962.

Chapter 10

1. Detergent/surfactant.
2. Water. The Class A additive just makes it penetrate better.
3. Class-A foam because most of what's burning is Class A. Class B foam would not be effective, just more expensive, because of its limited ability to penetrate packed combustibles.
4. High initial equipment cost, limited effectiveness, high maintenance costs, low GPM delivery.
5. It is not recommended. Flowing CAFS foam through a combination spray nozzle will strip a large portion of air from the stream, breaking down the foam's bubble structure.

Chapter 11

1. Stretching enough hose to reach and operate on the fire.
2. To protect occupants and ensure life safety.
3. The amount of storage present, the area involved, and the type of fire load such as cardboard, flammable liquids, or furniture.
4. To the side and a bit higher than the nozzle operator so they have a clear view of the operation.
5. When you can't see the floor while standing up.

6. The nozzle is opened and flowing the entire time while the crew is advancing.

7. Don't generate it in the first place, ventilation ahead of the fire, vertical ventilation.

8. The fire will be extinguished, another company will arrive in time to establish a water supply for continuous operation, the engine will run out of water and the attack will have to be abandoned.

Chapter 12

1. Intended flow rate, length, number of personnel available to lead it out and operate safely.

2. Preconnected beds, bulk unconnected beds, reverse or day-line bed with supply and attack line in the same bed.

3. Consideration when designing and packing hose loads, accurate size-up to determine size and length of hose line, shoulder loading hose rather than dragging or pulling with the arms.

4. A line that incorporates a smaller diameter initial attack line attached to a nozzle that is supplied by a larger diameter hose as a back-up.

5. A day line is a hose stretched from the water supply point to an engine positioned at the fire.

6. A forward lay is a hose stretched from the water supply point to an engine positioned at the fire, most effectively utilizing a large diameter hose.

7. Utilizing friction. For example, by wrapping an arm around the hose positioned under the armpit, or kneeling on a hose line to create friction with the ground.

Chapter 13

1. Flow rate, reaction forces generated, discharge patterns, quality of the stream, level of training needed, durability.

2. The force of a combination of the weight of the water being discharged and the pressure needed to propel that discharge.

3. Smoothbore

4. After doing deep research, consider these points: suitability for the intended task, personnel needed to operate, durability, price, and service support.

5. Include items such as scheduling inspections, having a check list for inspections, having a trained and qualified person perform the inspections and repairs, and describing an adequate supply of spare parts with which to make repairs.

Index

A
acceptance (proof test) pressure 218
adjustable gallonage, constant flow style 90, 96
adjustable gallonage nozzles 96
    advantages 100
    automatic nozzle vs. 100
    master stream device use 100
advanceable nozzles 92
aerial devices
    on engine companies 179
    heavy streams and 178
    nozzle reaction and failure 140
    placement of 185, 197
    setting outriggers for 209
    testing and recertification 232
AFFF (aqueous film-forming foam) 239
Aff, Ted (lieutenant) 327
agents. *See* Class A foam; Class B foam
air entrainment 76
Akromatic nozzle 114
Akron Brass 114, 178
alarms
    delay 41
    extra 188
    staging areas for multiple 194
American Insurance Association 143
American La France 85
amount of fire 256
anodized aluminum nozzles 320–321
apparatus
    accessories vs. practicality 27
    Class 125 series 106
    Class 135 series 106
    equipment included 27
    fog-pressure units 108
    mini-pumper 108
    positioning 53, 187
    pumper 23
    rescue-pumper 27
    tanker 26
    tender 26
    writing specifications for 27
aqueous film-forming foam (AFFF) 239
atmosphere 71, 82
    heat and 67
    pressure of 6
attack. *See also* direct attack; fire attack; indirect attack; interior attack; offensive attack; transitional attack
    crew 42
    high-volume 183
    pumper 281
Attacking and Extinguishing Interior Fires 61, 65
attack lines 8
    definition of 5, 285
    large-diameter 54, 56, 228
    length of 25
    low-flow 49
    medium size 50
    preconnected 25
    pressure requirements for 23
    two-person operation of 30
    universal 52
automatic extinguishing systems 28
automatic nozzles 52, 53, 88, 90, 96, 120, 164
    adjustable gallonage nozzle vs. 100
    Akromatic nozzle 114
    conventional nozzles vs. 120
    design 114–115
    effective flow evaluation 122
    efficiency vs. extinguishment 116
    evaluation of reaction 126

automatic nozzles (*continued*)
    fire pump relief valve vs. 114
    flow evaluation and 130–132
    flow meters and 118
    flow ranges 128–129
    flow variation and 115
    handline operations and 121, 122–125
    heavy streams and 180
    HTFT nozzle 114
    increasing flow and 118
    low operating pressure and 120
    master stream devices and 120
    mechanics of 113
    multiple from the same pumper 121
    nozzle reaction and 114, 126–128, 130
    operational hints 130
    pressure control device 85, 114
    price 120
    pump operations and 129–130
    pump operators and 116, 129–130
    reach vs. volume 121
    reserve power capabilities 124
    SM-10 114
    SM-100 114
    smoothbore tips vs. 122
    testing before purchasing 120
    theory 49
    variable flow style 90

B

back pressure 22, 149, 302. *See also* nozzle reaction
baffles 71
balanced-pressure injection metering systems 243
ball type shutoff 92
Baltimore City (MD) Fire Department 15
Barracks hazard 51
baseline flow 36, 39
    equipment for establishing 36
    high-heat fires and 42
battalion chiefs. *See also* captains; chiefs
    Creasey, Claude 162
    Dixon, Clarence 148
bayonet nozzle. *See* piercing nozzles
below-grade fires 102
bent discharge nozzle 105
bent loop design 179
bent loop devices 22
Big Brick theory 267
blitz attack 182–184
    lines 54
    second line of defense and 184
    water supply and 184

blowback effect 73
bomb lines 54
booster lines 36, 42, 46, 106, 109. *See also* 1" booster line
bootlegging 327
Boston (MA) Fire Department 15, 55
brass barrel strainer 6
brass couplings 9
brass nozzles 320
break-apart nozzles 92
breaking the back of fire 75
breathing apparatus 86
Bresnan nozzle. *See* distributor nozzles
budgets
    accountability with public funds 28
    cuts to 29
    nozzle price and 120
    purchasing apparatus and 27
building construction 46
building evaluations 56
bulk hosebeds 54, 293–294
Bureau of Land Management 235
burning, accelerated 40
burst pressure 219

C

CAF. *See* compressed air foam (CAF)
CAFS (compressed air foam system) 244–246
calculators 1, 18
calorific value 68
captains. *See also* battalion chiefs; chiefs
    Moreno, Gil 143, 162
career vs. volunteer departments 29
cellar nozzles. *See* distributor nozzles
centrifugal pump 106
charge pressure and large-diameter hose 23
Chicago belt 309
Chicago (IL) Fire Department 55, 108, 154
    fog-pressure apparatus 108
    Mystery nozzles and 166
    post-war apparatus replacement program 108, 109
    quick water attack program 163
chiefs. *See also* battalion chiefs; captains; Layman, Lloyd (chief); McMillan, Clyde (chief)
    Clark, William 148
    Dunn, Vincent (deputy) 81
    Fried, Emanuel 187
    Kozey, Gregory 182
    O'Donnell, Andrew (district) 129, 166
    Phipps, Ed 151
    Spahn, Edwin J. (deputy) 115

clappered Siamese 179
Clark, William (chief) 148
Class 125 series apparatus 106
Class 135 series apparatus 106
Class A foam 235
   advantages of 240
   balanced-pressure injection metering
      systems 243–244
   batch mixing 241–242
   on Class B fires 249–250
   Class B foam vs. 239, 250
   combination nozzles and 242
   compressed air and 238–239, 244–246
   development of 238–239
   discharge-side foam eductors
      (proportioners) 243
   firefighter safety and 248–249
   flashover and 249
   future of 250–251
   historical drawbacks of 238–239
   history of 237–238
   knockdown and 249
   mixing methods 240–241
   on-scene mixing 242
   overhaul and 272
   positive-pressure, discharge-side injection
      system 244
   premixing 241–242
   proportioners 243
   reignition and 249
   structural firefighting and 248
   surfactants and 239
Class B foam 237, 239, 250
clockwise vs. counterclockwise nozzle
   movement 70
Coast Guard nozzle 89
combination line 53
combination nozzles 22, 50, 64, 70, 83. *See
   also* peripheral deflection method
   adjustable gallonage, constant flow
      style 90, 96
   Class A foam and 242
   constant pressure, variable flow style 90
   drawbacks of 166
   estimating flow with 118
   heavy-caliber for protecting exposures 205
   high-flow handlines and 165
   impinging stream method 89
   increased airflow and 71, 73
   large-diameter lines and 56
   large fires and 198–200
   low-pressure 166–167
   low-pressure vs. standard pressure 76, 80, 84
   protection myth and 166
   reaction force formula 174–176
   shutoff device 92
   single gallonage, constant flow style 90, 96
   single gallonage, variable flow style 89
   standard operating pressure 84
   steam and 147
   stream pattern variety 84
   swirling deflection method 89
   teeth 84
   tips 93
   water exit velocity 73, 84
   water shape and 84
combination stream attack 151
combustion
   engineering 253
   process 60
compartment size 39
compound gauge 22
compressed air foam (CAF)
   disadvantages of 245–246
   qualities of 244–245
compressed air foam system (CAFS) 244–246
compressed air injection 244–247
concealed spaces 37
conflagration-type fires. *See* large fires
conserving water 49, 107–108
   extinguishment vs. 108
constant flow nozzle 87
constant pressure nozzles. *See* automatic nozzles
construction
   extruded (hose) 221
   methods and fire intensity 162
consumer awareness 29
conventional nozzles vs. automatic nozzles 120
Coordinated Fire Attack 67
counterbore 150
counterclockwise vs. clockwise nozzle
   movement 70
countering nozzle reaction 308–309
Creasey, Claude (battalion chief) 162–163
critical flow rate 42, 49, 68, 70
   importance for extinguishment 79
cross lay 289

D
Dacron 212
data gathering 30, 46
   to determine flow rates 39, 40
Davis, Larry 36
day lines 53, 154, 297
deck guns. *See also* prepiped deck guns
   removable mounted 179

defensive operations 187
   tactics 192–194
deflection methods 89. *See also* peripheral deflection method
   swirling and combination nozzles 89
deluge guns 22
Department of the Interior's Interagency Fire Center 235
departments. *See* fire departments
Des Moines warehouse experiment 69
Detroit (MI) Fire Department 50
direct attack 59, 82, 142
   operation of 60
   smoothbore nozzles and 148
   success of 62
discharge
   gate 17
   port 15
   pressure 17–18
discharge-side foam eductor 243
discharge-side injection system 244
discharge valves 2
   calculating flow and 18
disease vs. symptom 82
distributor nozzles 84, 85, 100
   additional lines used with 102
   description of 101
   operation of 101–102
District of Columbia Fire and Emergency Medical Services Department 50
Dixon, Clarence (battalion chief) 148
double-jacket hose 222–223
   mildew and 232
   protecting 223
downhill lay 22
"Down to Earth Talk About Nozzles" 115
drafting water 6–7
droplet size 170
dry hydrants 7
Dunn, Vincent (deputy chief) 81
DuPont 212

E

Eastford (CT) Fire Department 182
effective streams. *See* streams, effective
elbows 289
electronic calculators 1, 18
Elkhart Brass Manufacturing 85, 88, 94
   automatic nozzles 114
   low-pressure combination nozzles and 166
   nozzle survey 98–99
   SM-10 nozzle 114
   SM-100 nozzle 114

Elkhart S pistol grip booster nozzles 108
Encap 223
"Engineering Extension Service Bulletin #18" 67
engineer's lay 305–306
engine-mounted relief valves 133
engine pressure 23
   increasing 118
   reverse lay and 23
   selecting gallonage and 97
entrained air 74
evacuation 195
evaluation equipment 322–325
exit baffle 90
Exploratory Committee on the Application of Water 64
exposures 15, 25. *See also* protecting exposures
extension 47
exterior attack 65, 71
extinguishment
   conserving water vs. 108
   effective flow rate for 36
   faster 36
extruded construction 221

F

Fairfax County (VA) Fire Department 73, 77
FDIC (Fire Department Instructors Conference) 64
FDNY New York City Bureau of Research and Development 143
feast-or-famine water supply 113
feathers 139
finely divided water particles 61
fire. *See also* high-rise fires; large fires; ventilation
   amount of 256
   behavior 59
   below-grade 102
   Class A 238–239
   downwind spread 201
   flammable liquid 95
   flow vs. nozzle pressure 120
   heat release of 40
   heat vs. residual heat 42
   high-heat 42
   high-rise 102, 151–152
   intensity and construction methods 162
   intensity and flammables 161
   intensity and plastics 161
   intensity (increasing) 161–162
   larger-than-usual 45
   larger volume (out-of-control) 187
   liquefied petroleum gas 95
   load, historic vs. modern 40

location within a structure 255
loss 39, 45
lost causes 203
luck and 264
offensive attack vs. defensive operations 192
potential for 28
protection 29
reading the fire building 256
reports 39
routine 45
salvage vs. extinguishment 274
self-ventilated 71
spread 42
spread and stream placement 278
structures and 42
tactics for increased volume 162–163
time/severity chart 36
turbulence 95
two-handed 187
volume 39, 162
fire alarm delay 41
fire attack. *See also* blitz attack; direct attack; exterior attack; high-volume attack; indirect attack; interior attack; offensive attack; tactical fire attack
   combination stream 151
   defensive operations 187
   heavy streams and 180
   methods 142
   modern interior 142
   quick water 163
   transitional 263–264
   water and 262
Fire Department Instructors Conference (FDIC) 64
fire departments. *See also* Chicago (IL) Fire Department; Los Angeles (CA) Fire Department; New York City (NY) Fire Department; *See also* Osaka, Japan, Fire Department
   Baltimore City (MD) Fire Department 15
   Boston (MA) Fire Department 15, 55
   career vs. volunteer 29
   Chicago (IL) Fire Department 154
   Detroit (MI) Fire Department 50
   District of Columbia Fire and Emergency Medical Services Department 50
   Fairfax County (VA) Fire Department 73, 77
   Gary (IN) Fire Task Force 113
   Houston (TX) Fire Department 55
   insurance companies and lawsuits 29
   insurance companies ratings 29
   Las Vegas (NV) Fire Department 16
   Miami (FL) Fire Department 151–154
   Montgomery County (VA) Fire Department 77
   New Orleans (LA) Fire Department 28
   New York City (FDNY) Fire Department 143, 217
   New Zealand fire service 110
   Oakland (CA) Fire Department 327
   Parkersburg (WV) Fire Department 61
fire extinguishment. *See* extinguishment
Firefighter I 45
Firefighter II 45
firefighting. *See also* structural firefighting; tactical fire attack
   art of 253
   improving efficiency 47
   modernization of technology 29
   practical vs. impractical 95
   rural 36
   smoothbore tactics 142
Fire Fighting Principles and Practices 148
fireground
   efficiency 26
   operations 45
   overconfidence and failure 45
   safety 45
   water management plan 25
fire hydrants. *See* hydrants
fire intensity
   Iowa formula and 41
   National Fire Academy (NFA) formula and 41
Fireman [66]
fire pump
   operation of 5
   supplying water to 5
fire steam. *See* steam
fire streams. *See* streams
first-aid lines 110
fixed flexible teeth 94, 95
fixed fog nozzles 106
fixed-gallonage fog nozzles 114
   insufficient fire stream and 113
fixed-gallonage method of operation 125
fixed-gallonage nozzles 117, 118
   overpressurization of 118
fixed replaceable teeth 94, 95
fixed rigid teeth 94, 95
flame–fuel interface 60
flame turbulence 94
flammables
   increased fire intensity and 161
   liquid storage containers 37

flaps 297
flashover 82
flash point 106
floating strainer 6
floor below nozzle 105
flow. *See also* large-diameter flow systems
  baseline 39
  calculation 18, 41
  control 319–320
  control ring 90
  definition of 4
  delivering 42
  distribution 41, 47
  efficiency, 1½" vs. 1¾" lines 50
  evaluation 130–132
  laminar 3
  output 8
  path 80
  starting 113
  starting point 22
  supply flow, increasing 24
  tests 54
flow-and-go tactic 268
flowmeters 18, 23, 52, 118, 130, 323
  calibration of 322
  performing evaluations and 317, 322
  selectable gallonage nozzles and 97
flow minders. *See* flowmeters
flow range. *See* flow rates
flow rates 23
  automatic nozzles and 128–129
  critical 42
  effective 36, 41
  effects on fires 36
  faster extinguishment and 36
  ineffective 47
  personnel requirements for 36
  reach and 42
  reviewing past reports to determine 39
flush features 93
  inlet screen 94
  large debris removal 93
flying lay 305–306
FMC nozzles 108
foam. *See* Class A foams; Class B foams; compressed air foam (CAF)
foam nozzles 85, 93
fog guns (high-pressure) 89
fog nozzles 114. *See also* combination nozzles
  low-pressure 167
fog-pressure units 108
fog streams 62, 80, 86, 276–277
  advocates 106
  high-pressure 106–107

improper use of 40
issues with 66
nozzle reaction 107
forestry hose 223–224
  weeping 224
formulas (fire flow)
  Iowa formula 40, 41
  National FIre Academy (NFA) formula 41
forward lays 8, 300–305
  large-diameter hose 300
  multiple line 8
  relay valves and 304
  waterflow capability and 301
four-stage hydraulic power take-off pumps 108
four-way valve. *See* relay valves
Freeman, John R. 133
friction 308, 310
  Seattle Grip 308
friction loss
  definition of 3, 4
  excessive 21
  factors affecting 4
  large-diameter hose and 12
  multiple supply lines and 11
  rapid water additive and 50
  reducing 8
  in relay valves 23
  rubber hose and 108
  shutoff devices and 93
  supply lines 8
Fried, Emanuel (chief) 187
fuel 39
  heat release and 40
  historic vs. modern 40
  quantity of 36
  type of 36
fuel–flame interface 60

G

gallonage selector 96–97
garden apartment 42
garden hose metaphor 2
Gary (IN) Fire Task Force 113
gases 60, 74
gathering data. *See* data gathering
goal setting for water management plans 25
Grant multiversal 178
grips. *See also* pistol grips
  removable 311
  Seattle Grip 308
ground-attack monitors 201–202
  large fires and 201
ground monitors
  guns 54

positioning of 185
guns. *See* deck guns; prepiped deck guns
    deluge 22
    fog (high-pressure) 89
    ground monitors 54
    removable mounted 179
    split waterway 22
    top-mounted portable 178

H

handlines
    adjustable gallonage nozzle and 100
    automatic nozzles and nozzle
        reaction 126–128
    automatic nozzles, operation of 121
    beds 25
    combination nozzles and 165
    high-flow 164–166
    nozzles 114
    offensive attack and 189
    safety 169, 171
handling the line 306–313
handwheel discharge controls 182
hard suction hose 7
hardware (modern) 29
hazards 30
    application tactics for mitigating 35
    baseline flows for mitigating 35
    evaluating 25
    flow distribution and 47
head loss 22
heat
    atmosphere and 67
    balance 143
    overhead danger 275–276
    oxygen and 40
    residual vs. fire 42
heavy streams
    aerial ladders and 178
    automatic nozzles and 180
    bent loop nozzle design and 179
    criteria for use of 182
    devices 39
    directing 206–208
    effective use of 180–182
    equipment 178
    fire attack and 180
    high-volume attack and 182, 189
    initial attack and 180
    nozzles 179–180
    offensive attack and 180–182, 185
    pace of application 189
    portable guns and 178
    preconnected lines and 182
    prepiped deck guns and 178
    protecting exposures 204–206
    removable, mounted guns 179
    safety and 208–209
    second line of defense and 196
    situations requiring 177–178
    smoothbore nozzles and 180
    Snorkel units and 178
    top-mounted portable guns and 178
    water supply and 180
    water towers and 178
HEN nozzle 157
high-flow handlines 164–166
    combination nozzles and 165
    smoothbore nozzles and 165
high-heat
    fire 42
    release 40
high-pressure fog guns 89
high-pressure/maximum volume method of
        operation 123–125
    disadvantages of 123
high-pressure pumps 108
high-rise buildings 54
    firefighting tactics for 55
    fire suppression tactics formulation 54
    search and rescue operations in 55
    standpipe system for water and 54
    ventilation in 55
high-rise fires 102, 151–152
    bent discharge nozzles 105
high-rise hose 225, 226–228
    2" line vs. 2½" line 227
high-volume attack 183
    heavy streams and 182, 189
    high-caliber streams and 189
    preconnected and portable guns and 182
high-volume streams 121
hit-and-move tactic 268
holding tanks, portable 26
horizontal ventilation 278
horseshoes 297
hose 2. *See also* back pressure; high-rise hose;
        forestry hose; large-diameter hose;
        lightweight-lined; nozzle reaction;
        rubber-covered hose; yarn
    advantages of rubber-covered 222
    advantages of traditional 222
    attack line 8, 222
    components 212–213
    construction 212–213
    damaged 23
    diameter 4
    double-jacket 222–223

hose (*continued*)
    effective flow of 162–163
    equipment for evaluation 322–325
    first-aid lines 110
    handling 306–313
    hard suction 7
    heat and wear 219–220
    heat damage and water flow 220
    history of 211–212
    inner liner 212
    inspection and maintenance 232–234
    interior surface 4
    jackets 214
    kinking 3, 4, 7, 224
    laying multiple 8
    length 4
    length and pressure 23
    lightweight lined 224–226
    line placement 260–262
    loading 291–293
    mildew and 220, 232
    minimum personnel to operate 29
    minuteman load 293
    number of 8
    nylon 216
    outer jacket 219
    pigtail 56
    pistol grip and 311
    preconnected 17, 25, 39, 286
    pressure ratings 218–219
    reel line 110
    removable grip and 311
    repair 232–233
    rope hose tool 310
    safe control and operation of 162
    selection of 39
    shoulder-loading 292–293
    size and pressure 23
    size and purpose of 8, 21
    sizes 217–218
    soft suction 7
    static 25
    synthetic 214–215
    thermoplastic-lined 224–225
    tools to help control 310
    2" line vs. 2½" line for high-rise fires 227
    unsafe pressure and 52
    weeping 224, 232
    woven jackets 213
    yarns 214–217
hosebeds 13. *See also* preconnected hosebeds
    bulk 293–294
    cross lay packing 289
    forward lay 302–303
    leader line 297
    locating 288–291
    mattydale packing 289
    packing 312
    preconnected 286–293
    restraint devices 217
    reverse lay 296–297
    types of 285
hoseline. *See* hose
Houston (TX) Fire Department 55
HTFT nozzle 114
Hughes, George (fire marshal) 268, 319
Humat valve. *See* relay valves
hydrants
    dead-end 17
    deteriorated 7
    disconnection from water main 7
    dry 7
    fire pump and 5
    frozen 6
    pressure 14
    pumper apparatus and 108
    relay valve and 15, 23
    using multiple 14
    water pressure 8
    water volume 8
Hydrassist valve. *See* relay valves
hydraulics 2
    calculations 17
    fire stream 120
Hypalon 223

I

ideal rate of flow theory 40
impinging stream 89
Improvement in Nozzles (device patent) 85
incident command systems 194
indirect attack 59, 82, 142
    effective applications of 79
    fire victims and 71
    interior 65
    issues with 65
    life-safety and 70
    operation of 60
    origins 61, 64
    smoothbore nozzles and 142–143
    teeth and 94
    ventilation and 62, 70
    when to use 63
ineffective fire streams 113
initial attack and heavy streams 180
inlet gauge 22

insurance companies
    fire departments ratings from 29
    lawsuits against fire departments and 29
intake port 15
interior attack 95. *See also* Iowa State University research
    combination streams 151
    crew rotation 269
    low-pressure nozzles and 163
    smoothbore nozzles for 151
    training 151
    ventilation and 255
invisible fireman 182
Iowa formula 40, 41, 68
    applications of 40
    fire intensity and 41
Iowa State University research 40, 67, 87
    Des Moines warehouse experiment 69
    indirect attack and 66

K

kickback 149, 318. *See also* nozzle reaction
knockdown 268
Kozey, Gregory (chief) 182
Krupa, Steve 143

L

labor agreement 29
ladder hook 135
laminar flow 3
large-caliber streams. *See* heavy streams
large-diameter flow systems 230–232
    discharge components 230–231
    large-diameter hose and 230
    relief valves 231–232
    supply components 230
large-diameter hose 9, 179, 228–230
    attack lines 228
    construction types 229
    forward lays 300
    friction loss 12
    historic construction 9
    increasing capacity of 16
    initial charge pressure 23
    manageability 13
    maneuverability 38
    modern construction 10
    popular sizes 13
    pressure ratings 228–229
    reverse lays 300
    rubber-covered 220
large fires
    aerial devices and 197
    combat tactics 190–194
    combination nozzles and 198
    evacuation and 195
    ground-attack monitors and 201
    heavy streams and 206–208
    incident command systems and 194
    lost causes and 203
    offensive attack vs. defensive operations 192
    safety officers and 197
    spray streams and 199
    staging areas and 194–197
    20 minute rule and 191
    withdrawal actions and 192
Las Vegas (NV) Fire Department 16
lawsuits 28
    preventing 29
    recordkeeping and 28
Layman, Lloyd (chief) 59, 70–71, 87
    indirect attack and 61, 62, 63–65
    "Little Drops of Water" FDIC paper 64
lays. *See also* forward lays; reverse lays; supply line lays
    downhill 22
    engineer's 305–306
    flying 305–306
    straight 8
leader line bed 297
liability 29
life cycle of a fire 37
life safety 29, 42, 76, 256–259
    steam and 78
lightweight lined hose 224–226
    kinking and 224
    washboarding and 224
lines. *See also* attack lines; day lines; hand lines; preconnected lines; supply lines
    combination 53
    first-aid 110
    gauges 323
    handling 306–313
    placement 260–262
    reel 110
liquefied petroleum gas fires 95
    "Little Drops of Water" FDIC paper 64
location of fire within a structure 255
Los Angeles (CA) Fire Department 51, 55, 162
    nozzle maintenance and 327
Los Angeles City nozzle 163
low-pressure combination nozzles
    Elkhart Brass Manufacturing and 166
    history of 166–167
low-pressure nozzles 163
    background of 166–169

low-pressure nozzles (*continued*)
    combination 166–167
    droplet size and 170
    evaluation of 169–171
    fog 167, 171
    interior attack and 163
    Los Angeles City nozzle 163
    nozzle reaction and 163
    safety and 171
    steam and 171
luck 264

M

maintenance of nozzles 327–328
    preventative 327
    procedures 328
    safety and 328
master stream devices 114
    adjustable gallonage nozzle and 100
    automatic nozzles and 120
    evaluation of 325
    single inlet 25
master streams. *See* heavy streams
mattydale 289
McMillan, Clyde (chief) 49, 88
    automatic nozzle operation theory and 113–114
Miami (FL) Fire Department 151–154
mildew 220, 232
minuteman load 293
mission. *See* goal setting
misting 170
monitor gun 114
Montgomery County (VA) Fire Department 77
Moreno, Gil (captain) 143, 162
mounting portable guns 183–190
moving in 265–267
    cellars 266
municipal water department 8
mutual aid program 46
Mystery Fog Nozzle Company 166
Mystery nozzles 94, 166
    design 85

N

narrow-cone fog pattern 71
National Board of Fire Underwriters 64
National Fire Academy (NFA) 41
    fire intensity formula 41
National Institute of Standards and Technology 77, 147, 239
Navy all-purpose nozzles 86–87, 89
    drawbacks 87

Nelson, Bill 40, 66–67, 87
    indirect attack 69
New Orleans (LA) Fire Department lawsuit 28
Newton's Third Law of Physic 174–176, 306
New York City (NY) Fire Department 50, 55, 143, 217
    bulk hose beds 54
    standard handline history 50
New Zealand fire service 110
NFA (National Fire Academy) 41
NFPA 1901: Standard for Automotive Fire Apparatus 217
NFPA 14: Standard for the Installation of Standpipe and Hose Systems 153
NFPA 1961: Standard on Fire Hose 218
NFPA 1962: Standard for the Care, Use, Inspection, Service Testing, and Replacement of Fire Hose, Couplings, Nozzles, and Fire Hose Appliances 218, 232
NFPA 1964: Standard for Spray Nozzles and Appliances 172
North Richland Hills (TX) 268
nozzle crew 116
nozzle evaluation 316–322
    durability and service 320–321
    equipment for 322–325
    flow range 317
    physical and mental firefighter stress and 318–319
    pricing factors 326–327
    training and 321–322
    warranty services and 327
nozzle maintenance 327–328
    Los Angeles Fire Department and 327
    procedures 328
    safety and 328
The Nozzleman 67
nozzle operator 96, 97
nozzle pressure 1, 22, 90
    definition of 4
    fire flow vs. 120
    fire stream reach and 116
    fluctuations in 118
    maintaining constant 100
    measuring 4, 17
    operating 117, 118
    overpressurization 117
    variation in 100
nozzle reaction 38, 52, 122, 149, 150, 306–307
    aerial device failure and 140
    automatic nozzles and 114, 130
    back-pressures 150
    calculation of 149–150

combination nozzle vs. smoothbore
    nozzle 306–307
combination reaction force formula 174–176
countering 308–309
definition of 4
effects of 53
explanation of 173–175
fog stream and 107
friction and 308
handlines and automatic nozzles 126–128
increasing flow and 118
master nozzle flow rate and reaction
    chart 174
measurement of 4
overpressurization and 117
pistol grips and 108, 310–312
playpipes and 135
reducing 52, 149–150
removable grip for 311
75 PSI vs. 100 PSI 163
single person and 42
smoothbore nozzles and 150
smoothbore reaction force formula 174–176
nozzles 50, 315. *See also* adjustable gallon-
        age nozzles; automatic nozzles; combi-
        nation nozzles; distributor nozzles; *See
        also* fixed-gallonage nozzles; flush features;
        heavy stream nozzle; low-pressure nozzles;
        piercing nozzles; *See also* shutoff devices;
        smoothbore nozzles; teeth; tips
    advanceable 92
    anodized aluminum 320–321
    automatic pressure adjusting 88
    baffle 71
    bent discharge 105
    brass 320
    break-apart 84, 92
    cellar 84
    clockwise vs. counterclockwise movement 70
    clogging and 148
    Coast Guard 89
    comparison of 136–137
    constant flow 87
    conventional 120
    converting 118
    converting existing to low-pressure 172
    distributor 84–85
    Elkhart S pistol grip booster 108
    evaluation of reaction 126
    exit baffle 90
    exit pressure 71
    fixed fog 106
    fixed-gallonage fog 114

fixing 83
flame turbulence and 94, 95
FMC 108
foam 85, 93
four main functions 316
friction loss and 21
functions 83
gallonage selector 96–97
handline 114
HEN 157
high-flow 53
importance of 83
Improvement in Nozzles (device patent) 85
Los Angeles City 163
low-pressure fog 167, 171
manufacturers 21
master nozzle flow rate and reaction
    chart 174
matched 122
mechanics of 116
Mystery 166
Navy all-purpose 86, 89
operation 316
opposed discharge 103
over-pressurized 117, 139
pattern selector 93
piercing 85, 92
preventative maintenance and 321
purchasing 325–326
reaction force 52
repair of 329–330
selection 315
selector ring 96
shutoffs 133
sidewall 71
single-gallonage 118
spray 87, 88–89
stream-shaping devices 84, 133
underwriter's playpipe 133–134
Vortex 157
when to open 268
nylon in hose yarn 216

O

Oakland (CA) Fire Department 327
occupancy 259–260
O'Donnell, Andrew (district chief) 129, 166
offensive attack 161
    blitz 184
    converting to defensive position 192–194
    directing heavy streams and 206
    handlines and 189
    heavy streams and 178, 180–182, 185

offensive attack (*continued*)
   low-pressure nozzle tests 162
   surgical use of nozzle streams 268
1½" line 50, 217
   flow efficiency 50
   personnel requirement to stretch 50
1¾" line 217
   flow efficiency 50
   history 50
   larger fires and 51
   personnel requirement to stretch 50
   rapid water additive 50
1" booster line 48
   flow 49
   friction loss 49
1.88" line 50, 218
operations
   automatic nozzles and 129–130
   fixed-gallonage method 125
   high-pressure/maximum volume
      method 123–125
   hints 130
   predetermined pressure method 125
opposed discharge nozzle 103
Osaka, Japan, Fire Department 77, 78–79, 239
Our Lady of Angels school fire 257
outer jacket 219
   heat resistance 220
   protective coatings 223
output flow 8
output pressure 8
outriggers 209
overapplication of water 70, 80, 267
overconfidence and fireground failure 45
overhaul 272–273
   Class A foam and 272
   water and 272–273
overhead heat danger 275–276
overpressurization 139
   nozzle reaction and 117
overspray 1
Oyston, Charles 85–86

P
Parkersburg (WV) Fire Department 61
pattern selector 93
patterns of water distribution 80
peel-away 170
pencilling 276
penetration 143
   combination vs. smoothbore nozzles
      and 144–145
   fire stream 108

peripheral deflection method 87, 88, 89
personnel 46
   available for fire attack 262
   considerations for limited 47
   flow distribution and 47
   maximum flow rate and 36
   nozzle reaction and 38, 42
   reduction of 29
   staffing minimums 29
petrochemical industries 13
Phipps, Ed (chief) 151
piercing nozzles 85, 89, 92, 100
   operation of 103
   origin of 102
pigtail hose 56
pistol grips 310–312
   nozzle reaction and 108
piston pump 106
pitot gauge 134, 322, 323
plastics and fire intensity 161
playpipes 84
   ladder hooks and 135
   nozzle reaction and 135
   underwriter's 133–134
plied construction 221
portable ground monitor 288
portable guns 56
   mounting and storage 182–189
portable holding tanks 26
portable master stream equipment 46
positioning of
   ground monitors 185
positive pressure 244
   fans 279
   improper use of fans 40
   injection metering systems 243
pre-burn time 41
preconnected hosebeds 286–293
   elbows and 289
   packing 291–292
preconnected lines 39, 44
   advantages of 2" and 2½" 287–288
   determining ideal length of 53
   ground monitors guns and 54
   heavy streams and 182
   insufficient length of 46, 53
   locating hosebeds for 288–291
   nozzles and 287
   operational problems 286
   size up and 286
preconnect hosebeds
   locating 288–290
   packing 291–292

predetermined pressure method of operation 125
prepiped deck guns 48, 178–179
    handwheel discharge controls and 182
        proper use of 182
pressure. *See also* charge pressure; discharge pressure; engine pressure; nozzle pressure; pump pressure
    acceptance 218
    atmospheric 6
    attack hose 23
    burst 219
    calculating proper 23
    calculating working 23
    dangerously high 52
    nozzle and 83
    output 8
    proof test 218
    pump discharge 2
    rated operating 96
    ratings for hoses 218–219
    service 219
    standard operating 84
    too much vs. too little 23
preventative maintenance 321
pricing factors 326–327
    bootlegging and 327
primary search 42
priming for drafting water 6
private dwelling 36
proof test pressure 218
proportioners 243
    backpressure and 243
    Class A foams and 243
protecting exposures 204–206
    heavy-caliber combination nozzles and 205
    tactics 205
protection myth and combination nozzles 166
protein foam 106
pumpers. *See also* supply pumpers
    attack 281
    placement of 185
    relief valves 231
    rescue 27
    screaming 141
    steam 133
pump operators 18, 24, 52
    attack hose and 8
    automatic nozzles and 116, 129–130
    bulk hose beds and 54
    onboard water tank operations 5
pump pressure 8
    calculating 17–18
    definition of 4
    efficient 21
    elevation and 22
    engine 17
    height of the fire and 22
    hydrant 14
    increasing 14, 19
    inlet (compound) gauge 22
    measurement of 4
    relay valves and 10, 15
    too little vs. too much 22
    zone residual 14
pumps 23
    centrifugal 106
    discharge pressure 2
    efficient operation of 24
    four-stage hydraulic power take-off 108
    high-pressure 108
    three-stage 110
purchasing equipment 325–326

Q
quick water attack program 163

R
rags 139
Rand Corporation 50
rapid water additive 50
rated operating pressure 96
reach 4
reaction. *See* nozzle reaction
reading the fire building 256, 265
recordkeeping and lawsuits 28
reel line 110
reignition 249
relaying water 23
relay valves 10, 23
    brush fires and 15
    flow and 304–305
    forward lays and 304
    friction loss in 23
    hydrants and 15, 23
    pump pressure and 10, 15
    supply pumper 15
relief valves
    engine-mounted 133
    large-diameter flow systems and 231–232
    pumper built-in 231
relighting 281
removable grip 311
removable mounted guns 179
repair parts for nozzles 329–330
rescue 77
rescue-pumper 27

residual heat vs. fire heat 42
response district 30
reverse lays 8, 17, 23, 294–300
    advantage of 296
    advantages and disadvantages of 299
    day line 297
    forward lays vs. 296
    large-diameter hose and 300
    leader line bed 297
    operations 297–300
    packing 296–297
    skids, horseshoes, and flaps 297
Rochna, Ron 235
rollover 73, 268
rope hose tool 310
rotation of interior attack crews 269
Royer, Keith 40, 66–67, 87
    indirect attack 69
rubber-covered hose 220–223
    advantages of 222
    extruded construction 221
    plied construction 221
    weep vents 222

S

safety 196. *See also* life safety
    Class A foams and 248–249
    crew rotation and 270
    fireground 45
    heavy streams and 208–209
    low-pressure nozzles and 171
    nozzle maintenance and 328
    pencilling and 276
safety officer 197
salvage 274
San Francisco Bureau of Equipment 143, 162
Schiobohm, Paul 235
Scientific American 85
search
    evacuation operations and 44
    primary 42
    rescue operations in high-rise buildings and 55
Seattle Grip 308
second line of defense 196
selectable flow nozzles 52
selector ring 96
self-ventilated fire 71
service pressure 219
shaping water 83–84
Shapiro, Paul 16
shoulder-loading hose 292–293
shutoffs 52, 84
    friction loss and 93
    integral vs. separate 92
    nozzles and 133
    nozzle tips 93
    twist-type 92
    universal 93
Siamese (clappered) 179
single-gallonage nozzles 118
    constant flow style 90, 96
    variable flow style 89
single inlet master stream device 25
size-up 253–263
    amount of fire 256
    considerations 254
    life safety 256–259
    line placement 260–262
    location of fire within a structure 255–256
    occupancy 259–260
    personnel available 262
    preconnected lines and 286
    water for fire attack 262
skid load beds 46
skids 297
sky shots 200, 206
slug tip 152
    advantages 154
SM-10 nozzle 114
SM-100 nozzle 114
smoke 60
    visibility and 42
smoothbore nozzles 3, 22, 50, 98, 99, 133, 323
    break-apart nozzles and 92
    clogging and 148
    combination line and 53
    counterbore 150
    deluge guns with 22
    description of 2
    discharge style 84
    distribution of water 68
    estimating flow with 118
    firefighting tactics with 142
    flowchart 323
    heavy streams and 180
    high-flow handlines and 165
    inefficient fire streams and 113
    large-diameter lines and 56
    large fires and 197–200
    low-pressure 76, 80
    nozzle reaction and 150
    operating pressure of 4
    operational advantages of 142–143
    operation of 2
    over-pressurized 139
    penetration and 144–145

pitot gauges and 134
reaction force formula 174–176
stacked 114
standard operating pressure 84
standpipes and 148
steam from 145
stream-shaping and 133
tip pressure and 138
tips 137–140
training and 322
variations on 157–158
water shape and 84
Snorkel units 178
soap-skim 237
soft suction hose 7, 299
solid gasoline 161
solid-stream nozzles. *See* smoothbore nozzles
Spahn, Edwin J. (deputy chief) 115
spinning replaceable teeth 94
protection myth 95
split waterway guns 22
spray nozzles 76, 87
spray streams 68
impinging 89
large fires and 199
patterns 74
peripheral deflection method 88
swirling deflection 89
sprinkler systems 23, 28
Squirt units 200
stacked tips 137
staffing
minimums 29
modern unit size 161
volunteer 162
staging areas 194–197
standardizing equipment 93
standard operating pressure 84
standpipes 54
dry 23
failure of building valves and 56
pigtail and 56
Siamese 23
smoothbore nozzles and 148
wet 23
starting flow 113
starting point flow 22
static hose 25
steam 62–63, 76–78, 80, 145–148, 267
combination vs. smoothbore nozzles and 147
expansion 61, 73
life-safety and 80
low-pressure nozzles and 171

pumpers 133
smoothbore nozzles and 145
unwanted 147–149, 274–277
Storz device 229, 230, 300
straight-bore nozzles. *See* smoothbore nozzles
straight lay 8
straight streams 68, 76
efficacy of 75, 80
strainer
brass barrel 6
dry hydrant 7
floating 6
strategic water flow plan. *See* water management plan
streams. *See also* fog streams; straight streams
best practices 269
devices for heavy 121
effective 1, 41, 137
efficient 23
feathers 139
fog 276
ground-attack monitors and 201
heavy hit vs. conserving water 49
high-pressure 108
high-volume 121
hydraulics 120
impact of shape 68
impinging 89
ineffective 113
loss of pressure 3
maximum fire flow 14
misting and 170
nozzle pressure and reach of 116
nozzles and 84
overpressurization and quality of 117, 118
peel-away and 170
penetration power 108
placement and fire spread 278
poor 24
pressure 2
rags 139
reach 84, 96
reach vs. volume 121
safe 23
selection for maximum penetration 76
shaping devices 25, 84
shaping devices for smoothbore nozzles 133
shaping devices for threaded tips 155
spray vs. smoothbore 143
volume for effective extinguishment 118
working pressure of 2
street layout 46
street pipes 133

stress reduction 143
structural firefighting 87
    Class A foam and 248
suction hose 6–7
    adapters 7
    dry hydrants and 7
supply flow, increasing 24
supply line lays
    downhill 22
    reverse 8, 23, 294–300
    straight (forward) 8
supply lines 1, 295
    calculating flow of 17
    definition of 5
    downhill lay 22
    effectiveness 11
    fire pump and 5
    friction loss and 8
    increasing available water flow 10
    increasing hose diameter on 12
    increasing pressure for 14
    increasing the size of 19
    laying multiple 10
    length of 14, 25
supply pumpers 6, 8
    relay valves and 15
    water pressure 8
    water volume 8
surfactants 239
swirling deflection 89
symptom vs. disease 82

T

tactical fire attack
    cooling exposed compartments before ignition 278
    crew rotation and 269
    decision process 253–263
    flow-and-go tactic 268
    hit-and-move tactic 268
    moving in 265–267
    overhaul and 272–273
    size-up 253–263
    thermal imaging cameras and 269
    thermal injuries and 276
    ventilation and 271–272, 278
    vent then enter 280–281
tactics
    defensive operations 192–194
    flow-and-go 268
    high-rise buildings 54, 55
    hit-and-move 268
    increased volume 162–163
    large fire 190–194
    protecting exposures 205
    smoothbore 142
tankers, purchasing 26
tank water 281
Task Force shutoff model 92
Task Force Tips (company) 114, 157
teeth 94. *See also* spinning replaceable teeth
    broken 95
    fixed flexible 94, 95
    fixed replaceable 94, 95
    fixed rigid 94, 95
    indirect attack and 94
    narrow spray pattern and 95
tender (purchasing) 26
Texas Forest Service 237
Texas Snow Job 237
theories
    automatic nozzle 49
    ideal rate of flow 40
    of recency 189, 286
thermal imaging camera 269
thermal injuries 276
thermoplastic-lined hose 224–225
Third Law of Physic (Newton) 174–176, 306
threaded tips 154–156
3" line 51
three stage pump 110
time/severity fire chart 36
tips. *See also* nozzles
    discharge tables for 141
    pressure and smoothbore nozzles 138
    size 3
    size selection 140–141
    slug 152–153
    smoothbore 137
    stacked 137
    standardized sizes 137
    threaded 154–156
    twin 151–154
top-mounted portable guns 178
training 29, 46
    importance of 7
    nozzle evaluation and 321–322
transitional attack 59, 74, 82, 263–264
turbulence 3
20 minute rule 191
twin tips 151–154
    advantages 154
twist-type shutoff 92
2½" line 51, 217
    efficiency of 53
2¼" line 217, 227

two buckets metaphor 49
two-handed fires 187
2" line 50, 51, 217
   efficiency of 53
   friction loss 51
   high-rise attacks and 227
   uses for 51
   weight 51
two-person attack line operation 30

U

UL FSRI. *See* Underwriters Laboratories Fire Safety Research Institute (UL FSRI)
Ultrajet 157
Underwriters Laboratories Fire Safety Research Institute (UL FSRI) 59, 74, 80
underwriter's playpipe 133–134
Union Carbide 50
United Fire Training and Research Center 110
United States Steel 113
University of Illinois Fire Service Institute 60, 73, 275
U.S. Air Force 106
   high-pressure operations 111
U.S. Coast Guard 61, 63
u-shaped apartment 54
U.S. Navy 86, 106

V

valves. *See also* discharge valves; relay valves; relief valves
   building, failure of 56
   leaking 7
vaporization temperature 60, 106
variable flow nozzles. *See* automatic nozzles
velocity 9, 84
   definition of 3
   nozzle and 84
   turbulence and 117
ventilation 40, 80, 271–272
   below grade 42
   channel 102
   coordinated 76
   engine work and 278–281
   flow path 265
   high-rise buildings and 55
   horizontal 278
   indirect attack and 70
   interior attack and 255
   relighting and 281
   self-ventilated fire 71
   unventilated areas and fire movement 73
   vertical 278, 280

visibility and 272
victims
   indirect attack and 71
   rescuing 77
visibility 272
   poor 42
volume (fire) 8
volunteer vs. career departments 29
Vortex nozzle 157

W

wartime firefighting manuals 106
washboarding 224
water. *See also* conserving water
   art of applying 263–271
   cost of 235
   damage 36, 80, 274
   drafting 6–7
   droplet size 77–78, 80
   extinguishers 42
   finely divided particles 61
   fire attack and 262
   hammer 23
   mains 14
   overapplication of 70, 80, 267
   overhaul and 272–273
   penetration 76
   relaying 23
   shaping 83, 83–84
   steam and 267
water distribution patterns 80
   clockwise 70
   counterclockwise 70
   side-to-side 75
water flow 1, 8. *See also* flow; hydraulics
   baseline 36
   capability and forward lays 301
   distribution 70
   efficiency 2
   elevation impact and 14
   estimating 22
   extinguishment and 22, 79
   factors affecting quantity 8
   height of the fire and 22
   hose diameter and 12, 19
   increasing 301
   increasing with a single hose 19
   low pressure and high velocity 7
   nozzles and 83
   quantity vs. form 68
   rate 23
   restriction of 2
   safe 2

water flow (*continued*)
   science of 2
   velocity 9
Watergate 28
water management plan 25, 39
   documenting to prevent lawsuits 29
   enacting 25
   flexibility and 31, 45
   goal setting 26
   ineffective 45
   maximizing efficiency using 44
   reduced personnel and 29
   simple 44
   surveys and 30
   target hazard average and 30
   training and 47
   unwritten or informal 45
water supply 1, 46, 281–282. *See also* hydrants; water tanks
   above vs. below the fire 22
   drafting water 6–7
   effective management of 17
   feast-or-famine 113
   improper vs. proper management of 24
   ineffective use of 25
   insufficient 16, 113
   secondary 5
   solving problems with 24
   water towers 178
water tanks 281
   onboard 5–6
   portable 5
weeping 224, 232
weep vents 222
Where's The Water 67
wide-angle protection myth 95
wildland firefighting with Class A agents 235
withdrawal actions 192
World War II 9, 61, 110
   changes in firefighting and 86
   protein foam 106
   tree sprayers experiments 106
   wartime firefighting manuals 106
wyes 4, 56
   supply lines and 8

Y

yarns (hose) 214–217
   entangled 215
   filament 215
   nylon 216

About the Author

David P. Fornell held the position of Deputy Fire Commissioner, second in command of the Detroit Fire Department, with fire operations, buildings, public information, and logistics as his primary responsibilities.

Before coming to Detroit, Fornell was assistant chief of the Casstown Community Fire Company, protecting 88 square miles of rural area in and around Casstown, OH. He formerly held the rank of captain of the Beckerle & Company Hose Company, Engine No. 9 of the Danbury, CT, Fire Department, and was the career chief of the Westchester, IL, Fire Department, a suburb of Chicago.

Fornell is a certified fire service instructor specializing in engine company operations, company officer development, water flow, and fireground safety courses. He has taught nationwide, including at the Fire Department Instructors Conference; Notre Dame Michiana Fire School; University of Missouri; Connecticut Fire School; University of Akron; South

Carolina Fire Academy; Illinois Fire Service Institute; Monroe, WI, Annual Fire School; Firehouse Expo; International Association of Fire Chiefs' annual conference; and the California Fire Instructors' annual conference.

Fornell serves as a consultant on operations, management, and safety issues, and counts the departments of New York City, Detroit, Boston, Chicago, Grand Rapids, and Memphis as among his many clients. For his work on the bunker gear program in New York City, he holds the rank of honorary battalion chief of the New York City Fire Department.

Fornell served 7 years as the executive editor of *Fire Apparatus* and has written articles for *Fire Engineering* and *Firehouse* magazines in the United States, and *Fire and Rescue* and *Military Firefighter* in the United Kingdom. He has hosted the rural water supply, engine company operations, and line officer training segments of "The First Line Supervisor" for the Fire Emergency and Training television network.